Cell-Free Analysis of Membrane Traffic

CELL-FREE ANALYSIS OF MEMBRANE TRAFFIC

Proceedings of a Conference on Cell-Free Analysis of Membrane Traffic Held at the European Molecular Biology Laboratory, Heidelberg, Federal Republic of Germany, October 1–4, 1986

Editors

D. James Morré
Department of Medicinal Chemistry
and Pharmacognosy
Purdue University,
West Lafayette, Indiana

Kathryn E. Howell
European Molecular Biology Laboratory
Heidelberg, Federal Republic of Germany

Geoffrey M.W. Cook
Department of Anatomy
University of Cambridge
Cambridge, England

W. Howard Evans
Laboratory of Protein Structure
National Institute for Medical Research
London, England

ALAN R. LISS, INC. • NEW YORK

Address all Inquiries to the Publisher
Alan R. Liss, Inc., 41 East 11th Street, New York, NY 10003

Printed in the United States of America

LIBRARY OF CONGRESS
Library of Congress Cataloging-in-Publication Data

Conference on Cell-Free Analysis of Membrane Traffic (1986: Heidelberg, Germany)
Cell-free analysis of membrane traffic : proceedings of a Conference on Cell-Free Analysis of Membrane Traffic, held at Heidelberg, Federal Republic of Germany, October 1–4, 1986 / editors, D. James Morré . . . [et al.].
p. cm.—(Progress in clinical and biological research; v. 270)
Includes index.
ISBN 0-8451-5120-7
1. Cell membranes—Congresses. 2. Cell fractionation—Congresses.
3. Cell organelles—Congresses. 4. Biological transport—Congresses. I. Morré, D. James, 1935- . II. Title. III. Series.
[DNLM: 1. Intracellular Membranes—metabolism—congresses. 2. Subcellular Fractions—analysis—congresses. W1 PR668E v. 270 / QH 601 C748c 1986]
QH601.C656 1986
574.87'5—dc19
DNLM/DLC 88-3005
for Library of Congress CIP

Contents

Contributors . . . . **xi**

Preface
D. James Morré, Kathryn E. Howell, Geoffrey M.W. Cook, and
W. Howard Evans . . . . **xvii**

Introduction
Cell Fractionation: Importance to Cell-Free Systems Development
George E. Palade . . . . **xix**

Section I. NEW DEVELOPMENTS IN SEPARATION TECHNIQUES

A. FREE-FLOW ELECTROPHORESIS

Introduction **3**

Free-Flow Electrophoresis: Preparative Applications to Cell-Free Analyses of Exocytotic Membrane Traffic
D. James Morré **7**

Co-Fractionation of BHK-21 Cell Endosomes and Lysosomes by Free-Flow Electrophoresis
Mark Marsh, Hildegard Kern, Erik Harms, Sandra L. Schmid, Ira Mellman, and
Ari Helenius **21**

Isolation of Functionally Distinct Endosome Subpopulations by Free-Flow Electrophoresis
Sandra L. Schmid and Ira Mellman **35**

Continuous Flow Electrophoresis Applied to the Separation of Cells, Membranes, and Membrane Functional Domains
N. Crawford **51**

Plasma Membranes From Amphibian Epidermis: Subfractionation by Preparative Free-Flow Electrophoresis
M.I. Burón, Placido Navas, G. García-Herdugo, and D. James Morré **69**

B. IMMUNO-ISOLATION AND DEVELOPMENT OF ORGANELLE-SPECIFIC ANTIBODIES

Immuno-Isolation of Subcellular Components
Kathryn E. Howell, Jean Gruenberg, Akio Ito, and George E. Palade 77

Immunoaffinity Techniques for the Purification and Functional Assessment of Subcellular Organelles
J. Paul Luzio, Barbara M. Mullock, W.J. Branch, and Peter J. Richardson 91

Characterization of the Trans-Golgi Network in BHK Cells
Ivan de Curtis and Kai Simons 101

Immuno-Isolation of Hepatocyte Membrane Compartments Using *S. Aureus* Cells
Ann L. Hubbard, William A. Dunn, Susette C. Mueller, and James R. Bartles . . 115

Organelle-Specific Antibodies: Production of Antibodies to Golgi Subcompartments
Jaakko Saraste, Mary Bronson, George E. Palade, and Marilyn Gist Farquhar . . 129

Analysis of Coated Vesicle Structure and Function Using Anti-Clathrin Monoclonal Antibodies
Frances M. Brodsky 141

C. DENSITY MODIFICATION, PHASE PARTITION, AND OTHER COMPLEMENTARY TECHNIQUES

Hepatic Endosome Subfractions: Preparation, Biochemical Properties, and Functional Roles
W. Howard Evans and Carlos Enrich 155

The DAB-Induced Density Shift: Principle, Validity, and Applications to Endosomes
Pierre J. Courtoy, Joël Quintart, Jean-Pierre Draye, and Pierre Baudhuin 169

The Isolation of Endosomes Using Colloidal Gold
Julia Beardmore and Colin Hopkins 185

Characterization of Intestinal Membrane Vesicles With Flow Cytometry
Jean-Pierre Gorvel, Zohair Mishal, and Suzanne Maroux 195

Purification of Membranes Enriched in Rough Endoplasmic Reticulum Domain by Using a Density Perturbant
Alain Amar-Costesec, Danièle Godelaine, and Michael Hortsch 211

Separation of Cell Organelles and Membrane Vesicles by Phase Partition
Per-Åke Albertsson 227

Isolation of Highly Purified Plasma Membrane From Hepatomas and Cultured Cells by Two-Phase Partition
Placido Navas, D. David Nowack, and D. James Morré 237

Use of Biotinylated Lectins and Immobilized Streptavidin for the Specific Isolation of Surface Glycoconjugates
J. William Buckie and Geoffrey M.W. Cook 243

Analytical Differential Centrifugation of Cell Organelles
Torgeir Flatmark 247

Section II. MEMBRANE TRAFFIC IN PLANTS

Introduction 253

Isolated Vacuoles and Tonoplast Vesicles as Experimental Tools to Study Solute Traffic to and From Vacuoles
A.M. Boudet and G. Alibert 255

Isolation by Free-Flow Electrophoresis and Identification of Tonoplast and Plasma Membrane for Studies of Plant Membrane Traffic
Günther F.E. Scherer and Barbara vom Dorp 269

Free-Flow Electrophoresis for Preparation of Plant Membranes
Anna Stina Sandelius and D. James Morré 281

Isolation of Plasma and Vacuole Membranes From Green Leaves by Preparative Free-Flow Electrophoresis
Guy Auderset, A.S. Sandelius, Claude Penel, Hubert Greppin, and D. James Morré 285

Association of a Peroxidase to Membranes of Pharbitis Cotyledons
Stefanie Kiefer, Hubert Greppin, and Claude Penel 289

NBD-PC: A Tool to Study Endocytosis and Phospholipase Activity in Plant Protoplasts
Lynne A. Dengler, Magaly Rincón, and Wendy F. Boss 291

Experimental Analysis of Membrane Traffic During Secretion of Scales in Green Algae
Michael Melkonian, Dieter Becker, Geoffrey I. McFadden, and Ingo B. Reize . . 297

A Method to Study the *In Vivo* Intermembrane Transfer of Lipids and Fatty Acids to the Plasma Membrane in Higher Plants
Patrick Moreau, Hélène Juguelin, René Lessire, and Claude Cassagne 303

Section III. RECONSTITUTION OF MEMBRANE TRAFFIC IN CELL-FREE SYSTEMS

Introduction 307

Reconstruction of Organelle Function in Cell-Free Systems: Introduction
James E. Rothman 311

Fusion in the Endocytic Pathway Reconstituted in a Cell-Free System Using Immuno-Isolated Fractions
Jean Gruenberg and Kathryn E. Howell 317

Reconstitution of Protein Transport From the Endoplasmic Reticulum to the Golgi Using a Cell-Free System
William E. Balch, Con J.M. Beckers, and David S. Keller 333

Cell-Free Reconstitution of the Insertion of a Newly Synthesised Glycoprotein Into the Plasma Membrane
J. Michael Edwardson and Philip G. Woodman 343

Reconstitution of an Endocytic Fusion Event in a Cell-Free System: Some Characteristics of Fusion *In Vitro*
Stella Hurtley and Graham Warren 355

Section IV. ANALYSES OF MEMBRANE TRAFFIC AND APPLICATIONS

Introduction **363**

Intracellular Transport of the Mannose-6-Phosphate Receptor
Suzanne R. Pfeffer, Yukiko Goda, and Nancy W. Chege **365**

Kinetics of Internalization and Recycling of Cell-Surface Glycoproteins
Lutz Thilo **377**

Functional Characteristics of the Endosomal Apparatus of Rat Liver Parenchyma
J.J.M. Bergeron, D.G. Kay, W.H. Lai, J.J. Doherty II, C.E. Smith, M.N. Khan, and B.I. Posner **391**

Identification of Antigen(s) Specific to a 'Late' Endosome Fraction
J.P. Ahluwalia, J.J. Doherty II, M. Troulis, B.I. Posner, and J.J.M. Bergeron . . **411**

Phosphomannosyl Receptor Binding is Localized in Vesicles Separated by Free-Flow Electrophoresis From Rough Endoplasmic Reticulum Preparations of Rat Liver
Nita Minnifield, Keri Safranski, and D. James Morré **417**

***In Vitro* Fusion of Microlipid Droplets From Mammary Gland**
H.M. Valivullah, D.R. Bevan, A. Peat, and T.W. Keenan **421**

Endocytosis and Subcellular Distribution of Transferrin in Rat Liver
H. Goldenberg, E. Wallner, R. Pumm, and M. Hüttinger **427**

A Mitotic Form of the Golgi Apparatus Identified Using Immuno-Electronmicroscopy
John M. Lucocq, James G. Pryde, Eric G. Berger, and Graham Warren **431**

Applications of Endocytosis Research to Drug Delivery
Glynn Wilson, Clive J. Dix, and Ian F. Hassan **441**

Index **447**

Contributors

J.P. Ahluwalia, Department of Anatomy, McGill University, Montreal, Quebec, Canada H3A 2B2 **[411]**

Per-Åke Albertsson, Department of Biochemistry, University of Lund, S-221 00 Lund, Sweden **[227]**

G. Alibert, Laboratoire de Biotechnologie et Amélioration des Plantes, ENSAT, 31076 Toulouse Cédex, France **[255]**

Alain Amar-Costesec, International Institute of Cellular and Molecular Pathology, ULC, B-1200 Brussels, Belgium **[211]**

Guy Auderset, Laboratory of Plant Physiology, University of Geneva, 1211 Geneva 4, Switzerland **[285]**

William E. Balch, Department of Molecular Biophysics and Biochemistry, Yale University, New Haven, CT 06510 **[333]**

James R. Bartles, Department of Cell Biology/Anatomy, Johns Hopkins University School of Medicine, Baltimore, MD 21205; present address: Department of Cell Biology and Anatomy, Northwestern University Medical School, Chicago, IL 60611 **[115]**

Pierre Baudhuin, Laboratoire de Chimic Physiologique, International Institute of Cellular and Molecular Pathology and Louvain University Medical School, B-1200 Brussels, Belgium **[169]**

Julia Beardmore, Department of Biochemistry, Imperial College of Science and Technology, London SW7 2AZ, England **[185]**

Dieter Becker, Botanisches Institut, Universität Münster, Schlossgarten 3, D-4400 Münster, Federal Republic of Germany **[297]**

Con J.M. Beckers, Department of Molecular Biophysics and Biochemistry, Yale University, New Haven, CT 06510 **[333]**

Eric G. Berger, Department of Pharmacology, Biocenter, University of Basle, CH-4056 Basle, Switzerland **[431]**

J.J.M. Bergeron, Departments of Anatomy and Medicine, McGill University, Montreal, Quebec, Canada H3A 2B2 **[391,411]**

D.R. Bevan, Department of Biochemistry, Virginia Polytechnic Institute and State University, Blacksburg, VA 24061 **[421]**

Wendy F. Boss, Department of Botany, North Carolina State University, Raleigh, NC 27695-7612 **[291]**

A.M. Boudet, Centre de Physiologie Végétale, UA CNRS 241, Université Paul Sabatier, 31062 Toulouse Cédex, France **[255]**

W.J. Branch, Department of Clinical Biochemistry, University of Cambridge, Addenbrooke's Hospital, Cambridge CB2 2QR, England **[91]**

Frances M. Brodsky, Becton Dickinson Immunocytometry Systems, Mountain View, CA 94043; present address: Department of Pharmacy, UCSF, San Francisco, CA 94143 **[141]**

Mary Bronson, Department of Cell Biology, Yale University School of Medicine, New Haven, CT 06510 **[129]**

The numbers in brackets are the opening page numbers of the contributors' articles.

J. William Buckie, Department of Pharmacology, University of Cambridge, Cambridge CB2 2QD, England **[243]**

M.I. Burón, Departamento de Biología Celular, Facultad de Ciencias, Universidad de Córdoba, 14071 Córdoba, Spain **[69]**

Claude Cassagne, Institut de Biochimie Cellulaire et Neurochimie, CNRS, 33077 Bordeaux Cédex, France **[303]**

Nancy W. Chege, Department of Biochemistry, Stanford University School of Medicine, Stanford, CA 94305 **[365]**

Geoffrey M.W. Cook, Department of Anatomy, University of Cambridge, Cambridge CB2 3DY, England **[xvii, 243]**

Pierre J. Courtoy, Laboratoire de Chimie Physiologique, International Institute of Cellular and Molecular Pathology and Louvain University Medical School, B-1200 Brussels, Belgium **[169]**

N. Crawford, Department of Biochemistry, Hunterian Institute, Royal College of Surgeons of England, London WC2A 3PN, England **[51]**

Ivan de Curtis, European Molecular Biology Laboratory, 6900 Heidelberg, Federal Republic of Germany **[101]**

Lynne A. Dengler, Department of Botany, North Carolina State University, Raleigh, NC 27695-7612 **[291]**

Clive J. Dix, Advanced Drug Delivery Research, Ciba-Geigy Pharmaceuticals, Horsham, West Sussex, RH12 4AB, England **[441]**

J.J. Doherty II, Departments of Anatomy and Medicine, McGill University, Montreal, Quebec, Canada H3A 2B2 **[391,411]**

Jean-Pierre Draye, Laboratoire de Chimie Physiologique, International Institute of Cellular and Molecular Pathology and Louvain University Medical School, B-1200 Brussels, Belgium **[169]**

William A. Dunn, Department of Cell Biology/Anatomy, Johns Hopkins University School of Medicine, Baltimore, MD 21205; present address: Department of Cell Biology and Anatomy, University of Florida, Gainesville, FL 32610 **[115]**

J. Michael Edwardson, Department of Pharmacology, University of Cambridge, Cambridge CB2 2QD, England **[343]**

Carlos Enrich, Laboratory of Protein Structure, National Institute for Medical Research, London NW7 1AA, England **[155]**

W. Howard Evans, Laboratory of Protein Structure, National Institute for Medical Research, London NW7 1AA, England **[xvii, 155]**

Marilyn Gist Farquhar, Department of Cell Biology, Yale University School of Medicine, New Haven, CT 06510 **[129]**

Torgeir Flatmark, Department of Biochemistry, University of Bergen, N-5000 Bergen, Norway **[247]**

G. García-Herdugo, Departamento de Biología Celular, Facultad de Ciencias, Universidad de Córdoba, 14071 Córdoba, Spain **[69]**

Yukiko Goda, Department of Biochemistry, Stanford University School of Medicine, Stanford, CA 94305 **[365]**

Danièle Godelaine, International Institute of Cellular and Molecular Pathology, B-1200 Brussels, Belgium **[211]**

H. Goldenberg, Department of Medical Chemistry, University of Vienna, A-1090 Vienna, Austria **[427]**

Jean-Pierre Gorvel, Centre de Biochimie et de Biologie Moléculaire, Centre National de la Recherche Scientifique, 13009 Marseille Cédex 9, France **[195]**

Hubert Greppin, Laboratory of Plant Physiology, University of Geneva, 1211 Geneva 4, Switzerland **[285, 289]**

Jean Gruenberg, European Molecular Biology Laboratory, 6900 Heidelberg, Federal Republic of Germany **[77, 317]**

Erik Harms, Kinderklinik und Poliklinic der Technischen, Universität München, 8000 München 40, Federal Republic of Germany **[21]**

Ian F. Hassan, Advanced Drug Delivery Research, Ciba-Geigy Pharmaceuticals, Horsham, West Sussex, RH12 4AB, England **[441]**

Ari Helenius, Department of Cell Biology, Yale University School of Medicine, New Haven, CT 06510 **[21]**

Colin Hopkins, Department of Biochemistry, Imperial College of Science and Technology, London SW7 2AZ, England **[185]**

Michael Hortsch, Cell Biology Program, European Molecular Biology Laboratory, 6900 Heidelberg, Federal Republic of Germany **[211]**

Kathryn E. Howell, European Molecular Biology Laboratory, 6900 Heidelberg, Federal Republic of Germany **[xvii, 77, 317]**

Ann L. Hubbard, Department of Cell Biology/Anatomy, Johns Hopkins University School of Medicine, Baltimore, MD 21205 **[115]**

Stella Hurtley, European Molecular Biology Laboratory, 6900 Heidelberg, Federal Republic of Germany **[355]**

M. Hüttinger, Department of Medical Chemistry, University of Vienna, A-1090 Vienna, Austria **[427]**

Akio Ito, Department of Biology, Faculty of Science, Kyushu University, Fukouoka 812, Japan **[77]**

Hélène Juguelin, Institute de Biochimie Cellulaire et Neurochimie, CNRS, 33077 Bordeaux Cédex, France **[303]**

D.G. Kay, Departments of Anatomy and Medicine, McGill University, Montreal, Quebec, Canada H3A 2B2 **[391]**

T.W. Keenan, Department of Biochemistry, Virginia Polytechnic Institute and State University, Blacksburg, VA 24061 **[421]**

David S. Keller, Department of Molecular Biophysics and Biochemistry, Yale University, New Haven, CT 06510 **[333]**

Hildegard Kern, Kinderklinik und Poliklinic der Technischen, Universität München, 8000 München 40, Federal Republic of Germany **[21]**

M.N. Khan, Departments of Anatomy and Medicine, McGill University, Montreal, Quebec, Canada H3A 2B2 **[391]**

Stefanie Kiefer, MSU-DOE Plant Research Laboratory, Michigan State University, East Lansing, MI 48824 **[289]**

W.H. Lai, Departments of Anatomy and Medicine, McGill University, Montreal, Quebec, Canada H3A 2B2 **[391]**

René Lessire, Institut de Biochimie Cellulaire et Neurochimie, CNRS, 33077 Bordeaux Cédex, France **[303]**

John M. Lucocq, Department of Biochemistry, University of Dundee, Dundee DD1 4HN, Scotland **[431]**

J. Paul Luzio, Department of Clinical Biochemistry, University of Cambridge, Addenbrooke's Hospital, Cambridge CB2 2QR, England **[91]**

Suzanne Maroux, Centre de Biochimie et de Biologie Moléculaire, Centre National de la Recherche Scientifique, 13009 Marseille Cédex 9, France **[195]**

Mark Marsh, Institute of Cancer Research, Chester Beatty Laboratories, London SW3 6JB, England **[21]**

Geoffrey I. McFadden, Plant Cell Biology Centre, School of Botany, University of Melbourne, Parkville 3052, Victoria, Australia **[297]**

Michael Melkonian, Botanisches Institut, Universität Münster, Schlossgarten 3, D-4400 Münster, Federal Republic of Germany **[297]**

Ira Mellman, Department of Cell Biology, Yale University School of Medicine, New Haven, CT 06510 **[21, 35]**

Nita Minnifield, Department of Biological Sciences, Purdue University, West Lafayette, IN 47907 **[417]**

Zohair Mishal, CTRSC, Service de Fluorométrie Analytique et Séparative, Centre National de la Recherche Scientifique, 94800 Villejuif, France **[195]**

Patrick Moreau, Institut de Biochimie Cellulaire et Neurochimie, CNRS, 33077 Bordeaux Cédex, France **[303]**

D. James Morré, Department of Medicinal Chemistry and Pharmacognosy, Purdue University, West Lafayette, IN 47907 **[xvii, 7, 69, 237, 281, 285, 417]**

Susette C. Mueller, Department of Anatomy, Georgetown University Medical Center, Washington, DC 20007 **[115]**

Barbara M. Mullock, Department of Clinical Biochemistry, University of Cambridge, Addenbrooke's Hospital, Cambridge CB2 2QR, England **[91]**

Placido Navas, Departamento de Biología Celular, Facultad de Ciencias, Universidad de Córdoba, 14071 Córdoba, Spain **[69, 237]**

D. David Nowack, Department of Biological Sciences, Purdue University, West Lafayette, IN 47907 **[237]**

George E. Palade, Department of Cell Biology, Yale University School of Medicine, New Haven, CT 06510 **[xix, 77, 129]**

A. Peat, Biology Department, Sunderland Polytechnic, Sunderland SR1 3SD, England **[421]**

Claude Penel, Laboratory of Plant Physiology, University of Geneva, 1211 Geneva 4, Switzerland **[285, 289]**

Suzanne R. Pfeffer, Department of Biochemistry, Stanford University School of Medicine, Stanford, CA 94305 **[365]**

B.I. Posner, Departments of Anatomy and Medicine, McGill University, Montreal, Quebec, Canada H3A 2B2 **[391, 411]**

James G. Pryde, Department of Biochemistry, University of Dundee, Dundee DD1 4HN, Scotland **[431]**

R. Pumm, Department of Medical Chemistry, University of Vienna, A-1090 Vienna, Austria **[427]**

Joël Quintart, Laboratoire de Chimie Physiologique, International Institute of Cellular and Molecular Pathology and Louvain University Medical School, B-1200 Brussels, Belgium **[169]**

Ingo B. Reize, Botanisches Institut, Universität Münster, Schlossgarten 3, D-4400 Münster, Federal Republic of Germany **[297]**

Peter J. Richardson, Department of Clinical Biochemistry, University of Cambridge, Addenbrooke's Hospital, Cambridge CB2 2QR, England **[91]**

Magaly Rincón, Department of Botany, North Carolina State University, Raleigh, NC 27695-7612 **[291]**

James E. Rothman, Department of Biochemistry, Stanford University, Stanford, CA 94305 **[311]**

Keri Safranski, Department of Medicinal Chemistry and Pharmacognosy, Purdue University, West Lafayette, IN 47907 **[417]**

Anna Stina Sandelius, Department of Plant Physiology, Botanical Institute, University of Göteborg, S-41319 Göteborg, Sweden **[281, 285]**

Jaakko Saraste, Department of Cell Biology, Yale University School of Medicine, New Haven, CT 06510 **[129]**

Günther F.E. Scherer, Botanisches Institut, Universität Bonn, Venusbergweg 22, D-5300 Bonn 1, Federal Republic of Germany **[269]**

Sandra L. Schmid, Department of Cell Biology, Yale University School of Medicine, New Haven, CT 06510 **[21, 35]**

Kai Simons, European Molecular Biology Laboratory, 6900 Heidelberg, Federal Republic of Germany **[101]**

C.E. Smith, Departments of Anatomy and Medicine, McGill University, Montreal, Quebec, Canada H3A 2B2 **[391]**

Lutz Thilo, Centre d'Immunologie INSERM-CNRS de Marseille-Luminy, 13288 Marseille Cédex 9, France **[377]**

M. Troulis, Department of Anatomy, McGill University, Montreal, Quebec, Canada H3A 2B2 **[411]**

H.M. Valivullah, Department of Biochemistry, Virginia Polytechnic Institute and State University, Blacksburg, VA 24061 **[421]**

Barbara vom Dorp, Botanisches Institut, Universität Bonn, Venusbergweg 22, D-5300 Bonn 1, Federal Republic of Germany **[269]**

E. Wallner, Department of Medical Chemistry, University of Vienna, A-1090 Vienna, Austria **[427]**

Graham Warren, Department of Biochemistry, University of Dundee, Dundee DD1 4HN, Scotland **[355, 431]**

Glynn Wilson, Advanced Drug Delivery Research, Ciba-Geigy Pharmaceuticals, Horsham, West Sussex, RH12 4AB, England **[441]**

Philip G. Woodman, Department of Pharmacology, University of Cambridge, Cambridge CB2 2QD, England; present address: Department of Biochemistry, The University of Dundee, Dundee DD1 4HN, Scotland **[343]**

Preface

In many areas of contemporary biology, rapid advances have been facilitated by the development of cell-free systems in which the individual steps and constituent processes can be isolated and studied under relatively simple conditions. Cell-free systems provide new opportunities and challenges for the analysis and dissection of membrane based events occurring inside the cell.

Intracellular trafficking is the consequence of the requirement to transport proteins and lipids from their sites of synthesis in the endoplasmic reticulum to their sites of functional residence. This involves the participation of various membrane systems, especially the Golgi apparatus, lysosomes, and specific domains of the plasma membrane. Superimposed on these predominantly exocytotic traffic routes is endocytotic traffic in which internalized receptor/ligand complexes of the cell surface may be rapidly internalized in clathrin-coated structures into the endocytotic compartment. Many of these processes can now be studied profitably in both animals and plants through the use of cell-free systems.

The possibility of reconstituting interactive exocytotic or endocytotic exchanges between two adjacent compartments in a cell-free system was demonstrated by Rothman and colleagues and others and has provided the major impetus for this conference on "Cell-Free Analysis of Membrane Traffic." As detailed by George Palade in a keynote address, the development of cell-free systems depends crucially on the foundations laid by pioneering cell biologists. It has been possible to isolate and characterize microsomes, lysosomes, and other cell particles from homogenates until, finally, every major cell component and most minor ones have been isolated from both animal and plant sources that are relatively free of contamination one from the other. The isolation and the characterization of the major subcellular components and organelles and membranes by methods that exploit size and density parameters have now been extended by application of new techniques that focus on charge (free-flow electrophoresis, phase partitioning), density modification, and specific surface epitopes (immunoisolation). These techniques have led, for example, to the subfractionation of the Golgi apparatus and the characterization of compartments within the endocytotic domain that are implicated in protein sorting, receptor recycling, and the fine control of membrane traffic.

This book collates the contributions of participants at a conference on "Cell-Free Analysis of Membrane Traffic" held at the European Molecular Biology Laboratory October 1–4, 1986. The multidisciplinary nature of the subject has resulted in a volume with four sections. The first concerns separation techniques. These new techniques involve among others, free-flow electrophoresis (Part IA), immunoisolation and development of organelle-specific antibodies (Part IB), and density modification and other isolation techniques (Part IC). Part II concerns membrane traffic in plants. Parts III and IV focus on specific applications to cell-free systems and other new and developing approaches to the study of membrane traffic. In addition to regular contributions, expanded abstracts were solicited from several participants to deal with subjects covered in workshops and poster discussions. These are grouped at the ends of the appropriate section.

We thank the advisors who assisted in program planning and development, especially Ira Mellman, New Haven, CT, USA; Hans Heidrich, Martinsried, FRG; Milan Bier, Tucson, AZ, USA; G. Weber, Bender and Hobein, Munich, FRG; and Neville Crawford, London, UK. We thank Dr. John Tooze and the organization of the European Molecular Biology Laboratory, Heidelberg, FRG, for graciously hosting the conference.

We thank especially Mrs. Janet Sweet for most capable assistance in conference organization and for producing this volume. Special appreciation is extended also to Mrs. Martha Harding for her role as copy editor and for the many corrections, revisions, and improvements that resulted from her painstaking efforts.

Financial support of the United States Air Force Office of Scientific Research, Bolling AFB, Washington, DC, USA; European Office of Aerospace Research and Development, London, UK; and Bender and Hobein, Munich, FRG, is gratefully acknowledged.

Additional support was provided by CIBA-Geigy, Horsham, UK; Protein Technologies, Tucson, AZ, USA; and the Purdue Research Foundation and the Purdue University Cancer Center, West Lafayette, IN, USA.

D.J. Morré
K.E. Howell
G.M.W. Cook
W.H. Evans

Introduction

Cell Fractionation: Importance to Cell-Free Systems Development

The study of membrane traffic in reconstituted, cell-free systems implies the availability of reasonably homogeneous populations of functionally competent subcellular components. It poses, in effect, additional exacting demands on cell fractionation, a technology that, over the years, has been one of the moving forces in the development of modern cellular and molecular biology. Cell fractions are no longer good enough if they provide only an inventory of specific chemicals and functions, however complex. They are expected to perform in vitro complex series of reactions, the final outcome of which is the transport of macromolecules from a donor to a receiving membrane-bound compartment, usually obtained from different cell fractions. This operation assumes the survival of intermembrane recognition systems, the activation of controlled membrane fusion reactions and the transport of content or membrane proteins from one compartment to another. Effective transport is assayed by monitoring specific modifications incurred by the transported protein upon its arrival in the receiving compartment.

Obviously the demands put on cell fractionation are unusually high, but researchers are venturesome spirits and, more important, subcellular components turn out to be unexpectedly sturdy pieces of biochemical equipment. They are still capable of performing the complex operations mentioned above, even after the severe trauma of cell disruption, after suspension for long hours in foreign media "non-iso" in any way, and after hours of being subjected to a variety of forces, most often centrifugal forces, required for their separation.

Given the importance of cell fractionation procedures as a starting technological base in this new research field, it is not surprising, and it is definitely appropriate, to find out that a substantial number of the papers included in this volume deal with current attempts to improve these procedures so as to obtain functional and, if need be, homogeneous populations of subcellular components. The reader is informed about the results obtained in separation procedures based on free-flow electrophoresis, isotachoelectrophoresis, countercurrent distribution, and centrifugal fields applied to particles of modified density. He is also informed about attempts to replace particle sorting based on nonspecific physical parameters (i.e., size, density, surface charge) by methods relying on specific biochemical interactions as done in immunoseparation or in fluorescence-activated particle sorting. The goal is evidently better cell fractions that can measure up to the stepped-up demands of current research. It is clear that there is a lot of ferment and promise in this field. Time will show to what extent that promise will be fulfilled.

Since many subcellular components are still poorly characterized, the volume understandably includes a few papers that identify and characterize markers for subcellular components that until recently were orphans of such attributes. And since a well organized symposium or concert needs it, the volume also includes a few attractive and interesting intermezzos and divertimentos.

After this preliminary but careful preparation, the reader is introduced to a series of remarkable papers that convincingly demonstrate that transport between membrane-bound compartments has been achieved in vitro in reconstituted systems for a variety of normal intracellular partners, such as endoplasmic reticulum (microsomes) and Golgi, different Golgi subcompartments, Golgi elements and plasmalemma, early endosomes and late endosomes, and endosomes and lysosomes.

We may be impressed by these remarkable achievements and we may admire the skill and the daring of the experimenters. Yet, beyond this first reaction, we should realize the real significance of these marvelous exercises. They are the necessary first steps in the construction of assay systems that will define in detail the environmental conditions, energy requirements, mutual recognition elements, and critical agonists that control these remarkable processes. The true excitement in the field is still to come, when delight and amazement will be replaced by true understanding.

George E. Palade
Department of Cell Biology
Yale University School of Medicine
New Haven, Connecticut

Section I
NEW DEVELOPMENTS IN SEPARATION TECHNIQUES
A. FREE-FLOW ELECTROPHORESIS

Cell-Free Analysis of Membrane Traffic, pages 3–6

New Developments in Separation Techniques

Advances in all areas of cell biology have depended on the availability of the appropriate methodology. In this first section of the book, the reader is introduced to the methods that have made possible the analysis of dynamic aspects of membrane traffic.

Classical studies on subcellular fractionation, based on differences in size and density, have enabled enormous advances in the development of cell-free systems for the study of membrane and organelle structure, composition and function. However, in examining the finer details of membrane traffic such studies have been frustrated by components of different origins having both overlapping sizes and densities. These difficulties have been largely overcome by resorting to techniques which utilize other inherent properties of the organelles under study. An example of this latter approach is the application of free-flow electrophoresis, which exploits differences in surface charge, to the analysis of membrane traffic. The investigator wishing to start using this approach may do so confident in the knowledge that reliable equipment is now commercially available, and in part 1A the reader will find practical details on how this technology has been successfully applied to the analysis of membrane traffic. Free-flow electrophoresis has the advantage of being both a preparative method and capable of continuous operation. In the opening chapter, uses of free-flow electrophoresis for separations involving exocytotic compartments are reviewed. With rat liver, for example, the Golgi apparatus can be successfully subfractionated into cis, trans and median compartments using this technique. Schmid and Mellman give details of the fractionation of endosomes from CHO cells and show that functionally and structurally distinct subpopulations can be obtained. Marsh et al. apply the method to cell lysates of BHK-21 cells and find that endosomes and lysosomes co-fractionate but that these components may be efficiently separated by Percoll density gradient centrifugation. In a review of continuous flow electrophoresis, Crawford gives examples of how the method has been applied to fractionating whole cell populations as well as the separation of subcellular organelles. The final article in this section details how plasma membranes from amphibian epidermis may be partially resolved according to activity gradients by free-flow electrophoresis.

An alternative approach to using physical parameters for subcellular fractionation is to exploit specific chemical differences, and in this respect immunoisolation appears to hold the greatest promise. On homogenization all vesicles formed from the same subcellular compartment do not necessarily have the same size, density nor possess the same surface charge. However, as long as all these vesicles have the same accessible epitope, immunoseparation provides an isolation method which circumvents the problems of polydispersity, and for the reader wishing to use such an approach, Part 1B of this volume provides a considerable amount of relevant practical detail. Howell et al. give an overview on the use of antibody-antigen interactions for cell fractionation, discuss the importance of the nature of the solid support in the isolation process and describe how their introduction of magnetic solid supports provides for a rapid and efficient retrieval of immunoisolated subcellular components. Luzio et al. stress the advantages of indirect immunoadsorption, where an antibody binding reagent is covalently bound to the solid phase, for subcellular fractionation. Practical advice is presented on the selection of antibodies and solid supports and on the optimization of the indirect immunoaffinity procedure. De Curtis and Simons give an example of how by combining density gradient centrifugation with immunoisolation procedures it is possible to obtain a fraction enriched in the trans-Golgi network. As these authors point out, the purification of a trans-Golgi enriched fraction is an important prerequisite for constructing an _in vitro_ assay to study the sorting and budding of vesicles destined for the plasma membrane of the cell. Hubbard et al. describe an immunoadsorption protocol applicable to relatively crude subcellular fractions, and useful in kinetic studies of endocytosis. In their chapter, Saraste et al. address the question of developing markers for the subfractions of the Golgi apparatus. Surveying previous attempts to produce anti-Golgi antibodies, these authors describe their strategy for raising polyclonal antibodies to material enriched in cis Golgi elements. They succeeded in generating an antibody recognizing a 58K protein that is restricted to cis Golgi cisternae. Up to now, there have been no known markers for cis Golgi elements. In the concluding contribution to this section of the book, Brodsky reviews the production of a panel of anti-clathrin monoclonal antibodies. Investigators will find this summary article on clathrin-coated vesicles useful as providing an example of how monoclonal antibodies may be used for studying organelles involved in intracellular traffic. Clearly it should be possible to employ the approaches reviewed by this author for the study of the structure, assembly and functions of other membrane organelles.

Part 1C of this volume concludes the discussion of isolation procedures with methods based on density modification, phase partition, and other complementary techniques. Evans and Enrich describe the isolation by centrifugation in isoosmotic Nycodenz gradients of membrane vesicular subfractions originating from the endocytic compartment of rat liver. On the basis of biochemical analysis these authors find appreciable compositional commonality between purified endosomes and plasma membranes, indicating that the endocytic compartment may comprise an "intracellular" domain of the surface membrane. The application of density modification methods for fractionating endosomes and rough endoplasmic reticulum is described by Courtoy et al., Beardmore and Hopkins, and Amar-Costesec et al. Endosome-specific isolation methods considered include the DAB shift procedure (Courtoy et al.), in which after internalization of ligand - HRP conjugates, the use of a DAB medium causes the selective accumulation of DAB-reaction product which results in a considerable increase in equilibrium density; and the use of colloidal gold monoclonal antibody complexes (Beardmore and Hopkins). Phase partition, a separation method based upon differences in the surface properties of the cell particles, is reviewed by Albertsson who notes that this method is especially efficient for the isolation of plasma membranes. Navas et al. give details of how this technique may be successfully applied as a rapid and facile alternative to gradient procedures for the isolation of plasma membranes. Flow cytometry has been applied by Gorvel et al. to the characterization of membrane vesicles. The sensitivity of the procedure is such that it is possible to detect an antigen present in small amounts on the membrane surface; however, immunoprecipitated vesicles cannot be analyzed by this method. Buckie and Cook show how contemporary interest in the avidin (Streptavidin) - biotin system for isolating biological molecules may be exploited for the specific isolation of cell surface molecules. Flatmark describes a simple and convenient method for the determination of the average sedimentation coefficient of subcellular particles which should prove useful to investigators wishing to design specific fractionation schemes for the optimal recovery of a particular organelle. Further progress in development of reconstituted cell-free systems is expected to be aided substantially by these several methodological advances in cell fractionation and analyses detailed in Part I.

Cell-Free Analysis of Membrane Traffic, pages 7-19

FREE-FLOW ELECTROPHORESIS: PREPARATIVE APPLICATIONS TO CELL-FREE ANALYSES OF EXOCYTOTIC MEMBRANE TRAFFIC

D. James Morré

Department of Medicinal Chemistry and Pharmacognosy, Purdue University, West Lafayette, IN 47907

INTRODUCTION

Beginning with the pioneering discoveries of Albert Claude, Philip Siekevitz and George Palade, methods of cell fractionation have been instrumental in the development of cell-free systems to study membrane and organellar structure, composition and function. Opportunities for decisive separations afforded by differences in size and density (s-rho space concept of Anderson and coworkers; Anderson, 1966), have been considerable. However, some applications, i.e. subfractionations of liver Golgi apparatus along its functional axis (Merritt and Morré, 1973) and endosome isolation (Evans and Enrich, this volume) proved frustrating due to components of different origins having both overlapping sizes and densities such that a need for additional separation parameters may be indicated. The use of imposed labels such as cytochemical products offered one approach (e.g. Courtoy, this volume). Differences in surface charge offered another.

Surface charges have been used to effect membrane separations in different ways including phase partitioning between solutions of different polymer content (Albertsson, this volume) and isoelectric focusing (Just et al., 1975). This paper reviews some recent applications of free-flow electrophoresis, a technique developed by Hannig, Heidrich and coworkers (Hannig and Heidrich, 1977), with emphasis on separations involving exocytotic compartments. Applications to endocytotic compartments are treated elsewhere in this volume in the chapter by Marsh et al.

In the technique of free-flow electrophoresis, the mixture of components to be separated is introduced as a fine jet into a separation buffer moving perpendicular to the field lines of an electric field (Fig. 1). Membranes bearing different electrical charge densities will migrate different distances across the separation chamber and thus may be resolved. The method is preparative and capable of continuous operation. The recoveries of membranes and proteins are high (about 90%). Separations are at 4° C, vesicle integrity is maintained (vesicles remain sealed), and most enzymatic activities are retained at or near initial levels (see Minnifield and Morré, this volume).

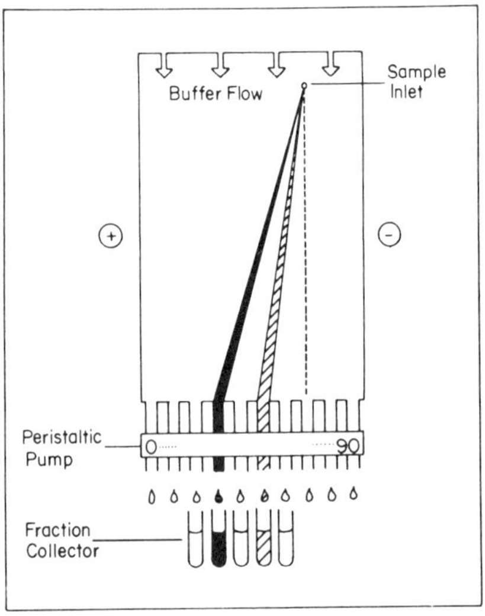

Figure 1. Principles of preparative continuous free-flow electrophoresis. Redrawn and reprinted with permission from Bender & Hobein, Munich, FRG.

For most separations involving animal membranes the electrophoresis medium (chamber buffer) contains 10 mM triethanolamine, 10 mM acetic acid, 5 mM glucose, 0.25 M sucrose and 0.5 mM $MgCl_2$, final pH 6.5 (1 N NaOH), osmolarity 270 mosm, conductivity 5.9 x 10^2 moho. The electrode buffer contains 100 mM triethanolamine and 100 mM acetic acid, pH 6.5. Electrophoresis conditions are 167 mA (constant current), 131 $\pm$ 10% V/cm, buffer flow 2.75 ml per fraction per h, sample injection 3.5 ml/h, temperature 6° C. The electrophoresis equipment used were VAP-5, VAP-21 or VAP-22 continuous free-flow electrophoresis units (Bender & Hobein, Munich, FRG).

SUBFRACTIONATION OF RAT LIVER GOLGI APPARATUS

Male Holtzman rats (ca. 250 g), provided a standard diet and drinking water ad libitum, were sacrificed by cervical dislocation and drained of blood. Livers (10-13 g) were excised and minced with razor blades.

To isolate intact Golgi apparatus as the starting material, the rapid procedure of Morré et al. (1972) was used. Livers were homogenized in 2 volumes of a grinding medium containing 37 mM Tris-maleate pH 6.4, 0.5 M sucrose, 1% dextran (Sigma; average molecular weight 225,000 d), 5 mM $MgCl_2$ and 5 mM mercaptoethanol (Morré et al., 1969; Morré, 1971). Homogenization was for 40 sec at 10,000 rev/min using a Polytron homogenizer (20 ST, Kinematica, Lucerne, Switzerland). The homogenates were centrifuged at 6,000 X g for 15 min, and the upper yellow-brown portion of the pellet was resuspended in a small amount of the electrophoresis chamber buffer. From a total volume of 6 ml, 2 ml aliquots were overlaid onto 3.4 ml portions of 1.2 M sucrose contained in 5.4 ml cellulose nitrate tubes, and centrifuged for 30 min at 100,000 X g in a swinging bucket rotor (Spinco SW 50.1). The supernatant was removed with a Pasteur pipette and the Golgi apparatus "rug" was carefully removed from the top of the 1.2 M sucrose layer, taking care to exclude, to the extent possible, any of the 1.2 M sucrose. The Golgi apparatus from the three gradient tubes were combined and the volume was brought up to 4-5 ml with fresh chamber buffer.

Prior to electrophoresis, the Golgi apparatus were unstacked as diagrammed in Figure 2. For this purpose, the resuspended intact Golgi apparatus were incubated for 45 min at 4°C with a mixture containing 3 mg each of crude α-amylase Type X-A from _Aspergillus_ _oryzae_ (Sigma; 132 units per mg protein) and α-amylase Type III-A from barley malt (Sigma; 1.7 units α-amylase/mg solid and 2.7 units β-amylase per mg solid). The function of the dextran added to the isolation medium is to prevent the Golgi apparatus from unstacking during their isolation. Therefore, one function of the unstacking enzymes is to aid in removing the added dextran so that unstacking may proceed. Unstacking is completed at the end of the incubation by drawing the suspension approximately 40 times in and out of a Pasteur pipette with an inside diameter of about 1 mm at the tip. The resulting mixture is used for the electrophoretic separation. Procedures for assay of marker enzymes and a more complete characterization of the fractions obtained are given by Morré et al. (1983, 1984), Navas et al. (1986) and Minnifield et al. (1986).

In the electrophoretic separation diagrammed in Figure 1, the most electronegative fractions migrate toward the anode located at the left. The samples after separation are collected through an assembly of 90 exit ports leading to a fraction collector with 90 different tubes. Normally, the Golgi apparatus subfractions are contained in about 10 tubes between fraction 30 (the most electronegative with the greatest migration toward the anode) and fraction 40 (the least electronegative with the smallest migration from the point of sample injection (Fig. 3)).

That separation of different membrane types has actually taken place is illustrated by the marker constituents presented in Fig. 3. Total sialic acid, a constituent that increases across the stacked Golgi apparatus from cis to trans (Rambourg et al., 1969), and sialyl- and galactosyl-transferase activities, both concentrated in cisternae of the trans face by enzyme

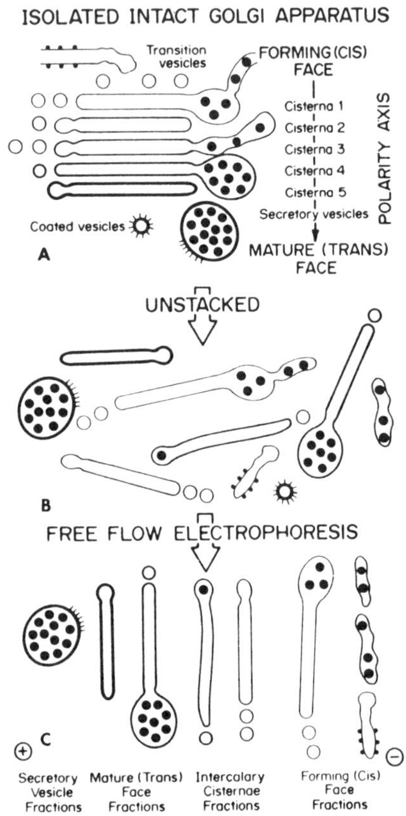

Figure 2. Diagrammatic representation of Golgi apparatus subfractionation by free-flow electrophoresis. A. The various components of the Golgi apparatus of the liver cell as displayed in cross-sectional representation from the cis to the trans face across the polarity axis. Differentiation of membranes from ER-like to plasma membrane-like is indicated by an increasing line width of the membranes. B. The same structure after unstacking into individual cisternal components. C. Resorting of the different structures by free-flow electrophoresis along the same polarity axis as seen in the starting material. By providing isolated Golgi apparatus subfractions derived

from membranes predominantly from one face of the structure or the other, the technique can help elucidate molecular details of Golgi apparatus function in membrane flux and the processing of secretory products and membrane proteins.

---

immunocytochemistry and/or cytochemistry (Roth and Berger, 1982; Matyas and Morré, 1983; Roth et al., 1985), are found associated with the most electronegative fractions to the left of the electrophoretic separation(Fig. 3). In contrast, nucleoside diphosphatase with UDP as substrate and NADH-cytochrome c reductase, putative cis face constituents, show the opposite distributions with greatest activities in the least electronegative fractions (Fig. 3). Thus, according to these activities and other markers such as the enzyme thiamine pyrophosphatase, localized to the trans Golgi apparatus in cytochemical analyses, the subfractionation of the Golgi apparatus achieved by free-flow electrophoresis is from cis to trans with the trans-most cisternae to the left and the cis-most cisternae to the right.

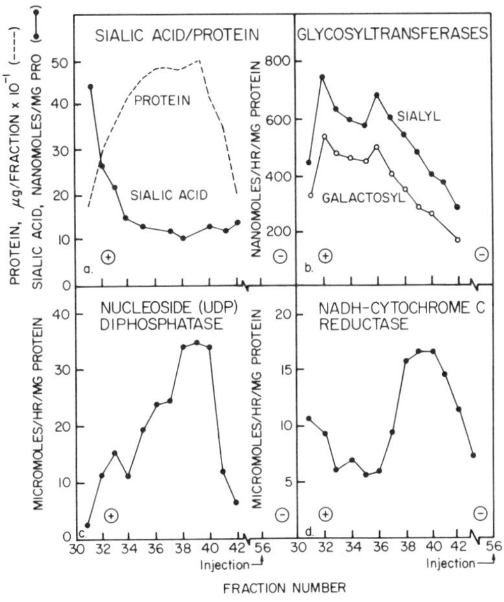

Figure 3. Electrophoretic separations of Golgi apparatus of rat liver comparing cis (fractions 36-39) to trans (fractions 31-33) separations of various marker constituents. A. Total sialic acid. B. Glycosyl-transferases. C. Nucleoside diphosphatase (UDP) as substrate. D. NADH cytochrome c reductase. From Morré et al. (1984)

To mark intermediate cisternae, we have used NADPase (Navas et al., 1986) which, under our conditions, is located in median portions of the stacked cisternae and low or absent from both the cis-most and trans-most Golgi apparatus elements. This activity is found associated with the mid-portions of the electrophoretic separations.

Not all Golgi apparatus activities show clear cis to trans gradients. With Golgi apparatus constituents involved in trafficking to lysosomes (β-hexosaminidase, the high mannose oligosaccharide processing enzyme, α-mannosidase II, the two enzymes involved in the biosynthesis of the phosphomannosyl recognition marker, and the phosphomannosyl receptor itself), a pattern was observed where specific activity or amount decreased from cis to intermediate (Minnifield et al., 1986). However, these activities once more increased in the most electronegative fractions, i.e. those cisternae, vesicles and tubules derived from the trans-most Golgi apparatus elements. These are exemplified by results for the phosphomannosyl receptor (Fig. 4). The distribution was characterized by two peaks, one cis and one trans. When these determinations were repeated with membranes pretreated with mannose-6-phosphate to displace bound lysoso-

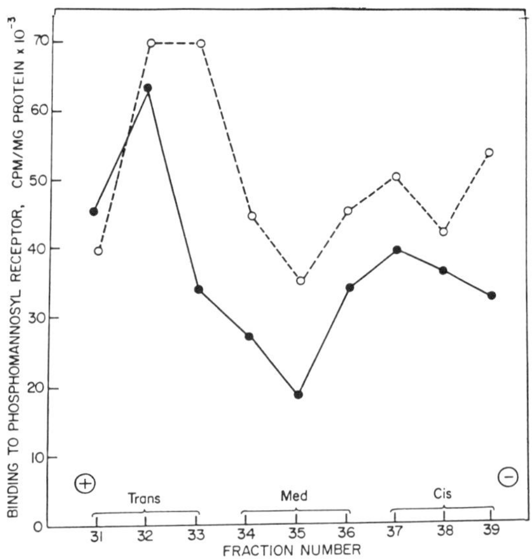

Figure 4. Binding of $[^{125}I]$-β-glucuronidase to measure phosphomannosyl receptor of Golgi apparatus subfractions prepared by free-flow electrophoresis for two different separations. Experimental details are in Minnifield et al. (1986).

mal enzymes prior to binding of the radioiodinated β-glucuronidase, similar results were obtained. Thus, degree of receptor occupancy was not a major factor contributing to the unusual pattern of receptor distribution observed with free-flow electrophoresis. These findings, too, are consistent with the cytochemical findings of both Brown and Farquhar (1984) and of Geuze et al. (1984) with hepatocytes. Thus, it appears that Golgi apparatus subfractions prepared by continuous free-flow electrophoresis provide a reliable basis for examination of the cis to trans disposition of various activities involved in membrane flux, membrane recycling and the processing of secretory products and membrane proteins across the stacked cisternae that characterize the Golgi apparatus of the intact cell.

OTHER APPLICATIONS TO EXOCYTOTIC COMPARTMENTS

Purification of Endoplasmic Reticulum (ER) and Isolation of Transition Vesicles

Frequent contaminants of ER fractions include mitochondria, putative endocytotic compartments, Golgi apparatus fragments and coated and/or transition vesicles and elements. Coats may be clathrin derived or nap-like and not clathrin derived (Croze et al., 1982). Free-flow electrophoresis allows both purification of ER and isolation of minor ER-derived subfractions such as transition elements and vesicles. A typical separation profile is given by Minnifield and Morré (this volume).

Fractionation of Cultured Cells

Homogenates of cultured cells tend to aggregate due to abundant cytoskeletal components. Free-flow electrophoretic separations tend to disrupt such aggregates and give separations (Figs. 5-8) amenable to further fractionation as described in Figure 9.

Preparation of Plasma Membrane Subfractions

Important applications of free-flow electrophoresis

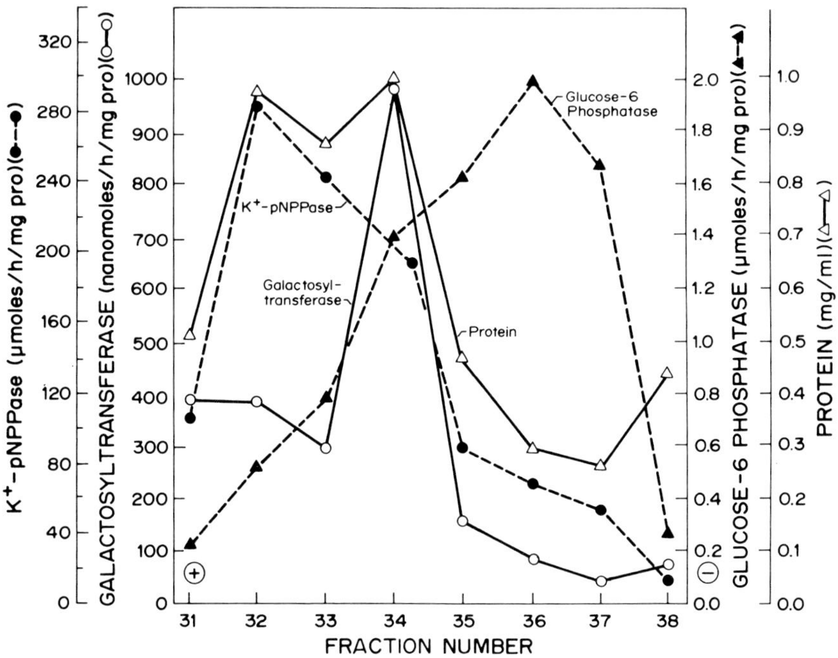

Figure 5. Electrophoretic separations of RLT-28 rat hepatoma cells homogenized with glass/Teflon in hypotonic medium. Glucose-6-phosphatase marks endoplasmic reticulum, galactosyltransferase marks Golgi apparatus and potassium stimulated-, ouabain inhibited-p-nitrophenylphosphatase (K^+-pNPPase) marks plasma membrane. Data of P. Navas.

are for separation of plasma membrane fractions of differing charge. Included are resolution of right side out and inside out red blood cell ghosts (Heidrich and Leutner, 1974), plasma membrane vesicles from amphibian epidermis apparently according to position in a developmental gradient (Buron et al., this volume) and for plasma membrane and tonoplast vesicles from plant cells (Sandelius et al., 1986, this volume; Auderset et al., 1986, this volume; Scherer et al., this volume). Both plasma membrane and tonoplast appear to be resolved into two populations depending on absolute orientation (right side vs inside out, Morré et al., in press).

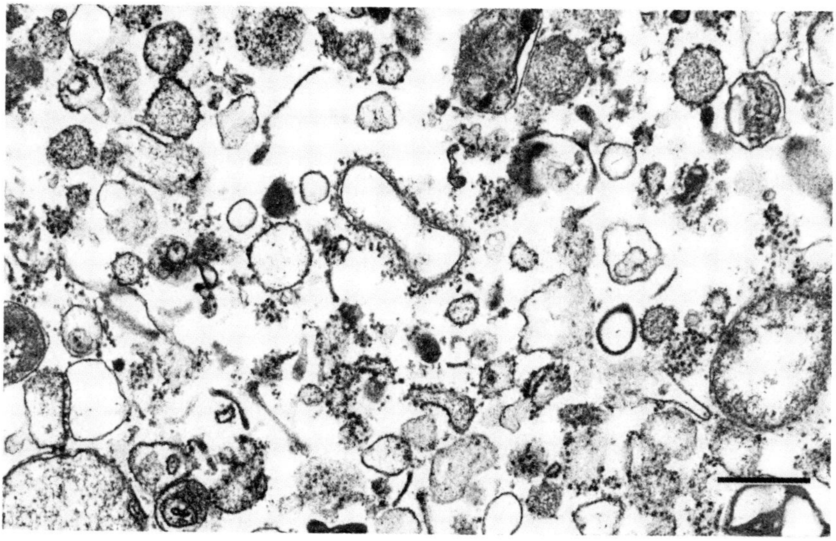

Figure 6. Electron micrograph of electrophoresis fraction 36 enriched in endoplasmic reticulum vesicles identified by polyribosomes of the outer membrane surfaces. Bar = 1 μm.

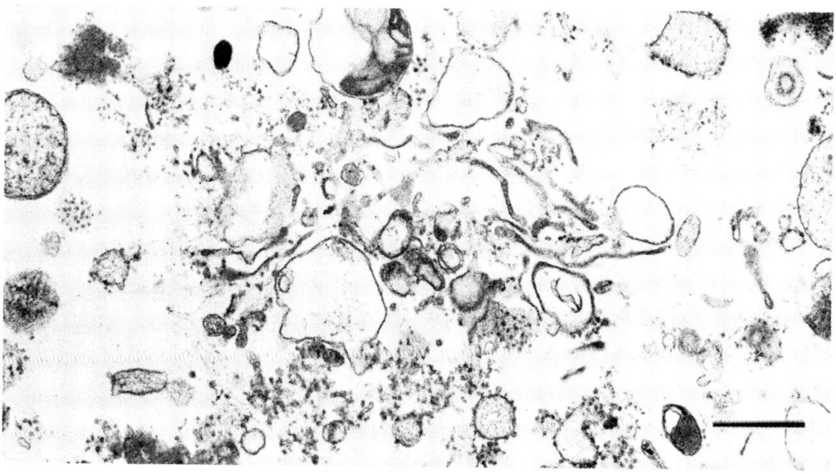

Figure 7. Electron micrograph of electrophoresis fraction 34. Golgi apparatus elements coincided with the peak in galactosyltransferase specific activity. Bar = 1 μm. Mitochondria and mitochondrial fragments occur throughout the electrophoretic separation while nuclei and nuclear fragments tend to occupy the more electronegative fractions.

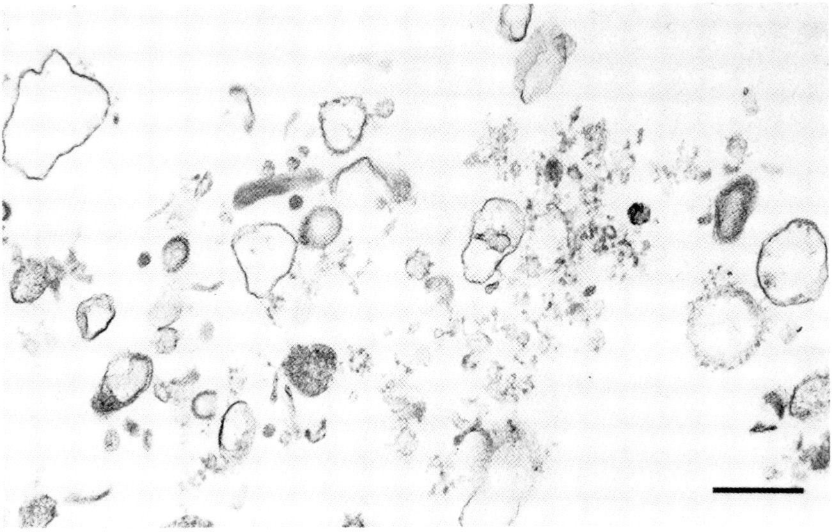

Figure 8. Electrophoresis fraction 32 contains principally vesicles of apparent plasma membrane origin. Scale marker = 0.5 μm.

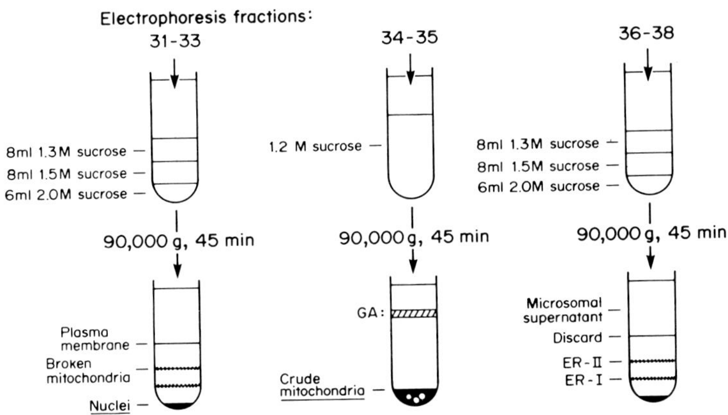

Figure 9. Procedure for subsequent purification of electrophoretic fractions from homogenates of cells in culture. Conventional sucrose gradients are employed as diagrammed. Golgi apparatus are enriched but not highly purified.

Purification of Dictyosome-like Structures (DLS) From Guinea Pig Spermatocytes

Germ cells of many mammalian species including man (Mollenhauer and Morré, 1984) contain structures that resemble the stacked cisternae (dictyosomes) of the Golgi apparatus but differ from Golgi apparatus in structural details and lack galactosyltransferase (Morré et al. 1985) as well as sulfate and sialic acid. The latter contributes to a lesser electronegativity for DLS to permit separation of DLS from Golgi apparatus by free-flow electrophoresis.

SUMMARY

Preparative free-flow electrophoresis provides a useful adjunct to conventional centrifugation methods with important applications to isolation of exocytotic and endocytotic compartments. With rat liver, Golgi apparatus subfractionation into cis, trans and median compartments is achieved. Simultaneous isolation of plasma membrane-, endoplasmic reticulum-, Golgi apparatus, mitochondria-, nuclear envelope- and possibly lysosome-enriched fractions is obtained with homogenates from cultured cells. Other applications include subfractionation of plasma membrane vesicles of differing charge (including separation of inside out and right side out vesicles), purification of endoplasmic reticulum, isolation of tonoplast (vacuole membranes) from plants and separation of Golgi apparatus membranes and dictyosome-like structures of spermatocytes. Of particular importance to cell-free systems for study of membrane trafficking are recent successes with preparative free-flow electrophoresis to obtain preparations of transition vesicles (those with nap-like but not clathrin-derived coats, coming from endoplasmic reticulum and thought to be responsible for delivery of materials to cis Golgi apparatus elements).

Work supported by grants from the NIH CA18801, CA 36761, and HD11908.

REFERENCES

Anderson NG (1966). The development of zonal centrifuges and ancillary systems for tissue fractionation and analysis, Bethesda: National Cancer Institute Monograph 21:253-269.

Auderset G, Sandelius AS, Penel C, Brightman A, Greppin H, Morré DJ (1986). Isolation of plasma membrane and tonoplast fractions from spinach leaves by preparative free-flow electrophoresis and effect of photoinduction. Physiol Plant 68:1-12.

Brown WJ, Farquhar MG (1984). The mannose-6-phosphate receptor for lysosomal enzymes is concentrated in cis Golgi cisternae. Cell 36:295-307.

Croze EM, Morré DJ, Morré DM, Kartenbeck J, Franke WW (1982). Distribution of clathrin and spiny-coated vesicles on membranes within mature Golgi apparatus elements of mouse liver. Eur J Cell Biol 28:130-138.

Geuze HJ, Slot JW, Strous GJAM, Hasilik A, von Figura K (1984). Ultrastructural localization of the mannose 6-phosphate receptor in rat liver. J Cell Biol 98:2047-2054.

Hannig K, Heidrich H-G (1977). Continuous free-flow electrophoresis and its application in biology. In Bloemendal H (ed): "Cell Separation Methods, Part IV. Electrophoretic Methods," Amsterdam: Elsevier-North-Holland Biomedical Press, pp 93-116.

Heidrich H, Leutner G (1974). Two types of vesicles from the erythrocytes ghost membrane differing in surface charge: Separation and characterization by free-flow electrophoresis. Eur J Biochem 41:37-43.

Just WW, Leon-V JO, Werner G (1975). Isoelectric focusing in continuous-flow electrophoresis. I. Separation of mixed red blood cells of different species. Anal Biochem 67:590-601.

Matyas GR, Morré DJ (1983). Coupling of uridine-5'-diphosphate (UDP) formation and nicotinamide adenide dinucleotide (NAD^+) reduction for cytochemical localization of glycosyltransferases. J Histochem Cytochem 31:1175-1182.

Merritt WD, Morré DJ (1973). A glycosyl transferase of high specific activity in secretory vesicles derived from Golgi apparatus of rat liver. Biochim Biophys Acta 304:397-407.

Minnifield N, Creek KE, Navas P, Morré DJ (1986). Involvement of cis and trans Golgi apparatus elements in the intracellular sorting and targeting of acid hydrolases to lysosomes. Eur J Cell Biol 42:92-100.

Mollenhauer HH, Morré DJ (1984). Dictyosome-like structures present in human testicular cells. Cell Biol Intern Repts 8:227-229.

Morré DJ (1971). Isolation of Golgi apparatus. Meth Enzymol 22:130-148.

Morré DJ, Merlin LM, Keenan TW (1969). Localization of glycosyltransferase activities in a Golgi apparatus-rich fraction isolated from rat liver. Biochem Biophys Res Commun 37:813-819.

Morré DJ, Cheetham RD, Nyquist SE, Ovtracht L (1972). A simplified procedure for isolation of Golgi apparatus from rat liver. Prep Biochem 2:61-69.

Morré DJ, Morré DM, Heidrich H-G (1983). Subfractionation of Golgi apparatus by free-flow electrophoresis. Eur J Cell Biol 31:263-274.

Morré DJ, Creek KE, Matyas GR, Minnifield N, Sun I, Baudoin P, Morré DM, Crane FL (1984). Free-flow electrophoresis for subfractionation of rat liver Golgi apparatus. Bio-Tech 2:224-233.

Morré DJ, Matyas GR, Mollenhauer HH (1985). Dictyosome-like structures from guinea-pig testes lack galactosyltransferase, a Golgi apparatus marker. Cell Tissue Res 240:35-40.

Morré DJ, Brightman A, Scherer G, vom Dorp B, Penel C, Auderset G, Sandelius AS, Greppin H (in press). Highly purified tonoplast fractions by preparative free-flow electrophoresis. Proc 1st Intern Workshop Plant Vacuoles, Sophia-Antipolis, France, July 6-11, 1986.

Navas P, Minnifield N, Sun I, Morré DJ (1986). NADP phosphatase: A marker in free-flow electrophoretic separations for cisternae of the Golgi apparatus midregion. Biochim Biophys Acta 881:1-9.

Rambourg A, Hernandez W, LeBlond DP (1969). Detection of complex carbohydrates in the Golgi apparatus of rat cells. J Cell Biol 40:395-414.

Roth J, Berger EG (1982). Immunocytochemical localization of galactosyltransferase in HeLa cells: Codistribution with thiamine pyrophosphatase in trans-Golgi cisternae. J Cell Biol 92:223-229.

Roth J, Tastjes DJ, Lucocq JM, Weinstein J, Paulson JC (1985). Demonstration of an extensive trans-tubular network continuous with the Golgi apparatus stack that may function in glycosylation. Cell 43:287-295.

Sandelius AS, Penel C, Auderset G, Brightman A, Millard M, Morré DJ (1986). Isolation of highly purified fractions of plasma membrane and tonoplast from the same homogenate of soybean hypocotyls by free-flow electrophoresis. Plant Physiol 81:177-185.

Cell-Free Analysis of Membrane Traffic, pages 21–33

CO-FRACTIONATION OF BHK-21 CELL ENDOSOMES AND LYSOSOMES BY FREE-FLOW ELECTROPHORESIS

Mark Marsh, Hildegard Kern+, Erik Harms+, Sandra Schmid*, Ira Mellman* and Ari Helenius*

Institute of Cancer Research, Chester Beatty Laboratories, Fulham Road, London SW3 6JB, England
+ Kinderklinik und Poliklinic der Technischen Universitat Munchen, Kolner Platz 1, 8000 Munchen 40, Federal Republic of Germany.
* Department of Cell Biology, Yale University School of Medicine, 333 Cedar Street, New Haven, Connecticut 06510, U.S.A.

INTRODUCTION

Endosomes are the primary sorting organelles of the endocytic pathway and important sites for the control of endocytic membrane cycling and receptor-mediated endocytosis. Markers, ligands and receptors internalized from the plasma membrane, by fluid phase and receptor-mediated endocytosis, are rapidly delivered to the endosomes and subsequently sorted to specific subcellular destinations, including the lysosomes, Golgi apparatus and domains of the plasma membrane (Goldstein et al., 1985). These functions are, at least in part, influenced by the acidification of the endosomal compartment through the activity of a H^+-ATPase (Mellman et al., 1986).

To understand the function of endosomes in more detail, and to characterise the composition, biochemical properties and biogenesis of these important components, we have attempted to isolate and purify the organelles to homogeneity. In attempting this isolation two major considerations were taken into account. Firstly, in order to study the functional properties of the isolated organelles we have attempted to modify their structure as little as possible. Secondly, efforts to isolate the organelles by conventional differential and gradient centrifugation techniques have been frustrated by the similar buoyant densities of endosomes with other smooth membraned organelles such as the Golgi complex and plasma membrane. We have therefore sought alternative physical parameters by which purification may be achieved. Here we describe the fractionation of baby hamster kidney (BHK-21) cell lysates by free flow electrophoresis and demonstrate the use of this technique in the isolation of functional endosomes.

MATERIALS AND METHODS

All reagents were from Sigma unless stated otherwise. Baby hamster kidney (BHK-21) cells were grown on 35mm tissue culture dishes or 850cm^2 roller bottles (Corning) in Glasgow's modified Eagle's medium (G-MEM) containing 5% foetal calf serum as previously described (Marsh et al., 1983).

Endosome Labelling

To label endosomes, confluent roller bottles of cells were incubated with fluid phase markers for 15min at 37°C, or 2hr at 20°C as previously described (Marsh et al., 1986). The markers used were dialysed fluorescein isothiocyanate (FITC) conjugated to dextran (FITC-dextran) at 10mg/ml or horse radish peroxidase (HRP) at 10mg/ml in Dulbecco's modified MEM.

Cell Surface Labelling with SFV

^{3}H-mannose-labelled SFV was bound to confluent BHK-21 cell monolayers for 1hr on ice in MEM (pH6.8). The cells were then washed with cold MEM and the cell-associated viruses fused with the plasma membrane by incubation in MEM at pH5.5 for 0.5min at 37°C (White et al., 1980). The cells were replaced on ice and washed with cold pH7.4 MEM. Unfused virus was removed by incubating the cells with 10mg/ml proteinase K in PBS for 10min at 0°C and the cells (which were dislodged from the dish by the protease treatment) were washed three times by centrifugation and resuspension in cold PBS containing 1% aprotinin.

Cell Homogenization

All manipulations and solutions were at 0-4°C unless otherwise indicated. BHK-cell monolayers were rinsed at least four times with cold Ca^{++}/Mg^{++}-free PBS and harvested using a rubber scraper. The cells were collected by centrifugation at 500 x g (5min). The cell pellet was carefully resuspended in 50ml of TEA buffer (0.25M sucrose, 10mM triethanolamine, 10mM acetic acid, 1mM EDTA, pH7.4) and collected by centrifugation at 750 x g for 10min at 4°C (Harms et al., 1980). The resulting pellet was resuspended in TEA-buffer (approximately 0.5ml per roller bottle) and left over ice for 5min. The cells were passed up and down through a 10ml glass pipette (1.5mm aperture) 10 times (Harms et al., 1980). The nuclei and non-lysed cells were removed by centrifugation at 750 x g for 10min and the pellet washed with 2ml TEA-buffer. The supernatants were pooled and centrifuged for 10min each at 850 and 950 x g to derive the postnuclear supernatant. Large aggregates were removed from the post nuclear supernatant by filtration through a double layer of prewashed Schleicher and Schull No. 574 filter paper.

The filtered post nuclear supernatants were applied to a Bender Hobein Elpho VAP 11 free flow electrophoresis machine either with or without TPCK-trypsin treatment (see text).

The chamber buffer was TEA buffer as above and the electrode buffer was 100mM triethanolamine, 100mM acetic acid, 10mM EDTA pH7.45 in 0.25M sucrose.

The machine was run at 5°C at a constant current of 210-220mA and field of 1400-1440 volts. The collector was set to fractionate the curtain at approximately 4.3ml/fraction/hour and the sample was perfused into the chamber at 3.3ml/hr.

Fraction Analysis

Fractions were analysed for HRP using o-dianisidine as described (Marsh et al., 1987). FITC-dextran and radio-labelled SFV were detected spectrofluorometrically and by scintillation counting respectively. Beta-hexosaminidase, INT-succinate reductase, cytochrome C-reductase, galactosyl transferase and 5'-nucleotidase activities were measured as previously described (Harms et al., 1980; Marsh et al., 1987). Protein was determined by the method of Bradford using Bio-Rad reagents (Bradford, 1976).

Acidification was assayed spectrofluorometrically, as described (Galloway et al., 1983). Endosomes (0.25ml) were diluted into 2.5ml buffer containing 10mM Hepes (adjusted to pH7.4 with KOH), 5mM $MgCl_2$, 2mM EDTA and 125mM KCl. Fluorescence was measured at emission and excitation wavelengths of 515nm and 485nm respectively. Carbonyl-cyanide p-trifluoromethoxy phenylhydrazome (FCCP) was used at a final concentration of 18μM and was dissolved in dimethyl sulphoxide (DMSO).

RESULTS

Previously we have used free flow electrophoresis (FFE) for analytical and preparative cell fractionation. Highly purified lysosomes have been isolated from rat liver, human lymphoblastoid cells and human fibroblast cell lines using FFE (Harms et al., 1980, 1981; Reggio et al., 1984), and the transport of endocytosed Semliki Forest virus (SFV) particles into the lysosome compartment has been followed in BHK-cells (Marsh et al., 1982). Here we determined the properties of BHK-21 cell endosomes subjected to FFE.

Co-fractionation of Endosomes and Lysosomes in Free Flow Electrophoresis

BHK-21 cell endosomes were labelled with a fluid phase marker, FITC-dextran, for 2hr at $20^{o}C$ and subsequently washed extensively with ice cold PBS. Under these conditions the label is taken into the endosome compartment exclusively (Marsh et al., 1983). It is not transferred to the lysosomes and appears to label almost the entire endosome compartment (Marsh et al., 1986; Griffiths et al., 1987). Labelled BHK-21 cells were harvested as described, and the cell pellet lysed by repeatedly (10 times) drawing the cells into, and expelling them from, a 10ml glass pipette, with a 1.5mm orifice. The procedure lysed 85-90% of the cells while >90% of the nuclei remained intact. In general approximately 80% of the FITC-dextran, and 85-90%

of the beta-hexosaminidase activities in the post nuclear supernatant remained in a 100KxG sedimentable form, indicating that only 20% of the endosomal marker and 10-15% of the lysosomal marker, respectively, were released from disrupted organelles during the lysis procedure. The nuclei, non-lysed cells and aggregates were removed by centrifugation, and the post nuclear supernatant used for electrophoresis.

We previously found that the separation of BHK-21 cell lysosomes from cell lysates was enhanced by briefly treating the post nuclear supernatant with trypsin (Marsh et al., 1982). For the initial separation of endosomes the samples were also trypsinized. The post nuclear supernatants were made 10μg/ml with TPCK-trypsin and incubated at 37°C. After 3 mins the samples were returned to ice and the trypsin activity inhibited by addition of soy bean trypsin inhibitor (SBTI) to a final concentration of 10μg/ml. The lysate was filtered once to remove aggregates and introduced into the FFE fractionation chamber through the injection port nearest to the cathode.

During transit through the chamber the sample, which appeared as an opaque stream, was deflected across the chamber towards the anode. Furthermore, the initial single stream was clearly observed to split into two streams, the peaks of which were separated by approximately 12 fractions (Fig.1). Analysis of the fractions indicated that the majority of the protein was present in the slower migrating, cathodal peak. Together with this major protein peak was the INT-succinate reductase activity, a marker for mitochondria. A second smaller protein peak was found further towards the anode. This second peak co-migrated with a major peak of beta-hexosaminidase activity and the bulk of the FITC-dextran, marking the lysosomes and endosomes respectively, and each was enriched 15-20 fold over the initial cell associated activity.

The faster moving peak appeared to contain only the endosomes and lysosomes. The markers for other organelles were found associated with the major protein peak. In addition to the mitochondrial markers, these included enzyme markers for the Golgi apparatus (galactosyl transferase) and endoplasmic reticulum (cytochrome c reductase), and radioactive markers for the plasma membrane. In the latter case we used the fusogenic properties of Semliki Forest

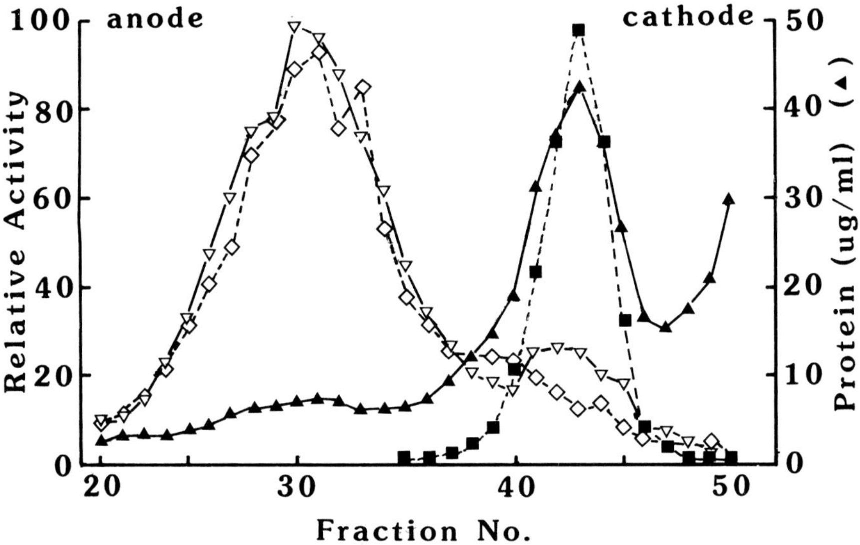

Figure 1. Free flow electrophoretic separation suspension of a BHK-21 cell post nuclear supernatant.

BHK-21 cells were labelled for 2hr at 20°C with 10mg/ml dialysed FITC-dextran. The cells were washed, lysed, and a post nuclear supernatant derived as described in materials and methods. The post nuclear supernatant was fractionated by free flow electrophoresis and collected through a 90 channel peristaltic pump. The fractions were analysed for FITC-dextran (▽—▽), protein (▲—▲), beta-hexosaminidase (◇—◇) and INT-succinate reductase (■—■).

virus (SFV) to insert an exogenous transmembrane protein into the plasmalemma (Marsh et al., 1983). During a low pH incubation of cells with bound virus, the virion membrane fuses with the cell surface, and, as the ^{3}H-mannose label is contained only in the viral spike glycoproteins, the radioactive SFV spike glycoprotein becomes a specific marker for the plasma membrane (White et al., 1980). The labelled cells were pooled with cells containing labelled endosomes and fractionated as described. The mannose label migrated with the major protein peak and, by SDS-polyacrylamide gel electrophoresis, was shown to be associated with intact viral spike glycoproteins (not shown).

We have also attempted to determine the distribution of components of the trans Golgi network (TGN) following FFE. The TGN can be labelled with newly synthesised spike glycoproteins of vesicular stomatitus virus (VSV) by incubating infected cells at 20°C (Griffiths et al., 1985). Preliminary experiments indicate that this label, as with the plasma membrane, Golgi, ER and mitochondria, is recovered in the major, cathodal peak and is excluded from the peak of endosome and lysosome activity (Ivan DeCurtis, Kai Simons and Mark Marsh, unpublished).

Endosome/Lysosome Separation is Dependent on Trypsinization

The separation illustrated in Figure 1 is achieved after the post nuclear supernatants have been subjected to a mild trypsinization. To determine the requirement for trypsin, we fractionated samples with and without trypsin treatment. In the absence of trypsin, the endosome, lysosome and mitochondrial activities were barely resolved from each other (Fig.2a) and migrated as components of a single peak of protein distributed between fractions 35 and 40. Following trypsinization for 3min as described, the endosomes and lysosomes separated from the mitochondria by approximately 12 fractions (Fig.2b). This separation is dependent not only on the presence of trypsin but on its enzymatic activity; no separation was observed in samples treated with SBTI and trypsin together.

Although the separation was dependent on trypsin activity, the latency of endosomes and lysosomes was not affected by the treatment (as determined by the ability of the FITC-dextran and beta hexosaminidase activities to sediment at 100KxG), nor was the buoyant density of the organelles on Percoll gradients affected.

The pool of endosomes and lysosomes derived from fractions 20 to 35 of the separation illustrated in Figure 2b, contain 28% and 34% of the cell associated FITC-dextran and beta-hexosaminidase activities, respectively, and the endosomes are enriched approximately 15 fold over the initial cell-associated activity. Within this pool more than 95% of the FITC-dextran can be sedimented by centrifugation at 100KxG for 30min and is judged to be contained within latent endosomes. A second peak of FITC-

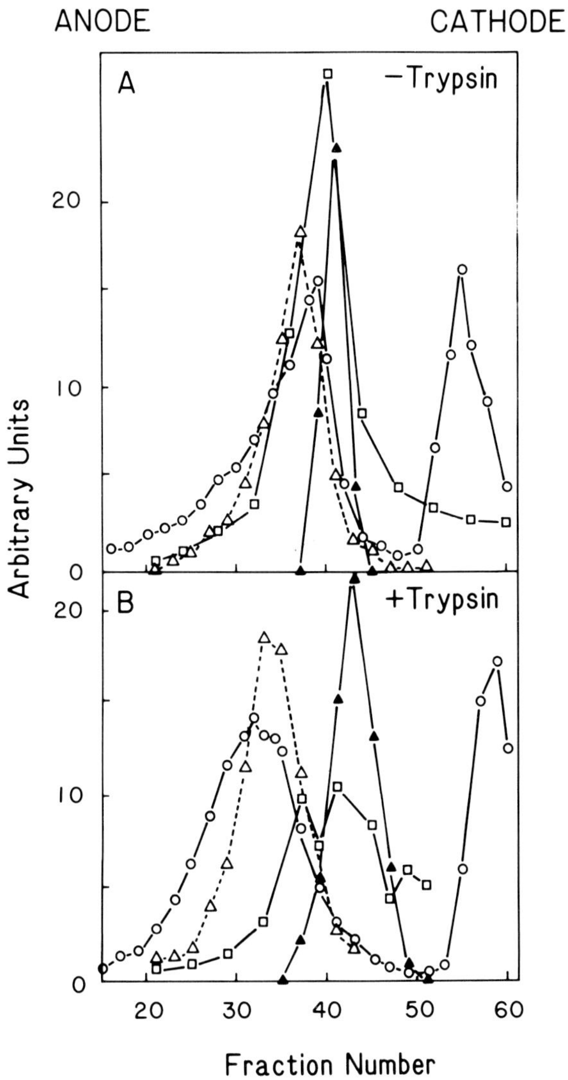

Figure 2. Effect of trypsin on the electrophoretic separation of BHK-21 cell endosomes and lysosomes.

A postnuclear supernatant was prepared from FITC-dextran labelled BHK-21 cells as in Fig.1. The supernatant was divided and one half treated with 10μg/ul trypsin for 3 min at 37°C. Following FFE the fractions were assayed as described. ○—○ FITC-dextran; △—△ beta-hexosaminidase; ▲—▲ INT-succinate reductase and □—□ protein.

dextran is found in a slowly migrating peak around fraction 60 (Fig.2). This activity does not sediment at 100KxG and is the label released from endosomal organelles damaged during the sample preparation.

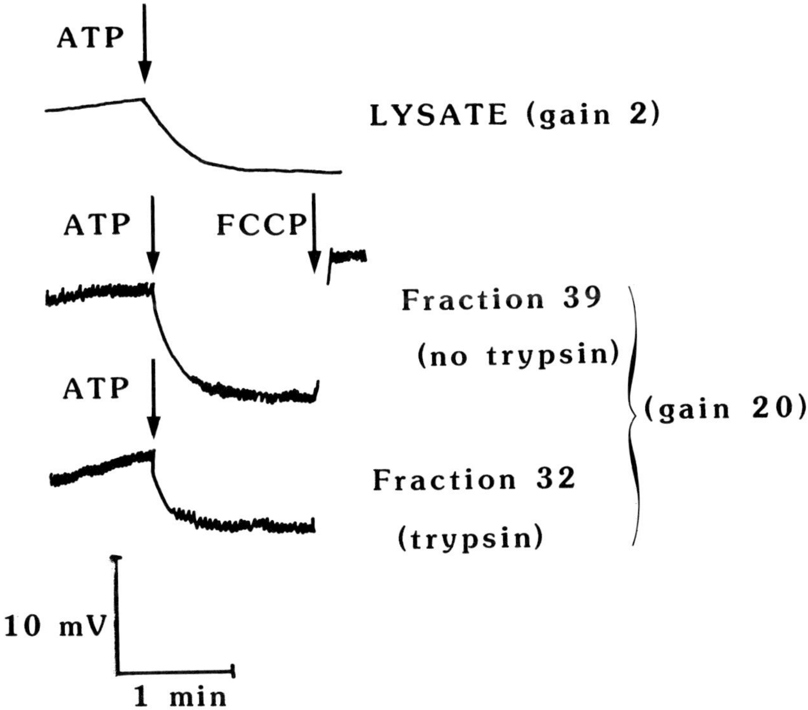

Figure 3. ATP-dependent acidification of isolated endosomes.

Aliquots of FITC-dextran labelled endosomes were taken from the cell lysate (250μg protein) and from the FFE fractions derived from the experiment illustrated in Fig.2. In the non-trypsinized sample (Fig.2a) aliquots were taken from fraction 38 (3μg protein) and in the trypsinized sample (Fig.2b) from fraction 32 (2.5μg protein). The samples were suspended into Hepes/KCl buffer (Galloway et al., 1983) and acidification assayed by fluorescence emission in the presence of ATP.

Acidification of Isolated Endosomes

We have previously demonstrated, using the pH sensitive marker FITC-dextran, that endosomes can undergo acidification in vitro on addition of ATP (Galloway et al., 1984). FITC-dextran labelled endosomes were tested for ATP-dependent acidification before and after trypsin treatment and after FFE. Non-trypsinized endosomes were taken from fractions 39 of the separation illustrated in Fig.2a and trypsinized organelles from fraction 32 of the same separation (Fig.2b). Figure 3 shows that in the presence of ATP, the fluorescence emission in both samples declined consistent with acidification of the organelles. Furthermore, the fluorescence emission was returned to the starting level, or slightly higher, by the addition of the proton ionophore FCCP.

DISCUSSION

Constitutive endocytosis is a fundamental property of virtually all cells and is responsible for the uptake and cellular processing of many physiological macromolecular ligands including nutrients, modified extracellular glyco-proteins, proteolytic enzymes and regulators of homeo-stasis, growth and division (Goldstein et al., 1985). In addition, this efficient internalization mechanism has been exploited by numerous viruses, toxins and other intra-cellular parasites to facilitate their entry into cells (Marsh, 1984). The endosomes are the primary organelles of the endocytic pathway to receive the internalized ligands and within these organelles many of the important events concerned with the regulation of endocytosis and processing of ligands and receptors occur (Helenius et al., 1983; Goldstein et al., 1985). Although a considerable amount is known of the functions of endosomes, many aspects of their properties remain unclear. In order to determine their composition, and characterise other aspects of their biology, we have sought to isolate endosomes. To this end we have attempted to use (1) procedures which do not sub-stantially modify the organelles, and (2) unique physical properties of the organelles which can be exploited to facilitate their isolation.

We have previously employed FFE in the preparative and analytical isolation of lysosomes from permanent cell lines and from whole tissue (Harms et al., 1980, 1981; Reggio et al., 1984; Marsh et al., 1983). The results presented here show that under conditions similar to those used to fractionate BHK-21 cell lysosomes, endosomes co-migrate with the lysosomes and are separated away from most other cellular membranes.

While it is clear that the electrophoretic mobility of endosomes and lysosomes is greater than that of most of the other membranes, the chemical natures of the groups responsible for conferring these charge differences are unknown. Recent experiments have demonstrated the presence of highly acidic sialoglycoproteins associated with the lumenal aspect of lysosome membrane (Lewis et al., 1985), however, it is not clear that these glycoproteins induce the increased electrophoretic mobility. Furthermore, the modifications induced by trypsin which increase the separation of endosomes and lysosomes from the other membranes is not understood. While the latency, buoyant density and acidification activity of the membranes are not affected by protease treatment, the enzymatic activity of trypsin is required. Comparison by SDS-polyacrylamide gel electrophoresis, of CHO-cell membranes before and after trypsinization has not revealed any changes in the polypeptide composition of the membranes (Marsh et al., 1987).

The free flow electrophoretic separation demonstrated here provides the basis for the rapid and efficient isolation of endosomes. To date the main problem in isolating these organelles has been that their buoyant density is very similar to those of other membrane systems. However, the fact that endosomes, together with lysosomes, can be separated from most other organelles by FFE indicates that differences in the net charge can be exploited for fractionation. The co-migration of endosomes and lysosomes in FFE is fortuitous as these organelles are efficiently separated by Percoll density gradient centrifugation (Marsh et al., 1983b) yielding purified samples of two endocytic compartments (Marsh et al., 1987). Thus endosome isolation can be achieved by a two step procedure involving (1) FFE and (2) Percoll density centrifugation. Experiments with Chinese hamster ovary (CHO) cells have produced highly purified (80-90 fold enriched), intact, acidification-competent endosomes in amounts appropriate for biochemical characterisation and immunization, in under 5hr by this two step procedure (Marsh et al., 1987).

ACKNOWLEDGEMENTS

We thank Brigitte Nade for technical assistance, Sally Swift for reading the manuscript and Marjorie Kipling for preparing the text. The work was supported by grants from the National Institutes of Health to AH and IM, from the Deutsche Forschungsgemeinschaft (Bonn-Bad Godesberg HA 757/7-6) to EH, and from the Cancer Research Campaign and Medical Research Council to MM. Collaboration was assisted by a Nato travel award.

REFERENCES

Bradford M (1976). A rapid and sensitive method for the quantitation of microgram quantities of protein utilizing the principle of protein-dye binding. Anal Biochem 72:248-254.

Galloway CJ, Dean GE, Marsh M, Rudnick G, Mellman I (1983). Acidification of macrophage and fibroblast endocytic vesicles in vitro. Proc Natl Acad Sci USA 80:3334-3338.

Goldstein JL, Brown MS, Anderson RGW, Russell DW, Schneider WJ (1985). Receptor-mediated endocytosis: concepts emerging from the LDL receptor system. Ann Rev Cell Biol 1:1-39.

Griffiths G, Bach R, Marsh M (1987). The endosome compartment in BHK cells. In preparation.

Griffiths G, Pfeiffer S, Simons K, Matlin K (1985). Exit of newly synthesised membrane proteins from the trans-cisterna of the Golgi complex to the plasma membrane. J Cell Biol 101:949-964.

Harms E, Kern E, Schneider JA (1980). Human lysosomes can be purified from diploid skin fibroblasts by free flow electrophoresis. Proc Natl Acad Sci USA 77:6139-6143.

Harms E, Kartenbeck J, Daria G, Schneider J (1981). Purification and characterisation of human lysosomes from EB-virus transformed lymphoblasts. Exp Cell Res 131:251-266.

Helenius A, Mellman I, Wall D, Hubbard A (1983). Endosomes. Trends in Biochem Sci 8:245-250.

Lewis V, Green SA, Marsh M, Vikho P, Helenius A, Mellman I (1985). Glycoproteins of the lysosomal membrane. J Cell Biol 100:1839-1847.

Marsh M (1984). The entry of enveloped viruses into cells by endocytosis. Biochem J 218:1-10.

Marsh M, Wellsteed J, Kern H, Harms E, Helenius A (1982). Monensin inhibits Semliki Forest virus penetration into culture cells. Proc Natl Acad Sci USA 79:5297-5301.
Marsh M, Bolzau E, Helenius A (1983). Penetration of Semliki Forest virus from acidic prelysosomes vacuoles. Cell 32:931-940.
Marsh M, Griffiths G, Dean G E, Mellman I, Helenius A (1986). Three dimensional structure of endosomes in BHK-21 cells. Proc Natl Acad Sci USA 83:2899-2903.
Marsh M, Schmid S, Kern H, Harms E, Male P, Mellman I, Helenius A (1987). Rapid analytic and preparative isolation of functional endosomes by free flow electrophoresis. J Cell Biol, in press.
Mellman I, Fuchs R, Helenius A (1986). Acidification of the endocytic and exocytic pathways. Anal Rev Biochem 55:663-700.
Reggio H, Bainton D, Harms E, Coudrier E, Louvard D (1984). Antibodies against lysosomal membranes reveal a 100,000-mol-wt protein that cross reacts with purified H^+K^+-ATPase from gastric mucosa. J Cell Biol 99:1511-1526.
White J, Kartenbeck J, Helenius A (1980). Fusion of Semliki Forest virus with the plasma membrane can be induced at low pH. J Cell Biol 87:264-272.

Cell-Free Analysis of Membrane Traffic, pages 35–49

ISOLATION OF FUNCTIONALLY DISTINCT ENDOSOME SUBPOPULATIONS BY FREE-FLOW ELECTROPHORESIS

Sandra L. Schmid and Ira Mellman

Department of Cell Biology
Yale University School of Medicine
New Haven, CT, USA, 06510

INTRODUCTION

Endosomes are a heterogeneous population of prelysosomal, acidic organelles which play a pivotal role in the sorting and targetting of internalized membrane and soluble content, directing their specific transport to appropriate intracellular destinations (Helenius et al, 1983; Pastan and Willingham,1983; Brown et al, 1983; Marsh et al, 1986; Mellman et al, 1986).

Internalized receptor-ligand complexes are first delivered to endosomal vesicles and associated tubules near the periphery of the cell. Membrane receptors are recycled to the cell surface for reutilization and ligands destined for the lysosome are carried into the perinuclear region of the cell where they are found in morphologically distinct endosomal vesicles and multivesicular bodies. Thus, at least two populations of endosomes can be distinguished morphologically in a variety of systems (Wall and Hubbard, 1985; Wileman et al, 1985; Marsh et al, 1986). Cell fractionation studies in rat liver and hepatocytes (Mueller and Hubbard, 1986; Baenziger and Fiete, 1986, Courtoy et al 1984) have suggested functional distinctions between two endosome subpopulations which have been defined biochemically as 'receptor-positive' and 'receptor-negative'. In addition, <u>in vivo</u> studies using pH-sensitive probes have shown that endocytic tracers encounter gradually decreasing pH as they traverse the endocytic pathway, demonstrating that endosomes are heterogeneous with respect to their internal pH.

To test whether this functional heterogeneity reflects the existence of distinct organelles, we applied the novel technique of free flow electrophoresis (FFE) to the separation of distinct endosome subpopulations. Here we report that functionally and structurally distinct subpopulations of endosomes can be resolved by FFE. Cell fractionation by FFE is rapid (requires <2 h), facilitating detailed kinetic analysis of ligand transport through endosome subpopulations en route to lysosomes. Since this procedure can be readily scaled up, large amounts of highly enriched early and late endosomes can be obtained for biochemical and functional analysis.

METHODS

Endosome labelling 1. Fluid phase markers Endosomes were specifically labelled by incubation of CHO cells in suspension (1-3 x 10^7 cells/ml) with FITC-dextran (5 mg/ml) in alpha-MEM for ≤15 min at 37°C as previously described (Galloway et al, 1983, Marsh et al, 1987). At the end of the labelling period cells were immediately cooled by adding >10 volumes of ice-cold PBS and washed extensively in PBS at 4°C. **2. Transferrin** Human diferric transferrin (Tfn), prepared as described by Klausner et al (1983) was iodinated in the presence of Iodogen (Pierce) to a specific activity of ~10^6 cpm/µg. After a 30 min incubation in serum-free alpha-MEM, to clear endogenous Tfn, CHO cells were resuspended at 1-3x10^7 cells/ml in ^{125}I-Tfn (10 µg/ml) and incubated for 30 min at 37°C. After incubation, cells are cooled to 4° C and surface-bound transferrin was removed by washing cells rapidly, twice in 25 mM acetic acid, 150 mM NaCl (to remove iron) and twice in PBS (to remove apotransferrin). **3. Semliki Forest Virus** CHO cells in monolayer were labelled with ^{35}S-methionine-labelled Semliki Forst Virus (SFV) by endocytosis of prebound SFV for 15 min at 37°C as previously described (Marsh and Helenius, 1980; Marsh et al,1987).

Preparation of CHO cell endosomes CHO cells labelled with endocytic tracers were pooled and combined with unlabelled carrier cells to a total of ~5x10^7 cells per free flow run. Pooled cells were washed in TEA-sucrose (10 mM triethanolamine, 10 mM acetic acid, pH 7.4, 1 mM EDTA and 250 mM sucrose) and resuspended in 5 volumes TEA-sucrose for homogenization. Optimum resolution of endosomes and lysosomes from the bulk of microsomal protein was highly dependent on the initial step in purification, cell homogenization. Cells were homogenized with 1-3 passes through a ball bearing homogenizer (Balch and Rothman, 1985) to obtain >80% lysis, leaving nuclei intact. Homogenization of cells was monitored by phase-contrast microscopy. We observed greater contamination of other microsomal fractions including Golgi and endoplasmic reticulum with increasing nuclear lysis, possibly due to negatively charged nucleic acids binding nonspecifically to microsomal membranes and enhancing their electrophoretic mobility (data not shown). Microsomes were prepared from a post-nuclear supernatant, resuspended to 1 mg/ml as previously described (Marsh et al, 1987) and incubated for 5 min at 37°C in 20-25 µg/ml TPCK-trypsin (Worthington). The digestion was stopped by addition of 100 µg/ml Soybean trypsin inhibitor (Sigma) and cooling on ice. Free flow electrophoresis (FFE) was performed using a Bender and Hobein ElphorVap 21 (EV21) instrument (Protein Technologies, Tucson, AZ) at 130 mA and 1700-1800 V as previously described (Marsh et al, 1987).

Acidification assays ATP-dependent acidification was assayed as previously described using endosomes isolated from cells which had been selectively labelled by endocytosis of FITC-dextran (Galloway et al, 1983, Roff et al, 1986). Organelles were equilibrated in a buffer containing 20 mM

Hepes-tetramethyl ammonium (pH 7.4), 5 mM $MgCl_2$, 145 mM KCl and 5 mM NaCl. ATP-dependent acidification was estimated from the characteristic quenching of FITC fluorescence emission which occurs as a function of decreasing pH.

RESULTS AND DISCUSSION

Separation of endosome subpopulations by free flow electrophoresis We have previously shown (Marsh et al, 1987, and these proceedings), that endosomes and lysosomes could be resolved from the bulk of microsomal proteins by FFE. To investigate whether distinct subpopulations of endosomes could be resolved by FFE, two different receptor-bound ligands marking distinct endocytic pathways were introduced into cells. Those endosomes involved in rapid membrane recycling were selectively labelled by incubating cells with ^{125}I-diferric transferrin (^{125}I-Tfn) which remains bound to its receptor and is recyled to the cell surface after releasing its iron content in endosomes (Klausner et al , 1983; Hopkins, 1983). Endosomes which target their content to lysosomes were selectively labelled by internalization of a well characterized membrane-bound endocytic tracer, ^{35}S-Semliki Forest virus (^{35}S-SFV) (Marsh and Helenius, 1980; Marsh et al, 1983). SFV, bound to the surface at 4°C is rapidly taken up at 37° by CHO cells ($t_{1/2}$ = 5 min) and delivered to lysosomes ($t_{1/2}$ = 30 min). After 15 min at 37°C, >90% of surface bound SFV is internalized into CHO cells, while <10% has been delivered to lysosomes (Schmid et al, manuscript submitted). Thus under these conditions, ^{35}S-SFV is predominantly located in endosomes en route to lysosomes.

As reported (Marsh et al, 1987; Marsh et al, these proceedings) the anodal migration of endosomes and lysosomes during FFE was enhanced following mild trypsinization of membranes. We therefore examined the effect of trypsin on the resolution of subpopulations of endosomes in order to optimize the resolution achieved. Cells labelled with either ^{125}I-Tfn or ^{35}S-SFV were

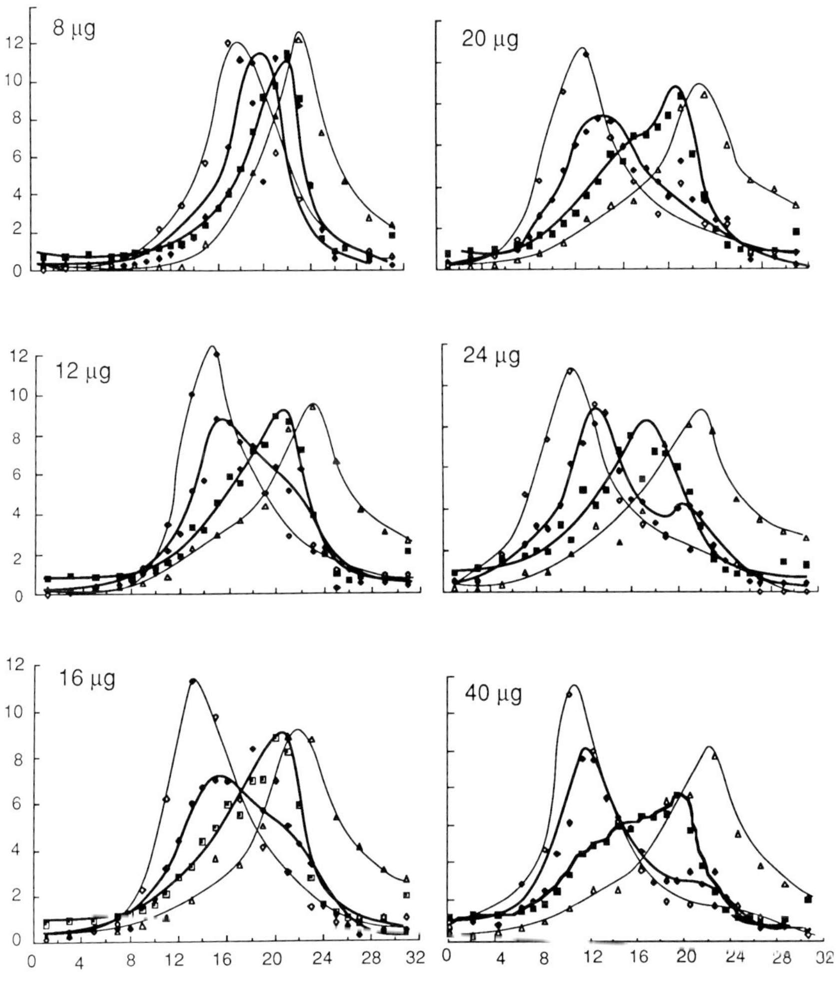

Figure 1 Titration of the effect of trypsin on the electrophoretic mobilities of early and late endosomes. (■) *early endosomes*, containing ^{125}I-Tf; (◆) *late endosomes*, containing ^{35}S-SFV; (◇) *lysosomes*, assayed by ß-hexosaminidase activity; and (△) *total protein*. Microsomal fractions at 1 mg protein/ ml were incubated as described in **Methods** with the indicated amount of trypsin (μg/ml) prior to electrophoresis.

pooled, washed and homogenized. The microsomal fraction was adjusted to 1 mg protein/ml in 0.25M sucrose-TEA and incubated for 5 min at 37°C with increasing concentrations of trypsin before being subjected to free flow electrophoresis. The elution profiles after FFE are shown in Figure 1. As can be seen, at low trypsin concentrations (≤8 μg/ml) lysosomes (detected by ß-hexosaminidase activity) can be resolved from endosomes and the bulk of microsomal protein by FFE. At higher levels of trypsin, the two endosome subpopulations shift anodally and are resolved from the bulk of the protein and from each other. Endosomes carrying SFV to lysosomes migrate more anodally than Tfn-containing endosomes involved in plasma membrane recycling. Optimum resolution of these two endosomal subpopulations is achieved following incubation of microsomal fractions in 20-25 μg/ml trypsin.

The reasons for the trypsin requirement are unclear, however, our results (Fig. 1; Marsh et al,1987) and that of others (Evans and Flint, 1985) indicate that there is very little or no separation of endosomes or lysosomes from the bulk of the input protein in the absence of a trypsinization step. Since our aim was to compare these two endosome subpopulations biochemically and functionally, we were concerned that this proteolysis step might alter the properties of either or both of these organelles. However, several control experiments described below strongly suggest that whatever the effect of trypsin digestion, there is little or no detectable alteration in the protein composition or in the functional integrity of either endosomes or lysosomes.

The effect of trypsin on endosome structure and function ATP-dependent acidification is at present the only available functional marker measurable in endosomes (Galloway et al, 1983, Merion et al, 1983). Figure 2 shows ATP-dependent H^+-translocation into endosomes labelled by incubating cells for 15 min with FITC-dextran. No inhibition of ATP-dependent acidification was observed after digestion of microsomes with 20 μg/ml trypsin indicating that by this criteria endosomes remain functionally active and intact.

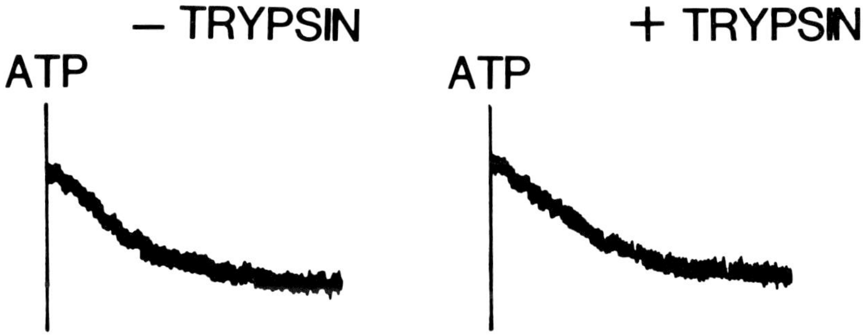

Figure 2 Effect of trypsin on ATP-dependent acidification of endosomes in a crude microsomal fraction. A microsomal fraction was prepared from CHO cells whose endosomes had been labelled with the pH sensitive fluorochrome FITC-dextran. ATP-dependent acidification of this fraction was determined with or without prior treatment with TPCK-trypsin (20 μg/ml) for 5 min at 37°C.

The effect of trypsin on the polypeptide composition of microsomes was analyzed after ^{35}S-methionine-labelled microsomes were incubated with 20 μg/ml trypsin and collected by centrifugation. SDS-PAGE analysis of microsomal pellets and of proteins released into the supernatant revealed only minor differences in protein composition after trypsinization (Fig. 3). Two-dimensional SDS-PAGE analysis of these fractions did not reveal any trypsin-induced increase in the acidic protein content of microsomes which would result in enhanced electrophoretic mobilities (data not shown). After digestion with 20 μg/ml trypsin, 13% of total

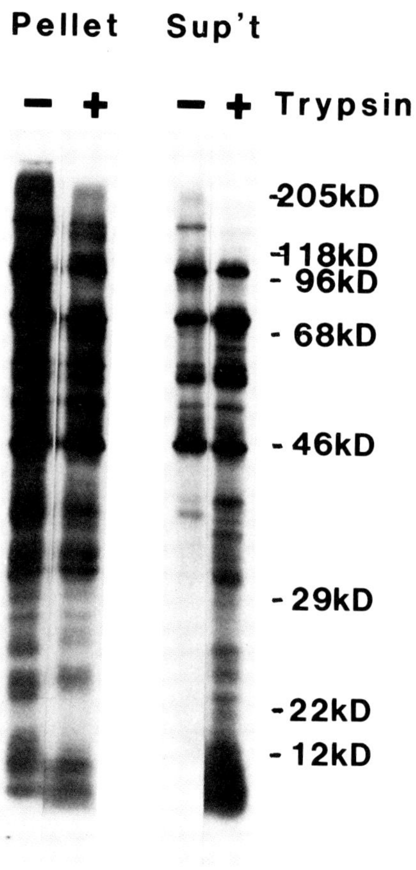

Figure 3 ^{35}S-labelled CHO microsomes before and after digestion with trypsin. CHO cells were labelled by overnight incubation on subconfluent plates with 50 µCi/ml ^{35}S-methionine. Microsomal fractions were prepared and incubated with trypsin as described in **Methods.** Microsomal membranes were pelleted for 15 min at 20 psi in a Beckman Airfuge and the supernatants and pellets were analyzed by SDS-PAGE on 10% polyacrylamide gels (Laemmli, 1970). The gel was visualized by fluorography using sodium salicylate (Chamberlin, 1979).

protein was released as TCA-soluble ^{35}S-methionine while 17% of total TCA-precipitable protein was released from microsomes in a trypsin-dependent manner. No major differences could be detected in the microsomal membrane fraction, however, the major effect of trypsin seen in Figure 3 is an increase of proteins released from the microsomal pellet (compare lanes 3 and 4). This data is in agreement with the suggestion of Beardmore and Hopkins (1987 and these proceedings) that trypsin results in release of microsomal membranes from a cytosolic matrix which promotes vesicular aggregation and therefore cofractionation by FFE and other methods.

In two additional experiments, a transmembrane protein, the Vesicular Stomatitis virus (VSV) G protein, was introduced into either microsomes by ^{35}S-methionine labelling of virus infected CHO cells or directly into endosomes by fusion of ^{35}S-VSV to the plasma membrane of CHO cells and subsequent internalization into endosomes at 37°C (Gruenberg and Howell, 1986). VSV-G protein has a 3000 kD cytoplasmic domain which is sensitive to proteolysis by trypsin (Katz et al., 1977). After incubation with trypsin under standard conditions, >95% of the VSV G protein in either microsomes (Fig. 4, panel a) or in endosomes (panel b, lanes 3 and 4) remained intact, indicating that extensive degradation of cytoplasmically-oriented microsomal or endosomal proteins did not occur.

Thus, we can find no evidence for deleterious effects of trypsin treatment or major effects on the protein composition of either microsomes or endosomes. In addition, the fact that endosomal acidification is also unaffected by trypsinization under these conditions, argues strongly that this treatment does not adversely effect the integrity of the endosomal membrane or of at least some of its functional components. Whether other endosome-associated functions are adversely affected will await identification of assays which measure these functions. Although, we have not determined why trypsinization facilitates detection of the electrophoretic mobility differences between endosomes and lysosomes, we have ruled out the possibility that these enhanced mobilities are a result of dramatic changes in protein composition of the organelles as a result of proteolysis.

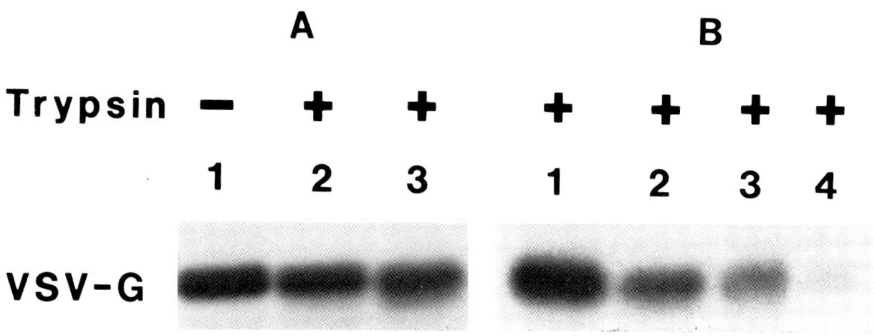

Figure 4 The cytoplasmic tail of the VSV-G protein is resistant to proteolysis under the conditions used for purification of endosomes. Panel a, microsomes prepared from VSV-infected CHO cells which had been labelled for 3 min and chased for 10 min so that the labelled G protein is present in both the endoplasmic reticulum and the Golgi (Balch, W.E., personal communication). Lane 1, undigested; lane 2, digested with 10 µg/ml trypsin; lane 2, digested with 20 µg/ml trypsin. Panel b, VSV G-protein was implanted into the plasma membrane of CHO cells by acid induced fusion with membrane-bound VSV. Cells were incubated for 15 min at 37°C to allow internalization of the transmembrane G protein into endosomes (Gruenberg and Howell, 1986). Lane 1, trypsin-digested microsomes;lane 2, unshifted fractions following FFE; lane 3, early endosome-corresponding fractions; lane 4, late endosome-corresponding fractions.

Resolution of kinetically-related early and late endosomes by FFE We have shown, (Schmid et al, manuscript submitted) that the membrane-bound ligand ^{35}S-SFV and fluid phase markers HRP and FITC-dextran are first delivered to early endosomes which comigrate with transferrin containing endosomes after FFE. At later time points, ligands and fluid phase markers appear in the more anodally-shifted late endosome fractions concomitant with their disappearance from early endosomal fractions. Thus the two endosome subpopulations resolved by FFE are kinetically related and can be defined as 'early' and 'late' endosomes. Early endosomes function in membrane recycling, while late endosomes deliver their content to lysosomes.

Free flow electrophoresis is a rapid procedure (total time < 2 h) for the enrichment of early and late endosomes and results in an ~20-fold purification of early endosomes and an ~50-fold enrichment of late endosomes with yields of 10% and 15% respectively over the post-nuclear supernatant. Early endosomes are less enriched as they are incompletely resolved from the bulk of microsomal protein and are mainly contaminated by Golgi fractions as detected by galactosyl-transferase activity (data not shown, Marsh et al, 1987). Late endosomes are chiefly contaminated with lysosomal fractions which could be removed by subsequent centrifugation in Percoll gradients (Marsh et al, 1987).

Early and Late endosomes are functionally distinct Data obtained _in vivo_ using pH-sensitive probes (Yamashiro et al, 1984; Tansugarin et al, 1984; Kielian et al, 1986) has indicated that endocytic tracers encounter a gradually decreasing pH gradient as they traverse the endocytic pathway. Consistent with this observation, we have shown that both the initial rate and extent of ATP-dependent proton influx measured _in vitro_ increases from early endosomes to late endosomes (Schmid et al, manuscript submitted).

In vitro acidification studies have also revealed an additional functional distinction between early and late endosomes. ATP-dependent acidification in early endosomes labeled either with FITC-conjugated transferrin or after incubation of CHO cells for 5 min with FITC-dextran (Fuchs and Mellman, Schmid et al, manuscripts submitted) is inhibited in buffers containing >25 mM Na^+. In contrast, the acidification of late endosomes and lysosomes is unaffected by Na^+-containing buffers. The inhibition of early endosomes under these conditions was not detected in the presence of vanadate. Together these data suggest the presence of a VO_4-sensitive Na,K-ATPase-like activity in early endosomes which inhibits acidification by generating a positive membrane potential opposing the electrogenic proton transport (Fuchs and Mellman, Schmid et al, manuscripts submitted). This activity is not detected in either late endosomes or lysosomes.

From these studies, two functional distinctions can be made between early and late endosomes. Early endosomes are less acidic than late endosomes, their pH being regulated at least in part, by the presence of a Na-dependent, VO_4-sensitive activity (possibly the Na,K-ATPase). In addition early endosomes are involved in the recycling of internalized membrane receptors, such as the Tfn-receptor, to the cell surface, while late endosomes deliver their internalized content to lysosomes.

SUMMARY

We have described a novel analytical procedure for the one-step purification of endosome subpopulations based on their differing electrophoretic mobilities when subjected to free flow electrophoresis. This procedure offers several advantages over existing schemes for endosome purification. Highly enriched endosomes are recovered in high yield and remain intact, facilitating biochemical and functional analysis of endosome subpopulations. It is rapid, requiring less than 2 hours to complete, and highly reproducible, thus facilitating detailed kinetic analysis of ligand transport through endosomal subpopulations. This technique can be easily scaled up to any degree enabling preparation of large quantities of early and late endosomes for further biochemical analysis and for antibody production. Our results have demonstrated the existence of at least two functionally and biochemically distinct, but kinetically related, subpopulations of endosomes.

ACKNOWLEDGEMENTS

We thank Drs. R. Fuchs and A. Helenius for helpful discussion and Philippe Male for excellent technical assistance with the FFE apparatus. This work was supported by an NIH grant to ISM and a Helen Hay Whitney Post-doctoral Fellowship to SLS.

REFERENCES

Baenziger JU and Fiete D (1986) Separation of two populations of endocytic vesicles involved in receptor-ligand sorting in rat hepatocytes. J Biol Chem **261**:7445-7454

Balch WE and Rothman JE (1985) Characterization of protein transport between successive compartments of the Golgi Apparatus: asymmetric properties of donor and acceptor activities in a cell-free system. Arch Biochem Biophys **240**:413-425

Beardmore JM, Howell KE, Miller K and Hopkins CR (1987) J Cell Sci **87**:495-506

Brown M S, Anderson RGW and Goldstein JL (1983) Recycling of receptors: a round trip itinerary. Cell **32**:663-667

Chamberlin JP (1979) Fluorographic detection of radioactivity in polyacrylamide gels with water-soluble fluor, sodium salicylate.Anal Biochem **98**:132-135

Courtoy PJ, Quintart J and Bauduin P (1984) Shift of equilibrium density induced by 3,3'-diaminobenzidine cytochemistry: a new procedure for the analysis and purification of peroxidase-containing organelles. J Cell Biol **98**:870-876

Evans WH and Flint N (1985) Subfractionation of hepatic endosomes in Nycodenz gradients and by free flow electrophoresis. Biochem J **232**:25-32

Galloway CJ, Dean GE, Marsh M, Rudnick G, and Mellman I(1983) Acidification of macrophage and fibroblast endocytic vesicles in vitro. Proc. Natl. Acad. Sci. (USA) **80**:3334-3338

Gruenberg J and Howell K (1986) Reconstitution of vesicle fusions in endocytosis with a cell-free system. EMBO J **5**:3091-3101

Helenius A, Mellman I, Wall D, and Hubbard A (1983) Endosomes Trends in Biochem Sci **8**:245-250

Hopkins, CR (1983) Intracellular routing of transferrin and transferrin receptors in epidermaoid carcinoma A431 cells. Cell **35**:321-330

Katz FN, Rothman JE, Lingappa V, Blobel G and Lodish HF (1977) Membrane assembly <u>in vitro</u>: Synthesis, glycosylation, and asymmetric insertion of a transmembrane protein. Proc Natl Acad Sci (USA) **74**:3278-3282

Kielian MC, Marsh M and Helenius A (1986) Kinetics of endosome acidification detected by mutant and wild-type Semliki Forest Virus. EMBO J **5**:3103-3109

Klausner RD, van Renswoude J, Ashwell G, Kempf C, Schechter AN, Deang A and Bridges KR (1983) Receptor-mediated endocytosis of transferrin in K562 cells. J Biol Chem **258**:4715-4724

Laemlli UK (1970) Cleavage of structural proteins during the assembly of the head of bacteriophage T4. Nature **227**:680-685

Marsh M and Helenius A (1980) Adsorptive endocytosis of Semliki Forest virus. J Mol Biol **142**:439-454

Marsh M, Bolzau E, and Helenius A (1983) Penetration fo Semliki Forest virus from acidic prelysosomal vacuoles. Cell **32**:931-940

Marsh M, Griffiths G, Dean GE, Mellman I, and Helenius A (1986) Three dimensional structure of endosomes in BHK-21 cells. Proc Natl Acad Sci (USA) **83**:2899-2903

Marsh M, Schmid S, Kern H, Harms E, Male P, Mellman I and Helenius A (1987) Rapid analytical and preparative isolation of functional endosomes by free flow electrophoresis. J Cell Biol **104**:875-886

Mellman I, Fuchs R and Helenius A (1986) Acidification of the endocytic and exocytic pathways. Ann Rev Biochem **55**:663-700

Merion M, Schlessinger P, Brooks RM, Moehring JM, Moehring TJ and Sly WS (1983) Defective acidification of endosomes in Chinese hamster ovary cell mutants "cross-resistant" to toxins and viruses.Proc Natl Acad Sci (USA) **80**:5315-5319

Mueller SC and Hubbard AL (1986) Receptor-mediated endocytosos of asialoglycoproteins by rat hepatocytes: Receptor-positive and receptor-negative endosomes. J Cell Biol **102**:932-942

Pastan I, and Willingham MC (1983) Receptor-mediated endocytosis: coated pits, receptosomes and the Golgi. Trends in Biochem Sci **8**:250-254

Roff CF, Fuchs R, Mellman I and Robbins AR (1986) Chinese hamster ovary cell mutants with temperature-sensitive defects in endocytosis. I. Loss of function on shifting to the nonpermissive temperature. J Cell Biol **103**:2283-2297

Tansugaran L, McNeil P, Reynold GT and Taylor DL (1984) Microspectrofluorometry by digital image processing: measurement of cytoplasmic pH. J Cell Biol **98**:717-727

Wall DA, and Hubbard AL (1985) Receptor-mediated endocytosis of asialoglycoproteins by rat liver hepatocytes: Biochemical characterization of the endosomal compartments. J Cell Biol. **101**:2104-2112

Wileman T, Harding C and Stahl P (1985) Receptor-mediated endocytosis. Biochem J **232**:1-14

Yamashiro DJ, Tycko B, Fluss SR and Maxfield FR (1984) Segregation of transferrin to a mildly acidic (pH6.5) para-Golgi compartment in the recycling pathway. Cell **37**:789-800

Cell-Free Analysis of Membrane Traffic, pages 51–67

CONTINUOUS FLOW ELECTROPHORESIS APPLIED TO THE SEPARATION OF CELLS, MEMBRANES AND MEMBRANE FUNCTIONAL DOMAINS

N. Crawford

Department of Biochemistry, Hunterian Institute, Royal College of Surgeons of England, Lincoln's Inn Fields, London, WC2A 3PN, U.K.

During the last few years there has been a rapidly growing interest in continuous flow electrophoresis (CFE) as a separative procedure in cell biology (Hannig et al., 1975; Hannig, 1978; Wagner et al., 1984; Wagner et al., 1984). Although many of the earlier studies were applied to subpopulation studies of peripheral blood lymphocytes and their precursors (Boltz et al., 1978; Shortman, 1977; Chollet et al., 1978), the field has now been extended widely to the study of many cell types which include other blood and bone marrow cells, sperm, parasites and tumour cells (Heidrich et al., 1979; Bol et al., 1981; Boltz et al., 1981; Schubert et al., 1973; Pretlow et al., 1979). The preparation of reasonably homogenous cell types from complex tissues such as liver and kidney have also been reported (Vandewalle et al., 1982; Heidrich and Dew, 1977; Oshiro et al., 1983).

Since the principles of CFE exploit differences in cell or particle surface charge, it was natural that cell biologists should also explore the value of CFE in subcellular fractionation and in the isolation and purification of organelles not readily possible by conventional separation procedures. Accordingly, CFE procedures have now been reported for plasma membrane vesicles, endoplasmic reticulum, sarcoplasmic reticulum, mitochondria, lysosomes and other storage organelles, and receptosomes, as also nuclear membranes and nuclear fragments (Crawford, 1985; Heidrich, 1971; Ryan et al., 1971; Heidrich, 1979; Morré et al., 1984; Stahn et al., 1970; Menashi et al., 1981; Stewart et al., 1983). To survey all of these whole cell and subcellular applications would be outside the scope of this review and therefore in the interests of compatibility with the aims and emphasis of this "Membrane Trafficking Workshop" my review will be rather selective and will focus attention on just a few areas, supplementing these with examples from interests with CFE techniques in the study of certain blood cells and their surface membranes and intracellular features in my own laboratory.

PRINCIPLES OF CFE

The electrokinetic property of cells or particles which is exploited in CFE is the "zeta potential." Most cells and organelles have a net electronegative charge expressed through surface-exposed polar groups of intrinsic membrane constituents and other extrinsic membrane-bound components. When suspended in an ionic medium the charged cell or particle attracts an ion cloud or electrical boundary layer which results in some of the surface associated anionic groups being neutralised by ions from the medium. Thermal fluctuations and other factors prevent total charge neutralisation and the zeta potential (still net negative), which determines the cell or particle's electrophoretic mobility is the potential at the plane of shear between the electrical double layer and the ionic environment surrounding the particle. Groups at the surface membrane of the cell contributing to the surface charge may be cationic such as protein -NH_2 groups, (either terminal or associated with intrachain basic amino acids) or cationic such as sialic acid -COOH, the carboxyl groups of glutamate and aspartate or lipid or protein associated phosphate groups. Sialic acid is a terminal residue in the carbohydrate chains of many membrane glycoproteins and is believed to be the major contributor to the surface charge characteristics of living cells and organelles. A particular consideration which is relevant in dealing with blood cells harvested from a protein-rich plasmatic environment and similarly with organelles separated from their colloid-rich cytosol, is the presence of loosely or non-specifically adsorbed proteins. This "proteinaceous halo" (Fig. 1), which of course may vary, dependent upon the initial isolation procedures, can significantly affect the electrokinetic behaviour of the cell or particle, and the reproducibility of fractionation depends upon maintaining some consistency in this property. Some of these protein-protein interactions, for example, are divalent cation dependent and can be affected by the use of chelating agents. Such surface binding has of course been exploited to advantage by the use of membrane domain specific antibodies to improve the resolution in mixed cell populations.

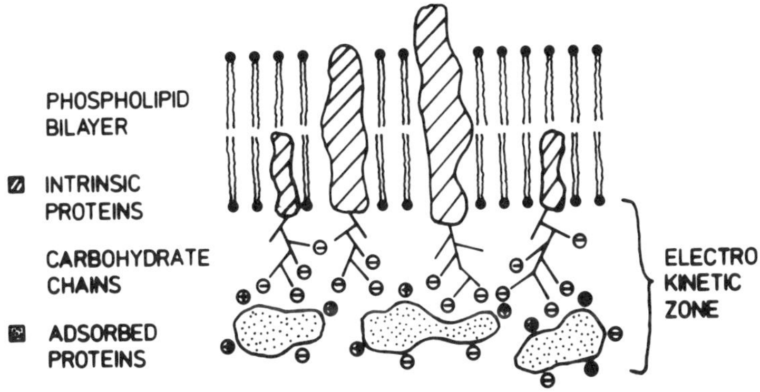

Figure 1. Diagrammatic representation of cell surface membrane showing negatively charged (sialic and sulphate residues) carbohydrate domains of intrinsic glycoproteins to which adsorbed proteins are associated.

Although other instrument configurations have been described, most commercially available CFE equipment includes a chamber, of width about 1-200 times the thickness, through which a downward flowing buffer film is passed. The consistency of flow is maintained by a multi-channel pump. An electrical field is applied perpendicular to the buffer flow direction and its stability is maintained by ion flow through semi-permeable membranes separating the electrode compartments from the chamber. Cells or subcellular components are pumped into the top of the chamber into the middle of the buffer stream, on the cathodal side, and separations thus depend upon their movement at different velocities towards the anodal side of the chamber whilst at the same time passing down the chamber with the buffer film. An angle of flow is thus described and separations depend upon differences in electrophoretic mobilities due to surface charge expression, and the

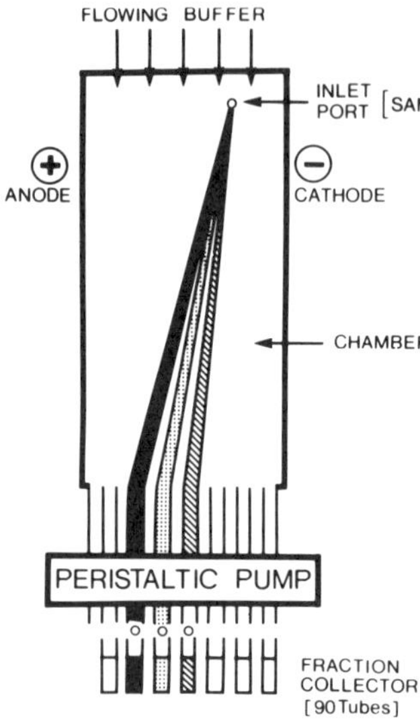

Figure 2. Diagram of CFE chamber into which cells or organelles are introduced into the flowing buffer film.

resolution depends upon residence time in the chamber (usually 2-4 min) and the pH and ionic strength of the flowing buffer, as also the temperature and potential difference applied across the chamber. Generally an outlet system at the bottom of the chamber (~90 fraction tubes) allows the collection of samples across the full width of the chamber (See Fig. 2).

WHOLE CELL SEPARATIONS BY CFE

Although analytical cell electrophoresis, using cytopherometers, has been used for cell surface characterisation for over 50 years it was really not until the middle 1960's that preparative procedures using continuous flow electrophoresis became established in cell biology research. Erythrocytes were the first cells to be harvested in high yields and relatively free from other blood cell contaminants using CFE. Subpopulation heterogeneity is less well expressed with red cells than with other cells, however, but since red cells fix well with glutaraldehyde or formalin without significant change in electrophoretic mobilities they have been used widely as electrophoretic reference markers for other cell separations. The apparent narrow separation profile (with respect to surface charge) that the circulating red cell pool displays has mitigated against their value in CFE studies in clinical disorders and disease states.

Much more attention however has been focussed upon white blood cells and particularly the circulating and tissue-associated lymphocytes since earlier studies by Zeiller and his colleagues (Zeiller et al., 1972; Zeiller, 1972) established that T and B lymphocytes possess significantly different electrophoretic mobilities. Following these earlier observations a great many papers have appeared concerned with the subdivision of these two major lymphocyte classes into subpopulations differing in surface markers and functional responses to PHA and other mitogens. The field was broadly reviewed in 1978 by Hannig (Hannig, 1978) and in 1979 by Pretlow and Pretlow (Pretlow and Pretlow, 1979) but it has now widened greatly into specialised areas of subpopulation profiling in, for example, viral transformation, rheumatoid arthritis and autoimmune disorders, athymic mice, immunosuppression and irradiation and lymphoproliferative disorders. In many of these areas, however, the functional and constitutional characterisation of the cell surfaces have lagged well behind the studies of their electrokinetic properties and much remains to be done before the value of such studies in disease states can be properly assessed.

Despite a wealth of literature on red cells and lymphocytes there has been comparatively less interest in CFE procedures applied to polymorphonuclear leucocytes, monocytes and platelets. Some of my own interests here have been focussed upon heterogeneity studies with circulating granulocytes and platelets. If the former cells are initially separated from blood by a conventional Ficoll-Hypaque procedure and resuspended in a suitable medium, when applied to the chamber of a CFE apparatus, they can be separated into a charge dependent population profile of roughly gaussion characteristics and extending over 20-30 fraction tubes (see Fig. 3). Studies with cell fractions taken from this profile have revealed significant differences in surface oriented constituents (e.g. sialic acid, -SH groups etc.) and in receptor status, but more interestingly they show an inverse relationship between electronegativity and functional properties mediated through their membrane/cytoskeletal axis. In a standardised quantitative phagocytic model the least electronegative subpopulations have around twice the phagocytic capacity of the most electronegative

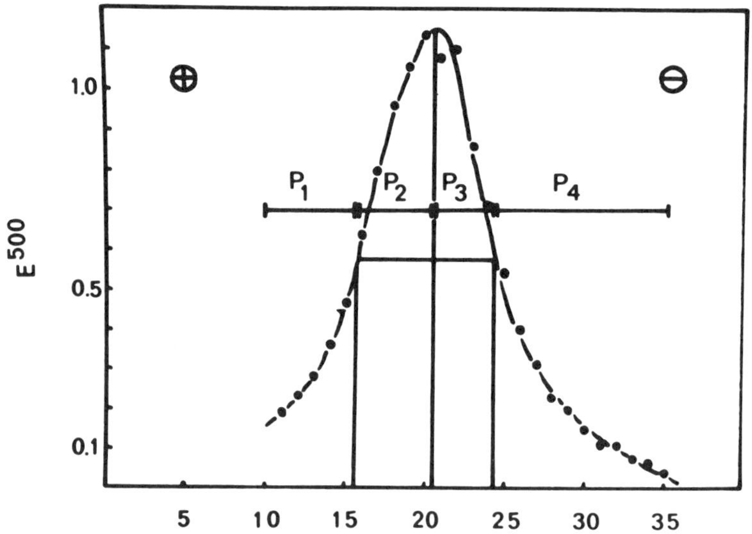

Figure 3. Free flow electrophoresis profile of neutrophils showing subdivision into subpopulation pools P_1-P_4 of decreasing electronegativity.

subpopulations (Fig. 4). Although the background literature to this concept is somewhat anecdotal, there have been previous reports suggesting a similar inverse relationship. For example Lichtman & Weed (Lichtman and Weed, 1972) have shown that bone marrow precursor granulocytes of high electrophoretic mobility do not spread well, are poorly phagocytic and chemotactic when compared with circulating granulocytes of lower surface electronegativity, which have highly attenuated motile behaviour. Other studies using agents to modify cell membranes such as the binding of chondroitin sulphate or neuraminidase desialyation seem to be supportive of this inverse relationship (Henricks et al., 1984). There have been few CFE studies applied to monocytes and macrophages, presumably because of their tendency to stick to foreign

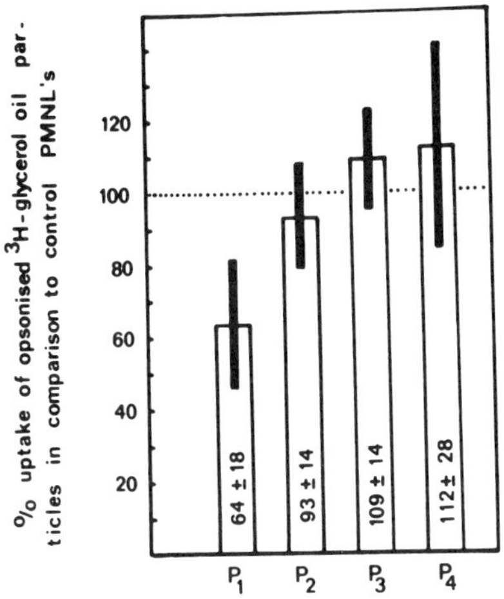

Figure 4. Quantitative measurement of neutrophil phagocytic capacity (uptake of ^{3}H glycerol in inert oil) in cell pools P_1-P_4. The 100% line refers to the phagocytic capacity of the cell pay load to the CFE chamber.

surfaces (i.e. the glass of the chamber) but this area is ripe for exploitation and particularly with respect to lymphokine receptor status and heterogeneity in cell lytic and tumouricidal properties. Similarly, most of the investigations with cultured or tissue-derived tumour cells have been at the analytical level with microscopic cytophotometry rather than at CFE preparative levels and detailed electrokinetic and molecular studies of metastatic variants seem worthy of attention. Electrophoretic profiling by CFE has been reported for mastocytoma ascites cells and a comparison with normal host cells has been made (Pretlow et al., 1981). There appeared to be little advantage in this CFE procedure over separations by rate sedimentation in centrifuges.

Tissues which have resulted in encouraging and potentially useful cell separations have been kidney (Heidrich and Dew, 1977; Vandewalle and Heidrich, 1980; Vandewalle and Heidrich, 1980; Heidrich, 1982), regenerating and neoplastic liver, spleen and thymus (Zeiller et al., 1975; Stein, 1976; Larrov et al., 1979; Zeiller et al., 1974), certain solid tumours (Pretlow et al., 1981; Jackson et al., 1977) and most recently pituitary (Plank et al., 1983), skin (Easty et al., 1986). One of the earlier successful procedures to be reported was the isolation of various cell populations from the renal cortex (Heidrich and Dew, 1977; Vandewalle and Heidrich, 1980; Heidrich, 1982). Using CFE to process cortical tissue which had first been passed through a fine press, these workers were able to separate, with good enrichment, cells from the straight portion of the proximal tubule and cells from the distal part of the nephron. The cells appeared to be intact and viable and could be used for a range of analytical and functional studies which included the identification of an active renin producing subpopulation (Heidrich, 1979). Many studies have been made with normal liver and hepatoma cells but almost without exception these have been directed towards "electrokinetic phenomenology" rather than towards composition or enzymatic differentiation. An interesting clinical application has been explored in the use of CFE in combination with a Ficoll/ sucrose density gradient for cytodiagnosis with cervical biopsy tissues (Boltz et al., 1976). These earlier studies were promising but a profile catalogue for normal and abnormal cells has not yet emerged, and it may be that the collagenase dispersion of the tissue cells results in an unacceptable variable in this technique. The interest in possible production of well characterised and highly purified pituitary cells for implantation stimulated some interest in the application of CFE techniques to these problems. Pituitary cells show much electrophoretic heterogeneity but viable growth hormone-producing subpopulations have been identifiable and can be maintained for some time in cultures (Plank et al., 1983). It is likely that further advances in this area will be overtaken by genetic engineering developments, although protein fractionation by CFE techniques may well have a role in the final stages of a commercial production process.

In the field of platelet research, despite earlier observations of electrokinetic heterogeneity in the circulating pool using cytopherometer techniques, there have been few attempts to explore subpopulations using CFE to provide yields adequate for surface membrane analyses. In our hands, subpopulation profiles extending over 20-30 fractions can be produced with formol fixed cells and although these are suitable for some analytical purposes [e.g. neuraminidase-labile sialic acid determinations] the identification of functionally attenuated "pace setters" within the pool perhaps responsive to low doses of agonists has not been possible with fixed cells. The use of taxol, a microtubule stabilising agent which appears not to inhibit agonist-induced activation (Crooke and Crawford, 1986) is being investigated and platelet subpopulation profiling in health and disease may now become feasible. Changes in the circulating profile in platelet consumptive coagulopathies, thrombocytopaenic states, extra corporeal circuitry and transplant rejection may shed some light on the dynamic aspects of bone marrow platelet production, life span kinetics and thrombopoietic regulation.

CFE SEPARATIONS OF SUBCELLULAR ORGANELLES

The limitations in resolution of subcellular particles from homogenates or subfractions using density gradient fractionation procedures, as also the potential damaging effects of sustained hydrostatic pressures on the less robust organelles, has led to considerable interest in the use of CFE techniques for organelle purification. One of the earliest applications was that of Heidrich (1971) in the study of membrane particles of rat liver mitochondria. Here the difference in electrophoretic mobilities of the outer and inner mitochondrial membranes was exploited for biochemical studies of the elementary particles of the inner mitochondrial matrix. Evidence for the sidedness of these membranes with respect to the localization of the elementary particles emerged from these studies. Techniques for the preparation of inside out and normally oriented membrane vesicles have been widely applied to red cells and ghosts but not generally. The experimental conditions required for controlling membrane orientation coupled with CFE techniques for resolving different vesicle populations are worthy of more detailed exploration since they would be highly applicable to the study of membrane vesicle trafficking, endo- and exocytotic events, and surface and intracellular ion-transport processes.

With respect to surface membrane domain studies, the elegant mitochondrial fractionations of Heidrich and his colleagues referred to earlier with their kidney proximal tubule cells enabled, by CFE techniques, to identify two different membrane domains from a brush border fraction, one of which contained only intact microvilli (apical membrane domain) and the other, characterised by membrane markers and the presence of functional complexes, to represent the basal membrane (interstitial domain) of the proximal tubule cell. An interesting feature of this work was the observation that EDTA-treated membranes were less electronegative than untreated membranes conflicting with earlier concepts that removal of bound Ca^{2+} increased electrophoretic mobilities. This may be an example of associated loosely bound proteins referred to earlier. Clearly there are Ca^{2+} dependent associations of proteins and other constituents ("proteinaceous halo") with cell surface membranes which influence their electrokinetic behaviour, and preparative procedures prior to the CFE will have to be examined in detail if consistent results are to be obtained.

One of the first membrane vesicle fractionations of direct interest to vesicle-trafficking enthusiasts was the study of Evans and his colleagues (Evans and Flint, 1985; Debanne et al., 1982) which was applied to the identification of processes involved in receptor interiorisation in the liver. They were able to isolate and characterise a population of endosomes (diacytosomes) believed to be involved in receptor mediated internalisation and subsequent processing of asialotransferrin and insulin. More recently, similar CFE purifications of endosome compartments from CHO (Schmid et al., 1986) and BHK cells (Marsh et al., 1986) have revealed, that in combination with conventional gradient sedimentation procedures, CFE has great potential for investigating the dynamic aspects of receptor mediated entry of ligands into cells as also for following the time course and direction of intracellular trafficking and receptor to surface membrane recycling processes. For a number of years Morré and his colleagues have been studying cell endomembrane compartments and have particularly focussed their attention upon the Golgi apparatus of liver cells (Morré, 1971; Morré and Ovtracht, 1977). More recently they have successfully exploited the resolving power of CFE to subfractionate cisternae and vesicles of the Golgi complex. Their biochemical data (sialic acid and membrane marker enzymes) and morphological characterisation suggest a progressive membrane differentiation across the polarity axis of the Golgi stacks. Since for a proper understanding of the relationship of Golgi membrane composition and functional heterogeneity to endoplasmic reticulum, lysosomes and endosomes as well as to the boundary membranes of secretory vesicles, knowledge of differential marker localisation and membrane domain segregation is a necessary prerequisite, such CFE fractionations could have a valuable and more general role in endomembrane characterisation.

An area of personal interest, involving the subfractionation by CFE of subpopulations of membrane vesicles of different cellular origin, has been in our investigations of platelet plasma membranes and intracellular (EC) membranes. Since the platelet surface membrane has its origin in the intracellular membrane of its precursor cell, the bone marrow megakaryocyte, the surface membranes and intracellular membranes present in a mixed membrane vesicle fraction harvested from density gradient fractionation of homogenates cannot be resolved cleanly in CFE without modification of the surface membrane at the whole cell level with neuraminidase. Surprisingly, and despite the rich content of sialoglycoproteins in the surface membrane of this cell, the EC membrane vesicles are substantially more electronegative than surface membrane from non-neuraminidase treated platelets (Crawford, 1985). However, the neuraminidase desialyation of the surface membrane before homogenisation allows a CFE separation of the two membrane species with exclusivity of marker localisation (Fig. 5). Similar covalent modifications to membranes to enhance CFE resolution may well have a place in the study of endosome and Golgi compartments. The use of phosphatases, certain lipases and charged probes for titrating -SH group status, for example, are all techniques worthy of further investigation here. In another field our interest in membrane segregatory events during membrane mediated cell motile phenomena led us to exploit the early events occurring during neutrophil phagocytosis (Spangenberg and Crawford, 1987). By using a quantitative phagocytic stimulus, involving a ^{3}H labelled and opsonised inert oil, neutrophil phagocytic vesicles can be cleanly separated from surface membrane domains not involved in the phagocytic process by first flotation in a density gradient and then purification by CFE (Fig. 6). This model system has led to the finding that during the phagocytic internalisation there is a selective segregation of a major sialoglycoprotein out of the forming vesicle and into the uninvolved membrane domains, whilst certain lectin receptors remain randomly distributed (Fig. 6). Although this membrane "reorganisation model" represents a

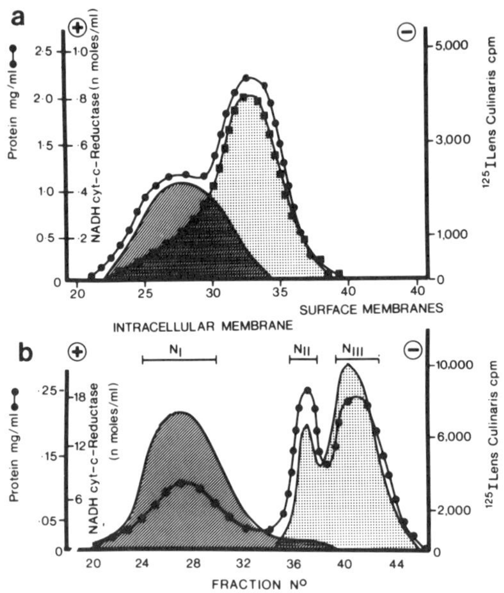

Figure 5. Electrophoretic profile of human platelet mixed membrane vesicles taken from a sorbitol density gradient of platelet sonicates. a) platelets untreated, and b) platelets treated with neuraminidase at the whole cell level before sonication. (Hatch area NADH cyt c reductase activity and dotted area ^{125}I lectin applied to the whole cells).

"frustrated phagocytic" event since it does not proceed to lysosomal fusion with the internalised vesicle, this has been advantageous in our work, and a similar approach to the study of endocytic vesicles perhaps using cytoskeletal poisons to prevent intracellular movement and fusagenic events together with CFE separations may well shed some light on the small-print molecular mechanisms involved in receptor mediated internalisation of important ligands.

In conclusion, although CFE techniques and equipment were introduced into cell biology many years ago, it suffered a "bad press" largely due to instrumentation limitations. It was the pioneering work of Hannig and his co-workers at the Max Planck Institute in Munich that

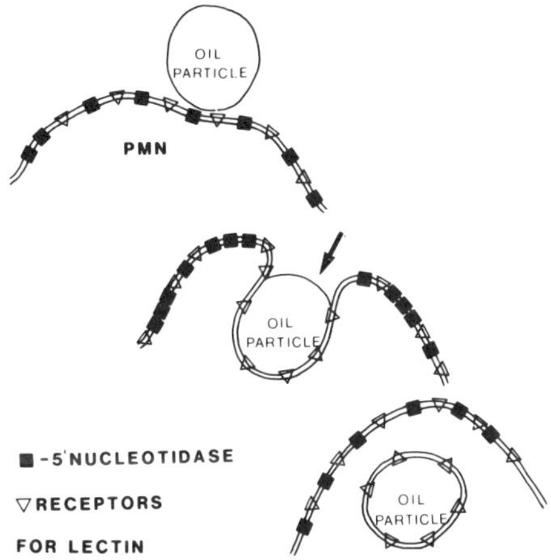

Figure 6. Diagrammatic representation of selective segregation of major sialoglycoprotein (5'nucleotidase) out of forming vesicle into uninvolved domain during phagocytosis. Lectin receptors remain randomly distributed.

generated a renaissance over the last 10-15 years and ensured a place for CFE technology in many areas of cell biology. Variations in approach using continuous isoelectric focussing and isotachophoresis complement zonal free flow electrophoresis and are beginning to widen its potential as a separative tool for cells, cell organelles and soluble materials either alone or in adjunct with more conventional separation procedures.

ACKNOWLEDGMENTS

I wish to thank the Royal Society, the British Heart Foundation and the Medical Research Council for financial sponsorship of our work in this field over the last few years and also Dr. G. Weber of Messrs. Bender & Hobein, Munich for many interesting discussions and much practical assistance with our problems.

REFERENCES

Bol S, van Vliet M, van Slingerland V (1981). Electrophoretic mobility of murine hemopoietic cells in different stages of development. Exp Haematol 9:431-433.

Boltz RC, Miller TY, Todd P, Kukulinsky NE (1978). A citrate buffer system for isoelectric focussing and electrophoresis of living mammalian cells. In Catsurpoolas A (ed): "Electrophoresis," Amsterdam and New York: Elsevier North Holland, Inc.

Boltz RC, Schmatz DM, Murray PK (1981). Evaluation of the electrophoretic separability of trypanosoma cruzi parasite stages. In Allen A (ed): "Electrophoresis '81," Berlin and New York: Walter de Gruyter & Co, pp 841-850.

Boltz RC, Todd P, Gaines RA, Milito RP, Docherty JJ, Thompson CJ, Notter MFD, Richardson LS, Mortel R (1976). Cell electrophoresis research directed towards clinical cytodiagnosis. J Histochem and Cytochem 24:16-23.

Chollet PH, Herve P, Chassagne J, Masse M, Plagne R, Peters A (1978). Application of cell electrophoretic methods (preparative and analytical) in demonstrating the surface heterogeneity of T-lymphocytes in man. Biomedicine 28:119-124.

Crawford N (1985). Use of free flow electrophoresis in studies of surface and intracellular membranes and specific surface functional domains in blood platelets and leucocytes. In Schuett L, Klinkmann H (eds): "Cell Electrophoresis," Berlin: De Gruyter.

Crawford N (1985). Structural and molecular properties of platelet membrane systems. In George JN, Neurden AT, Phillips DR (eds): "Platelet Membrane Glycoprotein," Chap. 2, New York: Plenum Press, pp 13-49.

Crooke M, Crawford N (1986). Unpublished observations.

Debanne MT, Evans WH, Flint N, Regoeczi E (1982). Receptor-rich intracellular membrane vesicles transporting asialo transferrin and insulin in liver. Nature 298:398-400.

Easty DJ, Crook MA, Crawford N, Evans DJ (1986). Proceedings of the International Electrophoresis Society. In Dunn MJ (ed): "Electrophoresis 86," London, pp 108-113.

Evans WH, Flint N (1985). Subfractionation of hepatic endosomes in Nycodenz gradients and by free-flow electrophoresis. Separation of ligand transporting and receptor enriched membranes. Biochem J 232:25-32.

Hannig K (1978). Continuous free-flow electrophoresis as an analytical and preparative method in biology. J Chromatog 159:183-191.

Hannig K, Wirth H, Bernd-Holger M, Zeiller K (1975). Free-flow electrophoresis I. Theoretical and experimental investigations of the influence of mechanical and electrokinetic variables on the efficiency of the method. Hoppe-Seyler's Z Physiol Chem 356:1209-1223.

Heidrich HG (1971). The sidedness of mitochondrial membranes and their subparticles as characterised by preparative free-flow electrophoresis. FEBS Lett 17:253-256.

Heidrich HG (1979). Electrophoretic separation of organelles and cells from rabbit kidney-cortex with respect to localisation of renin. Hoppe-Seyler's Z Physiol Chem 357:261-262.

Heidrich HG (1982). Cell populations from the renal proximal distal tubule isolated by free-flow electrophoresis. In "Cell Function and Differentiation (A)," New York: Alan R. Liss, Inc. pp 115-126.

Heidrich HG, Dew M (1977). Homogenous cell populations from rabbit kidney cortex. J Cell Biol 74:780-788.

Heidrich HG, Russman L, Bayer B, Jung A (1979). Free-flow electrophoresis for the separation of materia infected and uninfected mouse erythrocytes and for the isolation of free parasites (plasmodium vinckei). Z Parasitenkd 58:151-159.

Henricks PAJ, van der Tol ME, Verhoff J (1984). Sialic acid and cell function. Klin Wochen schv 62:487-488.

Jackson RJ, Stewart HD, Sachs G (1977). Isolation and purification of normal and malignant colonic plasma membranes. Cancer 40:2487-2496.

Larrov VF, Shestakova SV, Sitkowskii MV, Gulubeva NN (1979). Electrophoretic separation of lymphocytes from rat spleen and thymus and their response to mitogens in vitro. Bull Soc Exp Biol Med 88:1035-1037.

Lichtman MA, Weed RO (1972). Alteration of the cell periphery during granulocyte maturation. Relationship to cell function. Blood 39:301-316.

Marsh M, Kern H, Harms E, Schmid S, Mellman I, Helenius A (1986). Rapid isolation of function endosomes, this volume.

Menashi S, Weintroub H, Crawford N (1981). Characterisation of a human platelet surface and intracellular membranes isolated by free-flow electrophoresis. J Biol Chem 256:4095-4101.

Morré DJ (1971). Isolation of Golgi apparatus. Meth Enzymol 22:130-148.
Morré DJ, Creek KE, Matyas GR, Minnifield N, Sun I, Baudoin PH, Morré DM, Crane FL (1984). Free-flow electrophoresis for subfractionation of rat liver Golgi apparatus. Biotech, Sept/Oct. 224-233.
Morré DJ, Ovtracht JL (1977). Dynamics of the Golgi apparatus: Membrane differentiation and membrane flow. Int Rev Cytol Suppl 5:61-188.
Oshiro S, Takami M, Nakajuma H (1983). In Hirai H (ed): "Electrophoresis 83," Berlin: De Gruyter, pp 379-381.
Plank LD, Hymer WC, Kunze ME, Marks GM, Lanhan JW, Todd P (1983). A study of cell electrophoresis as a means of purifying growth hormone secreting cells. J Biochem and Biophys Meth 8:275-289.
Pretlow TG, Pretlow TP (1979). Cell Electrophoresis. Ann Rev Cyt 61:85-128.
Pretlow TP, Stewart HB, Sachs G, Pretlow II TG, Pitts A (1979). Free-flow electrophoresis of an ascites mast cell tumor. Proc AACR and ASCO 124.
Pretlow TP, Stewart HB, Sachs G, Pretlow TG, Pitts AM (1981). Free-flow electrophoresis of an ascites mast-cell tumour. Br J Cancer 43:537-541.
Ryan KJ, Kalant M, Llewellyn-Thomas E (1971). Free-flow electrophoretic separation and electrical surface properties of subcellular particles from guinea pig brain. J Cell Biol 49:235-239.
Schmid S, Fuchs R, Male P, Mellman I (1986). Isolation of functionally and structurally distinct endosome subcompartments from CHO cells by free-flow electrophoresis, this volume.
Schubert JC, Walther F, Holzberg E, Pascher G, Zeiller K (1973). Preparative electrophoretic separation of normal and neoplastic human bone marrow cells. Klin Wschr 51:327-329.
Shortman K (1977). The separation of lymphoid cells on the basis of physical parameters. Separation of B and T-cell subsets and characterisation of B-cell differentiation stages. Meth Cell Sep 1:229-249.
Spangenberg P, Crawford N (1987). J Cell Bio Chem, in press.
Stahn R, Maier K-P, Hannig K (1970). A new method for the preparation of rat liver lysosomes. J Cell Biol 46:576-580.
Stein G (1976). Separation of human lymphoid cells by preparative cell electrophoresis: Studies on cells of

lymphoid organs in haemotological normal persons. Biomedicine 23:5-11.

Stewart DIH, Emes CH, Crawford N (1983). Preparation of a highly purified surface membrane fraction from rabbit polymorphonuclear leucocytes using high voltage free flow electrophoresis. Biochim et Biophys Acta 733:154-162.

Vandewalle A, Heidrich HG (1980). Characterisation of different populations of rabbit kidney cortical cells separated by free-flow electrophoresis using hormone-stimulated cyclases. Int J Biochem 12:61-66.

Vandewalle A, Heidrich HG (1980). Morphological and biochemical heterogeneity of cortical kidney cell populations isolated by free-flow electrophoresis. Eur J Cell Biol 22:516.

Vandewalle A, Kopfer-Hobelsberger B, Heidrich HG (1982). Cortical cell populations from rabbit kidney isolated by free-flow electrophoresis: Characterisation by measurement of hormone-sensitive cyclase. J Cell Biol 92:505-513.

Wagner H, Mang V, Kessler R, Heydt A, Manzoni R (1984). Preparative free-flow electrophoresis of proteins, peptides and related compounds. In Hirai H (ed): "Electrophoresis '83," Berlin and New York: Walter de Gruyter & Co., pp 283-291.

Wagner H, Mang V, Kessler R, Speer W (1984). A free flow system for preparative separations. In Holloway J (ed): "Analytical and preparative isotachophoresis," Berlin and New York: Walter de Gruyter & Co., pp 347-356.

Zeiller K (1972). Separation of lymphocyte subpopulations by use of free flow electrophoresis: A comprehensive survey of recent research. Behring Inst Mitt 52:11-15.

Zeiller K, Holzberg E, Pascher G, Hannig K (1972). Free flow electrophoretic separation of T and B lymphocytes. Evidence for various subpopulations of B cells. Hoppe-Seyler's Z Physiol Chem 253:105-109.

Zeiller K, Loser R, Pascher G, Hannig K (1975). Free flow electrophoresis II. Analysis of the method with respect to preparative cell separation. Hoppe-Seyler's Z Physiol Chem 356:1225-1244.

Zeiller K, Pascher G, Wagner G, Leibich HG, Holzberg E, Hannig K (1974). Distinct subpopulations of thymus dependent lymphocytes. Immunol 26:995-998.

Cell-Free Analysis of Membrane Traffic, pages 69–73

PLASMA MEMBRANES FROM AMPHIBIAN EPIDERMIS: SUBFRACTIONATION BY PREPARATIVE FREE-FLOW ELECTROPHORESIS

M.I. Burón, P. Navas, G. García-Herdugo and D.J. Morré

Departamento de Biología Celular, Facultad de Ciencias, Universidad de Córdoba, Córdoba 14071, Spain and Department of Medicinal Chemistry, Purdue University, West Lafayette, Indiana 47907, U.S.A.

The amphibian epidermis is organized into four major strata that represent a developmental continuum of epidermal maturation and keratinization (Parakkal and Alexander, 1972). The outermost, mature and fully keratinized cells of the stratum corneum are derived from underlying stratum granulosum cells which in turn find their origins in the less differentiated stratum espinosum. The innermost stratum of proliferative cells, the stratum germinativum, is the ultimate source of the different epidermal layers. Plasma membrane of amphibian epidermis was isolated by aqueous two-phase partition and fractions of about 90% purity were obtained. These fractions were 40-fold enriched in K^+-stimulated, ouabain-inhibited, p-nitrophenylphosphatase activity (K^+-pNPPase) relative to the total homogenate and, based on activities of marker enzymes, contents of mitochondria, Golgi apparatus and endoplasmic reticulum were less than 0.15 that of the total homogenate (Table 1). Yields of plasma membrane were estimated to be 3 mg protein/100 mg of epidermis and represented a recovery of about 50%, according to methods described elsewhere (Croze and Morre, 1984).

Preparative free-flow electrophoresis separation is presumably based largely on surface charge (Hannig and Heidrich, 1977) such that plasma membrane vesicles carrying the greatest net negative charge should migrate the farthest toward the anode. When the two-phase purified membranes were subjected to free-flow electrophoresis, a gradient of KK^+-pNPPase, reminiscent of the activity gradient from base

TABLE 1: Marker enzyme composition of total plasma membranes prepared from frog skin by two-phase partition. Results are averaged from at least three different preparations ± SD.

Enzyme marker	Specific Activity[1] Total homogenate	Plasma membrane	Enrichment (PM/TH)
K^+-pNPPase (Plasma membrane)	6.2±2.4	266±8	43
Succinate-INT-reductase (Mitochondria)	0.066±0.001	0.004±0.0004	0.06
Glucose-6-Phosphatase (Endoplasmic reticulum)	0.15±0.002	0.006±0.001	0.04
NADPH-cyto c-reductase (Endoplasmic reticulum)	0.4±0.06	0.06±0.012	0.15
Galactosyltransferase (Golgi apparatus)	2.1±0.09	0.04±0.001	0.02

[1]Units of specific activity of K^+-pNPPase and galactosyltransferase are nanomoles/h/mg protein. All others are μmoles/h/mg protein.

to apex for frog epidermis from cytochemical investigations, was observed (Fig. 1).

A gradient of glycoconjugates with carboxyl groups or with sulfate esters from the base to the apex is well established for amphibian epidermis (Navas et al., 1985). Both sulfate and sialic acid showed a decreasing activity gradient from a maximum at about fraction 34 (Fig. 1) to a minimum at about fraction 41. The most electronegative fractions nearest the anode and to the left of the main protein peak were enriched in thick membranes similar to those of the stratum corneum.

Except for the stratum corneum and certain regions within the replacement cell layer, a gradient of increasing specific activity of the sodium pump would be expected among the different epidermal strata (Mills and Dibona, 1977). The concentration of the sodium pump in the plasma membrane of replacement cell layer appears to correlate with the concentration of sodium channels in the apical membranes (Zeiske and van Driessche, 1984). A gradient from base to apex of lectin binding to plasma membrane of amphibian keratinocytes recently has been shown (Villalba et al., 1986) and the same lectins induced a non-competitive inhibition of K^+-pNPPase activity of frog epidermis microsomes

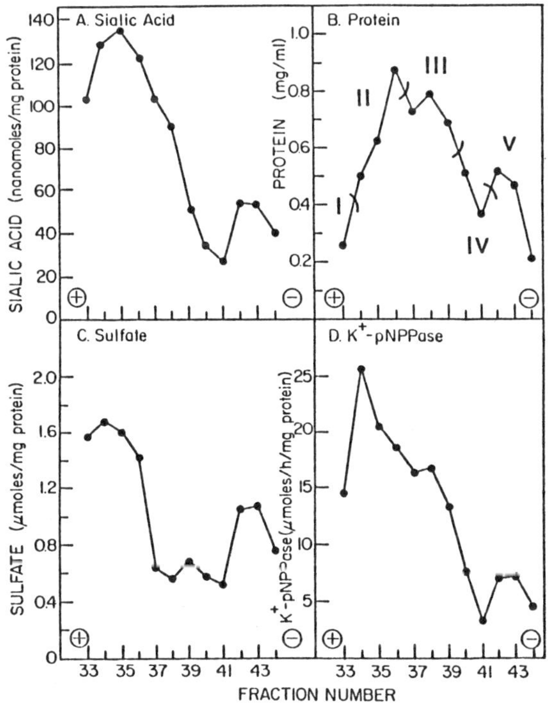

Figure 1: Electrophoretic separations of plasma membrane of amphibian epidermis comparing marker constituents from the most electropositive fraction (44) to the most electronegative fraction (31). **A.** Total sialic acid. **B.** Protein. **C.** Total sulfate. **D.** K^+-pNPPase.

(Burón et al., 1986). The specific sugars that bind these lectins are most representative carbohydrate residues in the B-subunit of the sodium pump (Shuurmans-Stekhoven and Bonting, 1981). Therefore, there is an apparent correlation between gradients of both lectin-binding and K^+-pNPPase, such that they may be considered as markers for plasma membrane differentiation of amphibian keratinocytes.

These results suggest that plasma membranes isolated from mixed cell populations, such as those from epidermis, are at least partially resolved according to activity gradients by free-flow electrophoresis. Additionally, the findings provide, for the first time, independent biochemical confirmation of a base to apex gradient of transport activity associated with the plasma membranes of cells of the different strata of amphibian epidermis.

REFERENCES

Burón I, García-Herdugo G, Navas P (1986). Lectin inhibition and kinetics of microsomal K^+-dependent p-Nitrophenyl phosphatase of frog epidermis. Comp Biochem Physiol (in press).

Croze EM, Morré DJ (1984). Isolation of plasma membrane, Golgi apparatus, and endoplasmic reticulum fractions from singles homogenates of mouse liver. J Cell Physiol 119:46-57.

Hannig K, Heidrich HG (1977). Continuous free-flow electrophoresis and its application in Biology. In Bloemendal H (ed): "Cell separation methods, part IV. Electrophoretic methods." Elsevier/North Holland: Biomedical Press, Amsterdam, pp 93-116.

Mills J, Dibona DR (1977). On the distribution of Na^+ pump sites in the frog skin. J Cell Biol 75: 968-973.

Navas P, Villalba JM, García-Herdugo G (1985). Carbohydrate cytochemistry on the cell surface coat of amphibian keratinocytes during differentiation. Z Mikrosk Anat Forsch, Leipzig 99, 35: 447-454.

Parakkal PF, Alexander NJ (1972). "Keratinization: A survey of vertebrate epithelia." Academic Press, pp 12-15.

Schuurmans-Stekhoven FMAH, Bonting SL (1981). Sodium-potassium-activated adenosine triphosphatase. Physiol Rev 61: 1-76.

Villalba JM, Navas P, García-Herduyo G (1986). Lectin binding patterns in amphibian epidermis. Acta Histochem 81: 51-57.

Zeiske W, van Driessche W (1984). The sensitivity of apical Na^+ permeability in frog skin to hypertonic stress. Plügers Arch 400: 130-139.

B. IMMUNO-ISOLATION AND DEVELOPMENT OF ORGANELLE-SPECIFIC ANTIBODIES

Cell-Free Analysis of Membrane Traffic, pages 77–90

IMMUNO-ISOLATION OF SUBCELLULAR COMPONENTS

Kathryn E. Howell, Jean Gruenberg,
Akio Ito* and George E. Palade**
European Molecular Biology Laboratory,
Postfach 10.2209, Heidelberg, F.R.G.
*Dept. of Biology, Faculty of Science,
Kyushu University, Fukouoka 812, Japan
**Dept.of Cell Biology, Yale Univ. School of
Medicine, New Haven, CT 06510

INTRODUCTION

Cell fractionation has been one of the most important methods cell biologists have used to establish structural-functional relationships for various subcellular components. The methods were generally designed to take advantage of differences in physical parameters, e.g., size, density and charge, among different organelles. Notwithstanding the essentially nonspecific character of these parameters, this first generation of methods was effective primarily because significant differences exist among the subcellular components of immediate interest (nuclei, mitchondria, microsomes, ribosomes and lysososmes). However, questions addressed more recently concern minority components, e.g., Golgi subcompartments, endosomes and vesicular carriers, among which there are only minor differences in physical properties. Functionally meaningful fractions representing these compartments are required to advance our understanding of protein sorting and membrane traffic. The purpose of this conference was to review and assess many of the new innovative approaches under development in this field.

Specific biochemical criteria for cell fractionation would be a desirable alternative to nonspecific physical parameters. Although other chemically specific interactions are available, immuno-isolation using an antibody-antigen recognition system appears to give the greatest promise. The antibody must recognize an epitope of the cognate antigen exposed on the outer surface of the subcellular component to be isolated. Other antigen epitopes may be inaccessible. Using domain specific antibodies to the same membrane-spanning protein, it is possible to distinguish between right-side-out and inside-out vesicles derived from a common source, e.g., the plasma membrane, (Gruenberg and Howell, 1985).

The antibody must be bound to a solid support in order to retrieve the specific component from the other vesicles of the input fraction. The separation of bound from unbound subcellular components is usually accomplished by pelleting the bound fraction together with the solid support. The ensuing pellet is washed by a few rounds of resuspension and pelleting in order to remove occluded or non-specifically bound material.

Advantages

Immuno-isolation circumvents polydispersity, a major problem encountered in traditional procedures. After homogenization of a cell, all the vesicles formed from the same subcellular compartment do not have the same size, density or charge. If these vesicles have the same accessible epitope exposed, they can all be immuno-isolated with the corresponding specific antibody, irrespective of their physical polydispersity. Even minor subcompartments of the cellular membrane systems involved in protein and membrane traffic can be efficiently isolated. In principle, the procedure can be applied using any antigen molecule or macromolecular assembly. Isolation is rapidly achieved, thereby reducing the chances of inactivation which often accompanies lengthy separation procedures (Howell et al., 1978). In addition, immuno-isolation is easily scaled up for preparative needs or down for experiments using tissue culture cell systems.

An added advantage of all immuno-isolation procedures is the possibility of running controls with pre-immune or non-immune antibodies. The blanks thus defined can be used to assess nonspecific binding and correct experimental data. A counterpart for this type of control is not possible in other cell fractionation procedures.

Solid Supports Used for Immuno-Isolation

The solid support used for immuno-isolation is of considerable importance. Its choice depends on the ultimate goal pursued.

Luzio and his collaborators introduced the use of a cellulose fiber meshwork (see review of Richardson and Luzio, 1986, and Luzio et al., this volume). The fibers provide a high concentration of antibody in a small reaction volume; moreover, they have great flexibility so that the antigen can be more efficiently contacted. These features account for the high binding efficiency of the system. However, the meshwork nature of the support may increase entrapment and nonspecific binding. This system has been used extensively by Luzio and his colleagues as well as by others (Devaney and Howell, 1985; DeCurtis and Simons, this volume).

Ito and Palade (1978) used as a solid support commercially available polyacrylamide beads, approximately 2µm in diameter. With this system, occlusion is minimized. The isolated fractions can be characterized by using all standard cell fractionation criteria such as recoveries and enrichments of marker enzymes or antigens (Beaufay and Amar-Costesec, 1976). Most biochemical assays can be carried out directly on the solid support. In addition, beads with immuno-isolated fraction can be processed directly for electron microscopy to obtain information on the morphology of the isolated subcellular components. Morphological surveys are important since the final goal is to recognize the chemistry and function of different structurally defined components. The procedure also provides a representative sample of the fraction under study. It obviates the need to ascertain this feature by collecting the fraction on filters or by the systematic examination of layered pellets. The procedure as originally described was a modification of a method previously used by Kawajiri et al., (1977) to subfractionate hepatic microsomes. The modifications introduced by Ito and Palade (1978) made possible the morphological survey of the isolated fractions. The original procedure has been used by others (Miljanich et al., 1982).

Another solid support, fixed Staphylococcus aureus cells, was introduced by Merisko et al., (1982). In addition to all the positive features of the polyacrylamide beads, it has the major advantage that Protein A, an immunoglobulin binding protein present at high density on the surface of the bacterial cells, provides a built-in, specific linker molecule. Staph A cells, like the cellulose immunoadsorbent, have a high antibody binding capacity per unit volume. However, with both Staph A and polyacrylamide beads, the same density of specific antibody bound per μm^2 surface area is achieved at saturation (~900 IgG molecules/μm^2), providing the same efficiency in immuno-isolation (expressed as µg protein isolated/μm^2 solid support surface area) (Gruenberg and Howell, 1985). Staph A can be processed for electron microscopy with less difficulty than polyacrylamide beads, (infiltration of the resin into the solid support is uniform and more rapid). In addition, Staph A cells are very inexpensive. Merisko et al., (1982) originally used Staph A cells for the immuno-isolation of coated vesicles and clathrin cages from crude vesicular preparations. Roman and

Hubbard (1984) have applied this procedure for the isolation of specific plasma membrane domains from a purified rat liver plasma membrane. Pfeffer and Kelly, (1985) and Schulze-Lohoff et al., (1985) used Staph A for the separation of different classes of coated vesicles. Additional examples of the Staph A procedure are presented in this volume by Hubbard et al.

The magnetic solid support, introduced by Howell et al. (1986), has the advantage of making possible rapid and efficient retrieval of immuno-isolated subcellular components. The magnetic beads, designed and synthesized by J. Ugelstad, University of Trondheim, Norway (Ugelstad et al., 1983; Kemshead and Ugelstad, 1985) are monodisperse spheres 4μm in diameter which offer many coupling possibilities for any antibody or linker molecule. The major favorable features of the magnetic bead system are an increased yield of immuno-isolated vesicles and facilitated removal of occluded or non-specifically adsorbed vesicles. Both are achieved because the washing of the (bead-

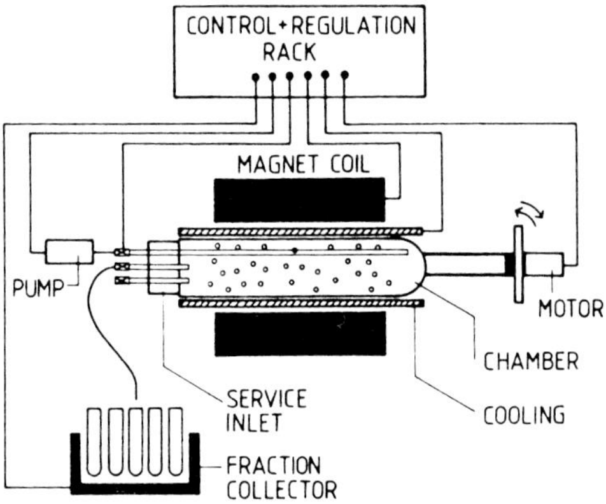

Figure 1. The magnetic beads are suspended as a dilute dispersion within a chamber placed within a magnetic field generated by an electromagnet. The beads and bound vesicles are retained within the chamber by the magnetic field while washing buffers are pumped through to remove the unbound and non-specifically bound components of the input fraction. The service inlet contains three ports; the buffer is pumped into the distal end of the chamber via the upper port; the middle port is connected to a fraction collector, and the lower port is used to withdraw samples during the experiment.
[Reprinted with permission from: Microspheres: Medical and Biological Application, (1987) A.R. Rembaum & Z. Tökes, Editors. Copyright, CRC Press Inc., Boca Raton, FL].

antibody-vesicle) complex is carried out under free-flow conditions thus eliminating the need for repeated pelleting and resuspending of the bound fraction and its solid support. This is an important consideration because vesicles bound to a solid support tend to break and reseal into smaller vesicles at successive pelleting and resuspension steps, and some of the small vesicles thus generated are lost during conventional washings. Free flow washing results in significantly increased yields and decreased contamination. The magnetic bead system is diagrammed in Fig. 1.

As with other solid supports, biochemical assays and a morphological survey of the isolated fractions can be routinely carried out directly on the magnetic beads.

Figure 2 is a composite of electron micrographs which illustrate the results obtained with the different immuno-isolation procedures described.

A. Immuno-isolation on cellulose fibers: Affinity purified sheep antibody raised against the F_c portion of mouse IgG was covalently coupled to a diazonium derivative of amino cellulose (Luzio, 1977). The specific antibody, a monoclonal raised against the cytoplasmic domain of the polymeric IgA receptor (Kühn and Kraehenbuhl, 1983; anti-SC 166), was immunoadsorbed. The input was a Golgi-enriched fraction isolated from rat liver at a sucrose gradient density <1.11 (Howell et al., 1978). Smooth membrane vesicles of various dimensions bound to the immunoadsorbent and were isolated. In thin sections the cellulose immunoadsorbent has an amorphous, diffuse appearance. Magnification x 22,000.

B. Immuno-isolation on Staph A cells (from Merisko et al., 1982): IgG from an anti-clathrin serum was first bound to the Staph A cells. A relatively crude smooth microsomal fraction from porcine brain was used as input preparation; its coated vesicles (cv) and clathrin cages (c) bound to the immunoadsorbent and were isolated. None of the smooth membrane vesicles of the input fraction were immuno-isolated. Magnification x 30,000.

C. Immuno-isolation on polyacrylamide beads: the beads are supplied (Bio-Rad, Immunobeads Catalog 170-5104, Richmond, CA) with an anti-rabbit IgG linker already bound to their surface. The specific antibody, affinity purified rabbit anti-cytochrome P_{450} reductase was immunoadsorbed. The input fraction was a smooth membrane fraction of density = 1.15 (Howell et al., 1978); a reasonable uniform population of smooth membrane vesicles (typical of smooth endoplasmic reticulum) was isolated.

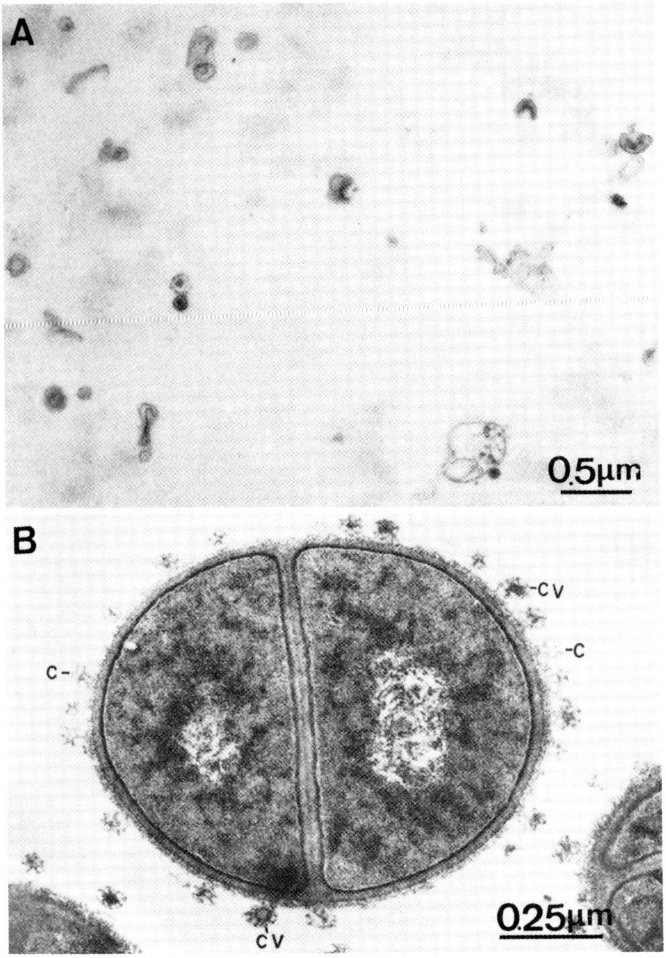

Figure 2.

These vesicles should be contrasted to those isolated on the magnetic beads, in Panel D (In order to section polyacrylamide beads, prolonged infiltration of the embedding resin (Epon or Spurr's) is required.). Magnification x 32,000.

D. Immuno-isolation on 4μm magnetic beads (Ugelstad et al., 1983) (Dynobeads M-450, Dynal Inc., Oslo, Norway and Fort Lee, NJ). The linker molecule, affinity purified sheep antibody raised against the F_C portion of mouse IgG, was covalently bound to the bead surface (Nustad et al., 1982). The specific

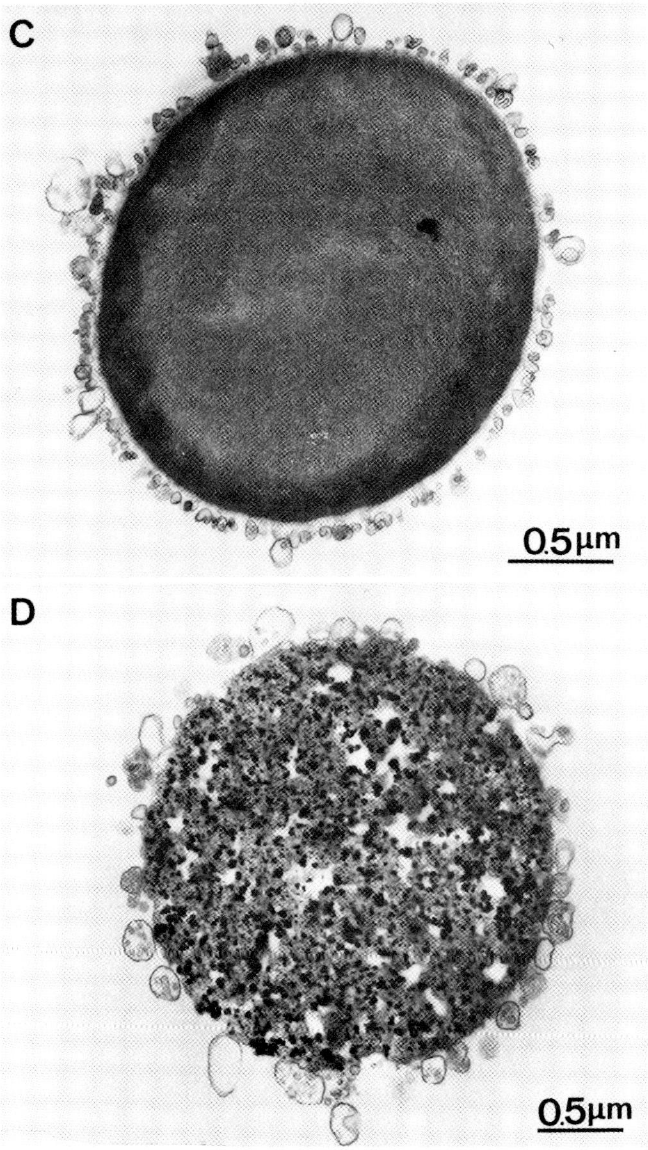

Figure 2. (Continued)

antibody, affinity purified rabbit anti-cytochrome P_{450} reductase was immunoadsorbed. The input was a Golgi-enriched fraction isolated from rat liver at a sucrose gradient density <1.11 (as in experiment in Table 1 and Fig. 2A). Vesicles filled with lipoproteins were immuno-isolated. Magnification x 25,000.

Selection of Antigens for Immuno-Isolation

An important element in the design of an immuno-isolation procedure is the selection of an appropriate antigen. The first requirement is that the antigen must be restricted in distribution to the subcellular component to be isolated. It must be absent or inaccessible for binding in the other vesicles of the input fraction. In order to obtain high yields of the compartment of interest, each vesicle formed from the compartment during cell homogenization must have at least one antigen. In one system, we have defined the critical density of antigen required for successful immuno-isolation as ~50 molecules/μm^2 membrane surface area (Gruenberg and Howell, 1985).

Two fundamentally different approaches to antigen selection have been followed in experiments reported: one relies on endogenous antigens and the other on foreign viral antigens. Endogenous antigens have been used for the isolation of subcellular components in which they are present as resident proteins. Foreign viral antigens have been applied for the isolation of time-defined compartments of the endocytic or secretory pathways.

Examples of Previous Applications Using Endogenous Antigens

In the experiments of Ito and Palade (1978) immuno-isolation made it possible to address the question of the location of a specific antigen, NADPH cytochrome P_{450} reductase, in a relatively purified Golgi fraction. This antigen had been traditionally used as a marker enzyme for the endoplasmic reticulum. Ito and Palade were able to show that the activity assayed in this fraction was not the result of its contamination by vesicles derived from the endoplasmic reticulum since the antigen was also present in vesicles with the morphological characteristics of Golgi elements (see example in Fig. 2, D). This line of investigation has been extended to the analysis of a wider spectrum of enzymatic activities found in microsomal, Golgi and other subcellular fractions and the data are presented in Table 1 (Howell et al., 1981).

The experiments were carried out so that an increased amount of the immunoadsorbent was added to a constant amount of enriched Golgi fraction until maximum immuno-isolation of the vesicles which contained the NADPH cytochrome P_{450} reductase was achieved. At that concentration of input Golgi fraction, protein and enzymatic activities were assayed on aliquots of the input fractions, the immuno-isolated fractions and the unbound fractions. Recoveries were always > than 90%.

A maximal amount of the NADPH-cytochrome b_5 reductase activity of the input fraction was recovered in the same immuno-isolate, suggesting that the two reductase activities were localized to the same vesicles. In contrast, only 20% of the protein and ~20% of the Golgi markers of the input fractions were found in the immuno-isolate. The immuno-isolated vesicles were found to contain almost all of the two microsomal reductase activities, without enrichment for Golgi transferases or thiamine pyrophosphatase. They were also found to be essentially negative for lysosomal or plasma membrane enzyme assays. Similar experiments were carried out by Moriyasu and Ito (1986).

This is an example of experiments in which a relatively purified fraction was used as the input preparation for immuno-isolation. Often, as in this case, prefractionation is a solution to deal with an antigen which has a wide subcellular distribution.

Until now, immuno-isolation has been applied most extensively for the improved separation of plasma membrane fractions (Luzio et al., 1976; Roman and Hubbard, 1984; Sztul et al., 1985; Devaney and Howell, 1985). This is because there are at present many more antibodies available to ectodomains of cell surface proteins than to cytoplasmic domains. Furthermore, antibodies against plasma membrane proteins can be easily raised using whole cells as antigen (Devaney and Howell, 1985).

Problems Encountered in the Selection of Endogenous Antigens

Generating useful antibodies for the immuno-isolation of certain intracellular compartments and subcompartments is a more difficult undertaking. For instance, with the secretory pathway (i.e., the endoplasmic reticulum- Golgi complex- plasma membrane pathway), every compartment has its own resident membrane proteins in addition to membrane proteins in transit to distal destinations. In all cases, antibodies for immuno-isolation will have to be raised against accessible epitopes located in the cytoplasmic domain (endodomain) of these proteins. The number of isolated and well characterized resident membrane proteins is relatively large for the endoplasmic reticulum, but considerably smaller for Golgi subcompartments. So far, specific membrane resident proteins have not been

Table 1. ANALYSIS OF IMMUNO-ISOLATION EXPERIMENT: CO-ISOLATION OF SELECTED MARKER ENZYMATIC ACTIVITIES

	Amounts or Activities **					
Antibody or Beads	**Observed**	**Corrected**	**Observed**	**Corrected**	**Observed**	**Corrected**
Endoplasmic Reticulum Markers (6)						
	Protein		**NADPH-cyt P_{450} reductase**		**NADH-cyt b_5 reductase**	
Non-Immune	91.3±6.1	-	98.7±4.9	-	98.2±2.9	-
Anti-Reductase*	72.8±9.2	80.0	15.1±5.0	16.4	17.1±6.3	19.0
Golgi Markers (6)						
	Galactosyltransferase		**Sialyltransferase**		**Thiaminepyrophosphatase**	
Non-Immune	85.3±8.7	-	82.1±8.3	-	82.9±6.1	-
Anti-Reductase*	68.2±8.2	80.0	66.9±9.1	81.5	70.1±8.7	85.4
Lysosomal Markers (3)						
	Aryl Sulfatase A & B		**N-Acetyl-β-D-glucosaminidase**			
Non-Immune	87.3±2.4	-	92.7±7.7	-		
Anti-Reductase*	83.9±6.2	96.1	91.4±7.7	98.6		
Plasma Membrane Markers (3)						
	5'-Nucleotidase		**Alkaline phosphodiesterase 1**			
Non-Immune	89.4±2.5	-	88.3±2.5	-		
Anti-Reductase*	83.5±3.4	93.4	85.5±3.9	96.8		

* Affinity purified anti-NADPH cytochrome P_{450} reductase

** Observed: Percent of total present in the unbound fraction. Corrected: Amount bound to non-immune beads added to unbound fraction of anti-reductase experiment identified in vesicular carriers. The only exception is represented by the integral and peripheral membrane proteins of a variety of secretion granules (Cameron et al., 1986).

The choice of useful antigens for immuno-isolation appears to be restricted to the endodomain of resident membrane proteins. Unfortunately, using the purified protein as antigen, it has proven to be more difficult to raise antibodies to the endodomains than to the ectodomains, although there are a few exceptions (glycophorin A: Cotmore et al., 1977 and synaptic vesicle membrane proteins: Pfeffer and Kelley, 1985). In cases in which the sequence of the cytoplasmic domain is known, synthetic polypeptides become promising antigens. An interesting alternative may be antibodies which recognize specific modifications (e.g., phosphorylation) of the cytoplasmic domain of transient membrane proteins.

Less is known about resident versus transient membrane proteins of the endocytic pathway. It is likely that the same problems will be encountered in the selection of endogenous antigens (or epitopes of antigens) for the immuno-isolation of the different compartments of the endocytic pathway.

In retrospect, the successes in the isolation of plasma membrane fractions were due to the use of antibodies recognizing epitopes in the ectodomains of resident plasma membrane proteins.

Examples of Previous Application Using Foreign Antigens

The use of an implanted viral glycoprotein can circumvent many of the problems discussed above (Gruenberg and Howell, 1986). The transmembrane glycoprotein of vesicular stomatitis virus, G-protein, was implanted into the plasma membrane of cultured cells by low pH mediated fusion (White et al., 1980). The cells were maintained at 4°C to insure that all antigen molecules are still in the plasma membrane at the onset of the experiment. When the cells were warmed to 37°C, normal membrane traffic resumed and the antigen molecules moved through the endocytic compartments as a synchronized wave. A plasma membrane fraction was immuno-isolated with antibodies against the ectodomain of the G-protein and time-defined endosomal fractions were immuno-isolated using antibodies raised against the endodomain of the same protein. The fusion capacity of these time-defined endosomal fractions has been assayed in a cell free system and is described in this volume (Gruenberg and Howell, 1987a & b).

Perspectives

In this overview we have presented the use of antibody-antigen interactions for cell fractionation. The basic principle in designing such procedures is to replace non-specific, physical parameters by specific, high affinity interactions in the isolation of subcellular components. Other high-affinity, complementary interactions could function equally effectively; for example: ligands and their cognate receptors, substrates or inhibitors and their enzymes, or lectins and specific carbohydrate groups of glycoproteins accessible at the surface of specific membrane compartments. When the location of the antigen used for immuno-isolation is restricted to, or highly concentrated in, a specific subcellular compartment, presumably one with a unique function, then immuno-isolation can provide a powerful means to isolate that component.

As our knowledge increases about minor compartments involved in membrane traffic along both exocytic and endocytic pathways, it will be possible to recognize or design better partners for immuno-isolation procedures. In the process, the analysis of immuno-isolated fractions may provide important information about the molecular mechanisms involved in protein sorting and membrane traffic control in different types of eukaryotic cells.

REFERENCES

Beaufay H, Amar-Costesec A (1976). Cell fractionation techniques. In Korn ED (ed): "Methods in Membrane Biology," NY: Plenum Press, 6: 1-99.

Cameron RS, Cameron PL, Castle JD (1986). A common spectrum of polypeptides occurs in secretion granule membranes of different exocrine cells. J Cell Biol 1103: 1299-1313.

Cotmore SF, Furthmeyer H, Marchesi VT (1977) Immunochemical evidence for the transmembrane orientation of Glycophorin A. Localization of Ferritin-Antibody conjugates in intact cells. J Mol Biol 113:539-553.

Devaney E, Howell KE (1985). Immuno-isolation of a plasma membrane fraction from the Fao cell. EMBO J 4: 3123-3130.

Gruenberg J, Howell KE (1985). Immuno-isolation of vesicles using antigenic sites either located on the cytoplasmic or on the exoplasmic domain of an implanted viral protein. Eur J Cell Biol 38: 312-321.

Gruenberg J, Howell KE (1986). Reconstitution of vesicle fusions occurring in endocytosis with a cell-free system. EMBO J 5: 3091-3101.

Gruenberg J, Howell KE (1987a). An internalized transmembrane protein resides in a fusion competent endosome for less than 5 min. Proc Natl Acad Sci USA 84: in press.

Howell KE, Ito A, Palade GE (1978). Endoplasmic reticulum marker enzymes in Golgi fractions- What does this mean? J Cell Biol 79: 581-589.

Howell KE, Ito A, Palade GE (1981). Heterogeneity of Golgi Membranes. J Cell Biol 91: 265a.

Howell KE, Ansorge W, Gruenberg J (1986). Immuno-isolation system using beads maintained in free flow within a magnetic field. In Rembaum A, Tökes Z (eds): "Microspheres: Medical and Biological Applications," Boco Raton, FL: CRC Press, pp 261-298.

Ito A, Palade GE (1978). Presence of NADPH-cytochrome P-450 reductase in rat liver Golgi membranes. J Cell Biol 79: 590-597.

Kawajiri K, Ito A, Omura T (1977). Subfractionation of rat liver microsomes by immunoprecipitation and immunoadsorption methods. J Biochem 81: 779-789.

Kemshead JT, Ugelstad J (1985). Magnetic separation techniques: their application to medicine. Mol & Cell Biochem 67: 11-18.

Kühn C, Kraehenbuhl J-P (1983). Monoclonal antibodies recognizing the secreted and membrane domains of the IgA dimer receptor. Ann N Y Acad Sci 409: 751-759.

Luzio JP, Newby AC, Hales CN (1976). A rapid immunological procedure for the isolation of hormonally sensitive rat fat-cell plasma membrane. Biochem J 154: 11-21.

Luzio JP (1977). Immunological approaches to the study of membrane features in adipocytes. In Reid E (ed): "Methodological Surveys in Biochemistry V.6," Chichester, UK: Ellis Horwood Limited, pp 131-142.

Merisko EM, Farquhar MG, Palade GE (1982). Coated vesicle isolation by immunoadsorption on Staphylococcus aureus cells. J Cell Biol 92: 846-857.

Miljanich GP, Brasier AR, Kelly RB (1982). Partial purification of presynaptic plasma membrane by immunoadsorption. J Cell Biol 94: 88-96.

Moriyasu M, Ito A (1986). Immunochemical subfractionation of a microsomal fraction of rat liver with antibody-coated *Staphylcoccus Aureus* cells. Cell Struct and Funct (Japan) 11: 351-358.

Nustad K, Ugelstad J, Berge A, Ellingsen T, Schmid R, Johansen L, Børmer O (1982). Use of monosized polymer particles in immunoassays. In Radioimmunoassay and related procedures in medicine 1982, International Atomic Energy Agency, Vienna pp. 45-55.

Pfeffer SR, Kelly RB (1985). The subpopulation of brain coated vesicles that carries synaptic vesicle proteins contains two unique polypeptides. Cell 40: 949-957.

Richardson PJ, Luzio JP (1986). Immunoaffinity purification of subcellular particles and organelles. Applied Biochem & Biotech 13: 133-145.

Roman LM, Hubbard AL (1984). A domain-specific marker for the hepatocyte plasma membrane. III. Isolation of bile canalicular membrane by immunoadsorption. J Cell Biol 98: 1497-1504.

Schulze-Lohoff E, Hasilik A, von Figura K (1985). Cathepsin D precursors in clathrin-coated organelles from human fibroblasts. J Cell Biol 101: 824-829.

Sztul ES, Howell KE, Palade GE (1985). Biogenesis of the polymeric IgA receptor in rat hepatocytes. II. Localization of its intracellular forms by cell fractionation studies. J Cell Biol 100: 1255-1261.

Ugelstad J, Söderberg L, Berge A, Bergström J (1983). Monodisperse polymer particles- a step forward for chromatography. Nature 303: 95-96.

White J, Kartenbeck J, Helenius A (1980). Fusion of Semliki Forest Virus with the plasma membrane can be induced by low pH. J Cell Biol 87: 264-272.

Cell-Free Analysis of Membrane Traffic, pages 91–100

IMMUNOAFFINITY TECHNIQUES FOR THE PURIFICATION AND FUNCTIONAL ASSESSMENT OF SUBCELLULAR ORGANELLES

J. Paul Luzio, Barbara M. Mullock, W.J. Branch and Peter J. Richardson

Department of Clinical Biochemistry, University of Cambridge, Addenbrooke's Hospital, Hills Road, Cambridge CB2 2QR, UK

INTRODUCTION

Over the past decade the obvious potential of immunoaffinity techniques for subcellular fractionation (Luzio et al., 1974, 1976; Ito & Palade, 1978) has been realised in a number of laboratories with the isolation of many organelles from a variety of cells and tissues (for review see Richardson & Luzio, 1986). Unlike most subcellular fractionation techniques which rely on differences in physical parameters between organelles to be separated, immunoaffinity techniques rely on biological differences ie. the expression of different antigens. They are rapid preparative techniques capable of high yield and purity, do not require the exposure of organelles to unusual media or physical shock and can be used when only a small amount of cell material is available. Immunoaffinity techniques for subcellular fractionation can be divided into the use of either direct immunoadsorbents where anti-organelle antibody is itself covalently bound to the solid phase or indirect immunoadsorbents where an antibody binding reagent is covalently bound to the solid phase providing an immunoadsorbent capable of isolating the antibody-organelle complex (Fig. 1).

In practice direct immunoadsorbents have rarely proved successful and it has been argued by analogy with affinity chromatography that an indirect technique is required either to provide a more flexible spacer arm or for steric reasons due to the difficulties inherent in the binding of two large particles. The antibody binding

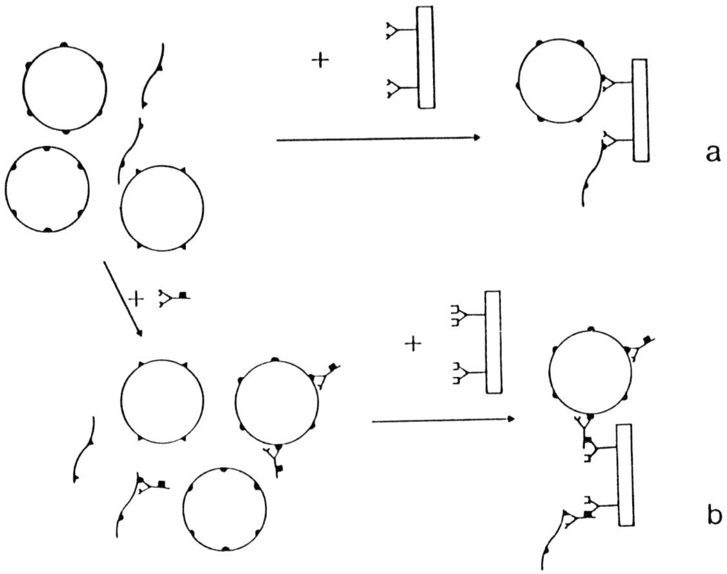

Figure 1. Schematic representation of direct (a) or indirect (b) immunoaffinity techniques for the preparation of subcellular organelles and membranes. The diagram shows that both techniques will select only those membranes and organelles showing the correct antigenic topography, ignoring those with other antigens or those that are "inside out." Reproduced with permission from Richardson and Luzio (1986).

reagent coupled to the solid support may be either an anti-immunoglobulin second antibody (polyclonal or monoclonal) or protein A from Staphylococcus aureus since this binds to the F_c portion of many immunoglobulin molecules (Kronvall et al., 1970). Fixed, heat-inactivated Staphylococcus aureus bacteria have also been used as the solid support (Merisko et al., 1982; Pfeffer and Kelly, 1985; Mueller and Hubbard, 1986). The use of the indirect technique makes it possible to standardise the solid phase for use in the preparation of many organelles with different first antibodies.

Criteria for selecting antibodies, solid phases and

optimal conditions for indirect immunoaffinity purification of subcellular organelles are discussed below.

ANTIBODY SELECTION

The choice of first and second antibodies may be dictated by availability though often it is possible to incorporate some element of design into the choice. In principle the first antibody should be specific to the organelle or membrane fragment of interest, and be of high avidity. In practice polyclonal antibodies prepared to isolated organelles are unlikely to be specific due to the impurity of the organelle preparations and because common antigens exist on different membranes. This may not prove an impossible obstacle, since, for example, in isolating plasma membrane a polyclonal antiserum can be reacted with intact cells before homogenisation so that only the plasma membrane is antibody labelled (Luzio et al., 1976). Performing such an operation in the cold prevents endocytosis of antibody and the kinetics of antibody binding with fast on rates and slow off rates militates against the subsequent redistribution of antibody.

Polyclonal anti-organelle antibodies may be prepared to specific antigens (protein, lipid or carbohydrate) and/or be immunoaffinity purified on the antigen before use. Such purification considerably improves the specificity of the reagent and removes unwanted immunoglobulin which will interfere with the use of the solid phase antibody binding immunoadsorbent. Removal of the non-bound immunoglobulin after the first antibody step normally necessitates washing, but the use of a suitable dilution of affinity purified antibody may preclude the need for this. The use of affinity purified anti-organelle antibodies also allows the possibility of a single purification step since they may be pre-bound to the anti-immunoglobulin immunoadsorbent before incubation with the cell homogenate or partially purified subcellular fraction (Ito and Palade, 1978). A modification of using a polyclonal antiserum directed at an endogenous organelle component is to alter the antigenic structure of the organelle. In the case of the plasma membrane and endosomes this has been achieved by incorporating a viral protein antigen and then isolating the membranes by reaction with anti-viral protein antibodies and anti-IgG immunoadsorbent (Gruenberg and Howell, 1985; 1986).

High affinity monoclonal antibodies are likely to be increasingly used as first antibodies for immunoaffinity subcellular fractionation. In the future antibodies of pre-determined specificity prepared against synthetic polypeptide epitopes or from a molecular cloning approach may prove to be the most useful assuming the antigen is present at sufficient density on the organelle of interest (Gruenberg and Howell, 1985; Devaney and Howell, 1985).

The anti-immunoglobulin antibody coupled to the solid phase for indirect immunoaffinity purification should be of high avidity, react with many immunoglobulin subclasses and be available in large amounts. Appropriately selected monoclonal antibodies are particularly useful since essentially unlimited quantities of antibody can be produced from a single hybridoma cell line. For some species of polyclonal first antibody (e.g. sheep and rabbit) suitable mouse monoclonal second antibodies are available, and if mouse monoclonal first antibodies are used, rat anti-mouse Ig monoclonal antibodies may be coupled to the solid support. Staphylococcus aureus protein A used on the solid support binds to the F_c portion of immunoglobulins, thus allowing directional single step immunoadsorbents to be prepared by reacting with affinity purified polyclonal antibodies and stabilising by chemical cross-linking (Schneider et al., 1982).

SOLID SUPPORT SELECTION

The choice of solid support to be used in the affinity purification of subcellular particles is determined by a number of criteria, the most important of which are high binding capacity, non-porous nature, adequate spacer arms, low non-specific binding, ease of separation from unbound particles and availability.

The binding capacity of any immunoadsorbent is governed largely by the number of attachment sites for the IgG binding molecule (usually the second antibody) and the degree of inactivation incurred during covalent bonding to the matrix. The most favoured matrices for affinity purification of subcellular particles have been non-porous (e.g. Sepharose 6MB, polyacrylamide, cellulose) since these give the greatest surface binding capacities. Although porous supports, e.g. Sepharose 4B, give high total binding capacities, most of the binding sites are within the Sepharose beads and so unavailable for the binding of relatively large subcellular organelles. The highest capacity (as determined by IgG binding) support reported (Richardson et al., 1984) was obtained with a monoclonal second antibody coupled to cellulose. Cellulose immunomatrices are easy to prepare (Hales and Woodhead, 1980), and to use in the separation of bound and free particles since this can be achieved by lowspeed centrifugation. Other solid supports may be separated by column based methods or magnetism. In the future, the latter may provide free flow systems maintaining the magnetic particles in suspension and thus giving a significant reduction in non-specific binding (Howell et al., 1986).

OPTIMISATION OF INDIRECT IMMUNOAFFINITY PURIFICATION

In any specific immunoaffinity protocol for isolating an organelle or membrane fragment the time course and dose response curves for binding of first antibody and immunoadsorbent must be experimentally determined to find the most appropriate conditions for specific isolation and high yield in the shortest possible time (Fig. 2). Incubation and washing conditions are also of importance and should be gentle to prevent disruption of organelles and vesiculation of membranes (Howell et al., 1986).

In contrast to classical subcellular fractionation methods, immunoaffinity purification may be assessed not only from the amount of a given contaminant which co-purifies with the organelle of interest but also by the non-specific binding of organelles to the matrix in the absence of the primary antibody. This is demonstrated in the case of fat cell plasma membrane purification by the data shown in Table 1.

Non-specific binding to matrices is determined both by the quantity of the matrix used and the amount of protein already coupled. Although there are a number of ways of calculating the efficiency of any given adsorption

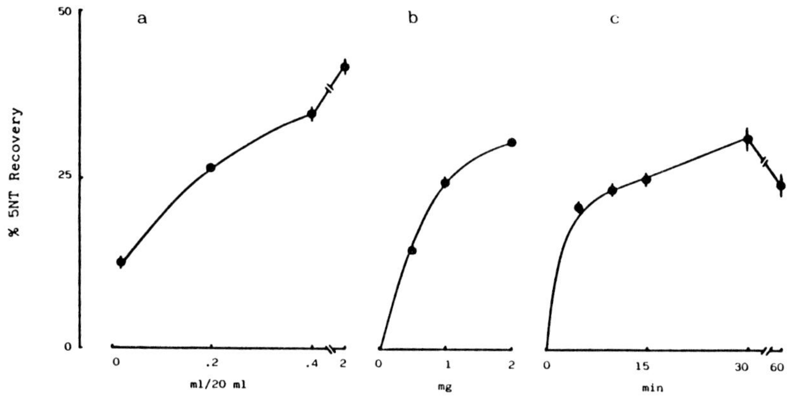

Figure 2. Factors affecting the yield of plasma membrane immunoaffinity purified on cellulose immunoadsorbent from collagenase-isolated rat fat cells. The fat cells from eight epididymal fat pads were incubated with (a) different amounts of anti-cell antiserum and then the homogenate incubated with 2 mg anti-Ig immunoadsorbent for 15 min, 4°C; (b) 1:100 anti-cell antiserum and then the homogenate incubated with different amounts of anti-Ig immunoadsorbent for 15 min, 4°C; (c) 1:100 anti-cell antiserum and then the homogenate incubated with 2 mg anti-Ig immunoadsorbent for different times at 4°C. Plasma membrane yield assessed by measuring 5'-nucleotidase (5NT) recovery. Data redrawn from Luzio et al. (1976).

process (Howell et al., 1986), the most informative is the ratio of specific:non-specific binding which illustrates the degree of purification achieved (Richardson and Luzio, 1986). This ratio falls with increases in the ratio of matrix to organelle used in the adsorption process (Miljanich et al., 1982). Consequently, high capacity immunoadsorbents tend to increase the efficiency of the process, by reducing the amount of matrix required (Richardson et al., 1984).

TABLE 1. Immunoaffinity purification of rat fat cell plasma membrane

Marker	% Homogenate Activity Recovered	
	Anti-IgG ImAd	NI ImAd*
5' Nucleotidase	35.3	4.2
Adenylate Cyclase (basal)	34.8	0.4
Adenylate Cyclase (NaF-stimulated)	36.4	1.6
Cytochrome oxidase	5.8	2.7
NADH-cytochrome c reductase	7.0	1.5
Lactate dehydrogenase	4.2	1.7

*NI ImAd, non-immune Ig coupled to cellulose (immunoadsorbent). Data taken from Luzio et al. (1976)

FUNCTIONAL ASSESSMENT OF IMMUNOAFFINITY PURIFIED ORGANELLES

Although it is difficult to separate membrane fractions from the immunoadsorbent without using extreme conditions, elution is rarely necessary since most immunoaffinity purified particles retain their normal biochemical functions while attached to the solid support. This was recognised in the earliest immunoaffinity purifications when it was shown that rat fat cell plasma membrane isolated on a cellulose immunoadsorbent retained adrenalin sensitive adenylate cyclase activity (Luzio et al., 1974; 1976).

A clearer example of the power of immunoaffinity purification in resolving biological problems was shown in the isolation of cholinergic nerve terminals from mammalian brain. It was suggested that extracellular adenosine might have a function in controlling acetyl choline release (Stone, 1981), but since transmitter-specific mammalian brain nerve terminals cannot be separated by physical methods it was not possible to determine the site of production of such adenosine. Using immunoaffinity purified rat brain cholinergic nerve terminals (Richardson et al., 1984) it was found that synaptic vesicle ATP was co-released with acetylcholine in response to membrane depolarisation and that this ATP could be broken down to ADP, AMP and adenosine by ecto-enzymes including 5' nucleotidase which is on the post-synaptic membrane (Richardson, 1983). The adenosine produced was sufficient to inhibit acetylcholine release in response to a second stimulation of the nerve terminals. Moreover the presence of anti-5'-nucleotidase serum could prevent this inhibition. Thus 5'-nucleotidase and associated ecto-phosphohydrolases located at the cholinergic synapse (Richardson & Brown, 1987) play a role in the feed-back modulation of synaptic transmission.

Immunoaffinity purified subcellular fractions are likely to play an increasingly important role in studying the dynamic interaction of organelles in vitro. Once an organelle is attached to the immunoadsorbent it can easily be added to and removed from an in vitro system. Thus, Gruenberg and Howell (1986) have used immunoaffinity purified subcellular fractions to reconstitute vesicle fusions occurring in the endocytic pathway in a cell-free system. Our own recent experiments have also been directed at establishing cell free systems to analyse events on endocytic pathways. Using density gradient isopycnic centrifugation as an analytical tool it has been possible to establish ATP- and cytosol-dependent, late endosome-lysosome interactions in a rat liver cell free system (Branch et al., 1987). The availability of immuno-affinity immobilised lysosomes as an acceptor fraction in such a cell free system should greatly facilitate the analysis of such endocytic pathway events.

CONCLUSIONS

In general it appears that indirect immunoaffinity techniques are most appropriate for subcellular fractionation. The first antibody should be specific, of high avidity and directed to an antigen present at high density. The immunoadsorbent should exhibit low levels of non-specific binding and be of high capacity using a high avidity second antibody. The availability of a greater variety of organelle-specific antibodies and better, general solid phase immunoadsorbents should allow the continued development of these techniques and particularly their use to investigate organelle interactions *in vitro*.

ACKNOWLEDGEMENTS

We thank the Medical Research Council, St. John's College, Cambridge and the Advanced Drug Delivery Research Unit, Ciba-Geigy Ltd., Horsham, Sussex for financial support.

REFERENCES

Branch WJ, Mullock BM, Shapira NA, Luzio JP (1987). Reconstitution *in vitro* of events on the endocytic pathway for asialoglycoproteins in rat hepatocytes. Biochem Soc Trans, in press.

Devaney E, Howell KE (1985). Immuno-isolation of a plasma membrane fraction from the Fao cell. EMBO J 4:3123-3130.

Gruenberg J, Howell KE (1985). Immuno-isolation of vesicles using antigenic sites either located on the cytoplasmic or the exoplasmic domain of an implanted viral protein. A quantitative analysis. Eur J Cell Biol 38:312-321.

Gruenberg JE, Howell KE (1986). Reconstitution of vesicle fusions occurring in endocytosis with a cell-free system. EMBO J 5:3091-3101.

Hales CN, Woodhead JS (1980). Labeled antibodies and their use in the immunoradiometric assay. Methods Enzymol 70:334-355.

Howell KE, Ansorge W, Gruenberg J (1986). Immuno-isolation system using beads maintained in free flow within a magnetic field. In Rembaum A, Tokes Z (eds): "Microspheres: Medical and Biological Applications," Boca Raton FL: CRC Press, in press.

Ito A, Palade GE (1978). Presence of NADPH-cytochrome P-450 reductase in rat liver Golgi membranes. Evidence obtained by immunoadsorption method. J Cell Biol 79:590-597.

Kronvall G, Seal VS, Finstad J, Williams RC (1970). Phylogenetic insight into evolution of mammalian F_c fragment of a γ G globulin using Staphylococcal protein A. J Immunol 104: 140-147.

Luzio JP, Newby AC, Hales CN (1974). Immunological isolation of rat fat-cell plasma membranes. Biochem Soc Trans 2:1385-1386.

Luzio JP, Newby AC, Hales CN (1976). A rapid immunological procedure for the isolation of hormonally sensitive rat fat-cell plasma membrane. Biochem J 154:11-21.

Merisko EM, Farquhar MG, Palade GE (1982). Coated vesicle isolation by immunoadsorption on Staphylococcus aureus cells. J Cell Biol 92:846-857.

Miljanich GP, Brasier AR, Kelly RB (1982). Partial purification of presynaptic plasma membrane by immunoadsorption. J Cell Biol 94:88-96.

Mueller SC, Hubbard AL (1986). Receptor-mediated endocytosis of asialoglycoproteins by rat hepatocytes : receptor-positive and receptor-negative endosomes. J Cell Biol 102:932-942.

Pfeffer SR, Kelly RB (1985). The subpopulation of brain coated vesicles that carries synaptic vesicle proteins contains two unique polypeptides. Cell 40:949-957.

Richardson PJ (1983) Presynaptic distribution of the cholinergic-specific antigen Chol-1 and 5'-nucleotidase in rat brain, as determined by complement-mediated release of neurotransmitters. J Neurochem 41:640-649.

Richardson PJ, Luzio JP (1986). Immunoaffinity purification of subcellular particles and organelles. Appl Biochem Biotec 13:133-145.

Richardson PJ, Siddle K, Luzio JP (1984). Immunoaffinity purification of intact, metabolically active, cholinergic nerve terminals from mammalian brain. Biochem J 219:647-654.

Richardson PJ, Brown SJ (1987). ATP release from affinity purified rat cholinergic nerve terminals. J Neurochem, in press.

Schneider C, Newman RA, Sutherland DR, Asser U, Greaves MF (1982). A one-step purification of membrane proteins using a high efficiency immunomatrix. J Biol Chem 257:10766-10769.

Stone TW (1981). Physiological roles for adenosine and adenosine 5'-triphosphate in the nervous system. Neuroscience 6:523-555.

Cell-Free Analysis of Membrane Traffic, pages 101–113

CHARACTERIZATION OF THE TRANS-GOLGI NETWORK IN BHK CELLS

Ivan de Curtis and Kai Simons

European Molecular Biology Laboratory
6900 Heidelberg, Federal Republic of Germany

INTRODUCTION

The trans-Golgi network (TGN) is a tubular network of membrane localized on the trans side of the Golgi stack which seems to play an important role in the sorting of the polypeptides in transit through the Golgi complex to their final destinations in the cell (Griffiths and Simons, 1986). This reticulum corresponds to the GERL structure, previously described by Novikoff and coworkers (1964; 1976) in numerous cell types. GERL was postulated to play a role in the sorting of lysosomal enzymes and secretory proteins. The role of the TGN in the sorting of proteins destined to the cell surface comes from studies using a temperature block. Matlin and Simons (1983) have shown that after infection of Madine Darby canine kidney (MDCK) cells with influenza virus, the transport of the viral spike glycoproteins to the cell surface can be blocked by incubating the cells at 20°C. Under these conditions the viral glycoproteins accumulated intracellularly in an endoglycosidase-H resistant form. Griffiths et al. (1985) have localized the site of accumulation of viral spike glycoproteins at 20°C in an analogous system using immuno-electronmicroscopy. Following infection of baby hamster kidney (BHK) cells with the temperature sensitive mutant ts O45 of the vesicular stomatitis virus (VSV) and incubation for

2 h at 20°C, the VSV G protein was shown to accumulate in the membranes of the TGN.

Our aim has been to obtain a fraction enriched in the TGN and to biochemically characterize this compartment. We have used the ts 045 VSV G protein as a functional marker for the TGN. A TGN fraction was obtained by combining density gradient centrifugation with immunoisolation techniques.

RESULTS AND DISCUSSION

G Protein Accumulates in the Golgi at 20°C

The VSV ts 045 mutant was used because it allows synchronization of the G protein transport along the biosynthetic transport pathway. By incubating BHK cells at the non-permissive temperature (39°C) after infection with VSV ts 045 (for 1 h at the permissive temperature (32°C)) the transport of the G protein was arrested at the level of the rough endoplasmic reticulum (RER), as shown by the diffuse intracellular staining of the cells with a polyclonal anti-G protein antibody (Fig. 1a). Shifting the infected cells from 39°C to 20°C for 2 h allowed transport of the G protein to the Golgi complex. Its distribution showed a typical Golgi pattern (Fig. 1c) and was indistinguishable from the staining obtained in the same cells with a monoclonal antibody against the 135 Kd Golgi protein (Burke et al., 1982). The 20°C block was reversible and during a further incubation at 32°C for 30 min most of the G protein was routed to the surface of the cells (Fig. 1e), while the distribution of the 135 Kd Golgi marker did not change (Fig. 1f).

Cell Fractionation of Infected Cells Incubated at 20°C

Based on these observations and on the previous immunoelectron microscopical localization of the G protein in the membranes of the TGN after

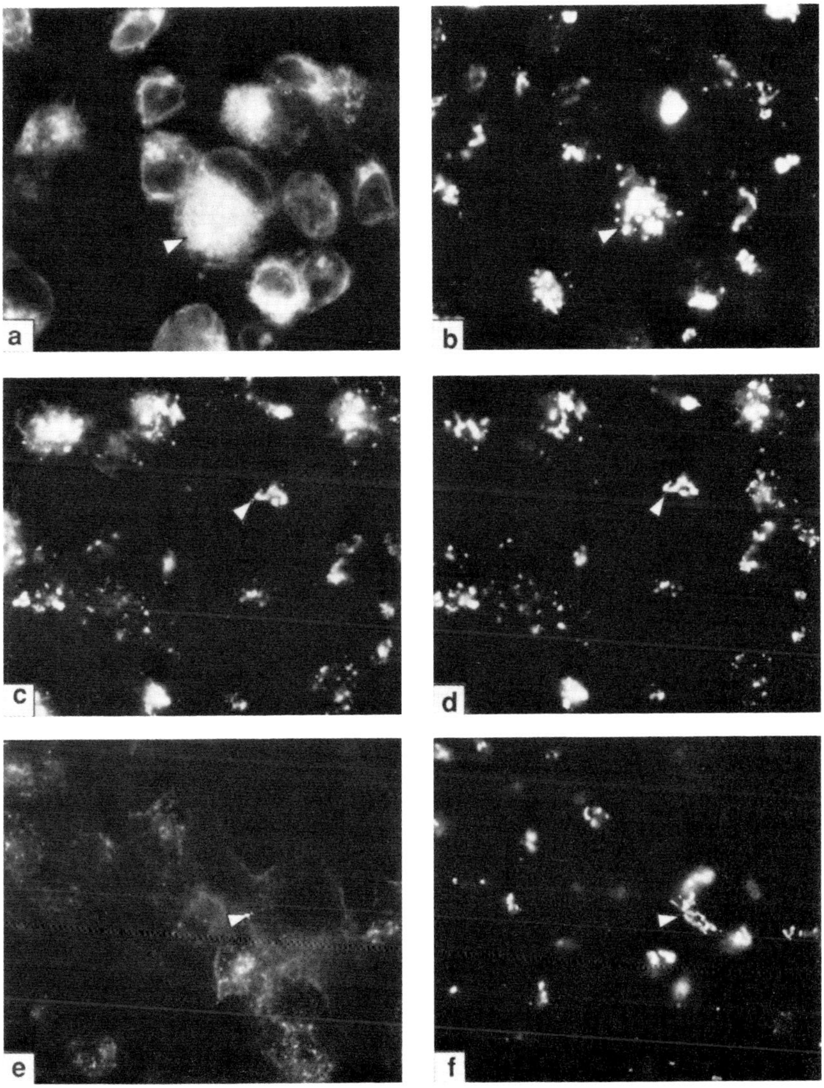

Figure 1. Double indirect immunofluorescence. BHK cells grown on coverslips were infected for 1 h at 32°C with 40 pfu/cell of VSV ts O45 in MEM containing 1% fetal calf serum (FCS), and incubated for 4 h at 39°C (a, b). A second sample was

incubated for 2 more h at 20°C in MEM containing 20 μg/ml of cycloheximide (c, d). A third sample was incubated for a further 30 min at 32°C after the 20°C incubation (e, f). The cells were then cooled on ice, fixed with 3% paraformaldehyde, permeabilized with 0.1% Triton X 100, and double stained with a mixture of a polyclonal anti-G protein antibody (a, c, e), and of a monoclonal antibody against the 135 Kd Golgi protein (b, d, f). Cells were then incubated with a mixture of fluoresceinated goat anti-rabbit IgG and rhodaminated goat anti-mouse IgG antibodies.

---

incubation at 20°C (Griffiths et al., 1985), we concluded that the G protein could be used as a functional marker of the TGN during the cell fractionation procedures. The infected cells were labeled at the end of the 39°C incubation either with ^{3}H-mannose or with ^{35}S-methionine. After the 20°C chase the cells were washed and homogenized as described in Fig. 2. The post-nuclear supernatant was subjected to sucrose density gradient centrifugation and the distribution of the labeled G protein along the gradient was determined by quantitation of the Triton X 114 extractable radioactivity (Bordier, 1981). The gradient fractions were analyzed by SDS-polyacrylamide gel electrophoresis (data not shown). The peak G protein fraction, having a density of approximately d=1.13 g/cm^3, contained only the endoglycosidase-H resistant form of the spike protein (G_2) while the endoglycosidase-H sensitive form (G_1) was concentrated at the bottom of the gradient, as expected for a protein fractionating with the RER (Knipe et al., 1977). Therefore by density gradient centrifugation we were able to remove most of the G_1 protein which had not reached the trans Golgi during the 20°C incubation. The activities of 2 Golgi markers, galactosyl-transferase (Roth and Berger, 1982) and sialyl-transferase (Roth et al., 1985), peaked at slightly lighter densities compared to the peak of the G_2 protein (Fig. 2).

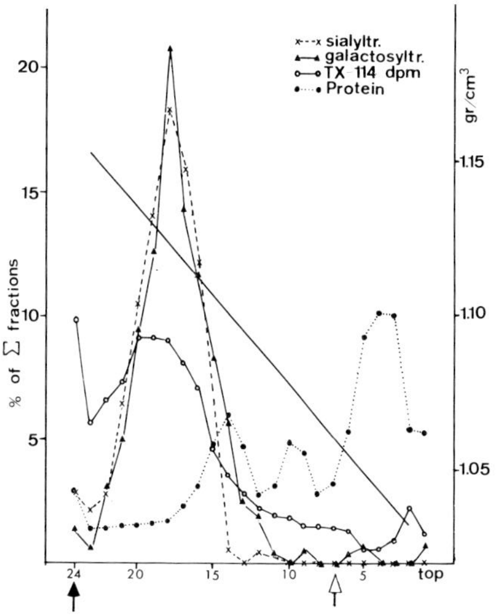

Figure 2. Equilibrium density gradient. BHK cells were infected with VSV ts 045 and incubated for 4 h at 39°C. Half of the cells were labeled during the last hour at 39°C with 50 µCi of ^{3}H-mannose/ml in glucose-free MEM. The cells were then incubated for 2 h at 20°C in presence of cycloheximide and excess unlabeled glucose. The cells were cooled on ice, washed 3 times with PBS, scraped, pelleted at 500 g_{max} for 5 min, washed once with homogenization buffer (0.25M sucrose, 10 mM Hepes-KOH, 1 mM EDTA, pH=7.4) and pelleted again at 500 g_{max}. The pellet was diluted 1:10 with the same buffer, and the cells were homogenized by passing them 5 times through a 22 gauge needle. A post nuclear supernatant (PNS) was obtained by spinning the cells for 10 min at 2000 g_{max}. 1 ml of labeled or unlabeled PNS were loaded on continuous linear sucrose gradients (16.5 ml of 0.3 to 1.3 M sucrose, 10 mM Hepes-KOH, 1 mM EDTA, pH=7.4) and run to equilibrium overnight at 24000 rpm in a SW 27.1 Beckman rotor. 0.7 ml fractions from the two gradients were collected. Radiolabeled fractions were analyzed for G protein (expressed as ^{3}H radioactivity, extractable into Triton X-114) and protein (Bradford, 1976). Unlabeled fractions from a second gradient were analyzed for sialyl-transferase (Bretz et al., 1980) and galactosyl-transferase (Pesonen et al., 1984) activities. The black and the white arrows indicate the location of the peak activity of the mitochondrial marker succinate-INT-reductase and of the lysosomal marker N-acetylglucosaminidase, respectively.

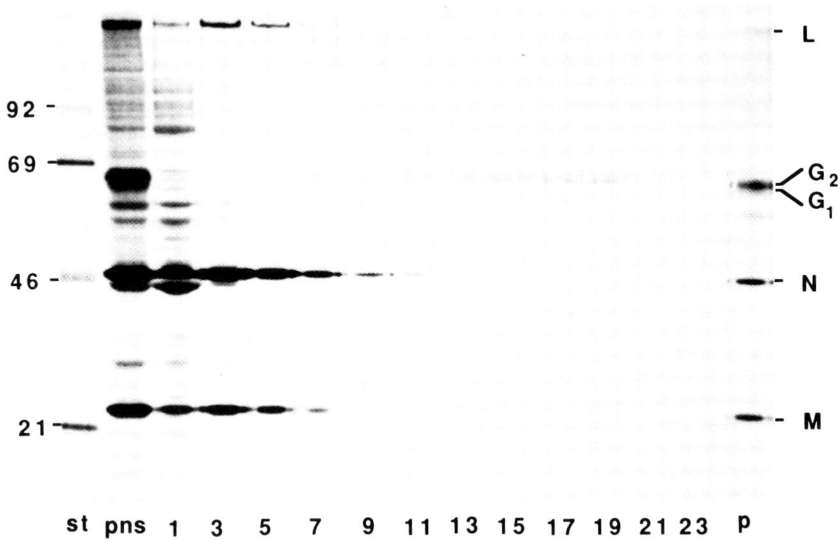

Figure 3. One hour sucrose density gradient. BHK cells infected as described in Fig. 2 were pulse-labeled for 15 min with 80 µCi of ^{35}S-methionine/ml of methionine-free MEM at the end of the 39°C incubation and then chased for 2 h at 20°C in presence of cycloheximide and excess unlabeled methionine. The cells were then cooled on ice, a PNS was prepared and loaded on a continuous sucrose gradient as described in Fig. 2. The sample was run for only 1 h at 24000 rpm. 0.7 ml fractions were collected and analyzed by SDS-PAGE (Laemmli, 1970). Fraction 1 represents the top of the gradient and p is the pellet.

Markers for lysosomes and mitochondria were separated from the G_2 protein (see arrows in Fig. 2). By performing a much shorter centrifugation (1 h) we observed that the bulk of the G_2 protein moved toward its equilibrium density (Fig. 3), whereas the RER form G_1 was still at the top of the gradient, suggesting that the G_2 form was contained in rather large size structures which had not vesiculated during homogenization. These data were confirmed by immunoelectron microscopy of the G_2 enriched fraction from the sucrose density gradient. The electron-micrographs showed well preserved and recognizable reticular structures containing G protein (data not shown).

Immunoisolation of a TGN-Enriched Fraction

Density gradient centrifugation resulted in only a 5-fold enrichment of G protein (G_2) over the starting homogenate. To obtain a higher enrichment of the TGN marker we tried several immunoisolation procedures using different solid supports to which a monoclonal antibody against the carboxyterminal cytoplasmic "tail" of VSV G protein (Kreis, 1986) was bound. The best results were obtained by using cellulose fibers to which a sheep anti-mouse IgG antibody was covalently linked (Luzio, 1977), followed by binding of the specific anti-G tail antibody. This immunoisolation procedure resulted in the recovery of 40% of the G_2 protein present in the peak gradient fractions used as starting material for the immunoisolation. Control fibers (lacking antibodies) bound less than 10% of the material recovered on the specific fibers. Fig.4 shows that the recovery of the activities of the trans-Golgi marker galactosyl-transferase (A) and of the medial Golgi marker N-acetylglucosaminyl-transferase I (B) (Dunphy and Rothman, 1983) were similar to the recovery of G_2 (C). The fact that the immunoisolation procedure was not able to separate the medial Golgi marker from the TGN marker suggests that cisternae of the Golgi stack also were present in the TGN fraction. In similar experiments the recovery of plasma membrane and endosomal markers after immunoisolation were respectively 10 and 30% as compared to G protein (data not shown). When the immunoisolated fraction was stained with a polyclonal antibody against the cytoplasmic tail of VSV-G protein (Fig. 5A) a strong fluorescent

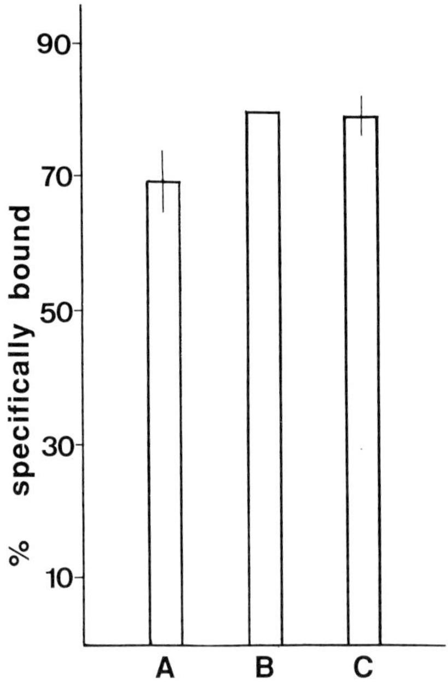

Figure 4. Immunoisolation of a TGN-enriched fraction. Four density gradient fractions corresponding to the peak fractions containing G_2 protein (Fig.2) were pooled, diluted to 0.25 M sucrose and centrifuged for 1 h at 100000 g_{max}. The pellet was resuspended in PBS/0.5% BSA together having the fibers with the specific anti-G-tail antibody, to a final concentration of 1 mg fibers/ml. The same amount of fibers without antibodies was used as a control. The immunoisolation was performed with rotation for 2 h at 4°C . At the end of the immuno-isolation the fibers were pelleted and washed twice with PBS/BSA. The pooled washes and the fibers were then analyzed for recoveries of G protein (A, n=3), galactosyltransferase (B, n=1), and N-acetylglucosaminyltransferase I (C, n=3). Bars represent standard errors. The ordinates express the percentages of the markers specifically recovered on the fibers.

signal was observed. However when this fraction was labeled with polyclonal antibodies against the luminal portion of the G protein very low immunofluorescence staining above the background was observed (5B). This observation suggested that the TGN was recovered on the fibers as sealed structures with the same orientation as in the intact cell. The immunoisolated fraction was analyzed by SDS-PAGE (Fig.6A). Its polypeptide composition was very different from the protein composition of the unbound fraction (B). G_2 was the

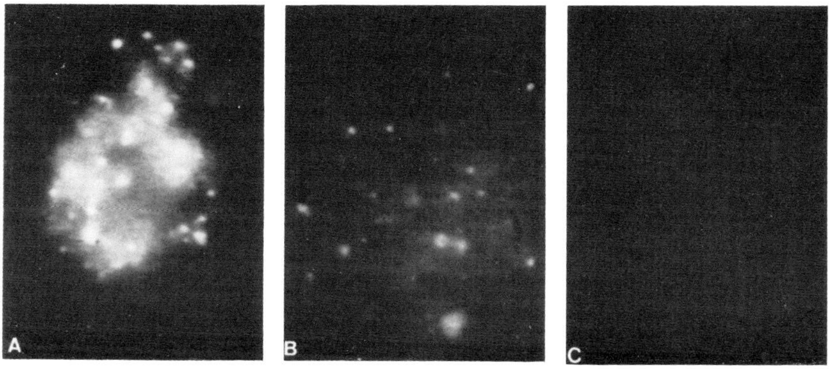

Figure 5. Orientation of the G protein in the immunoisolated TGN. Immunoisolation of the TGN fraction was performed as described in Fig. 4. Aliquots of the immunoisolated fraction were incubated with a polyclonal antibody against the cytoplasmic tail of G protein (A), with a polyclonal antibody against the luminal portion of G protein (B), and without any antibody (C). All three samples were subsequently incubated with a rhodamine labeled goat anti-rabbit IgG antibody.

most prominent band in the TGN fraction. A 25-fold enrichment was calculated for G_2 with respect to the total homogenate by densitometric analysis of autoradiograms. Control fibers (C) show a very low amount of non-specifically bound material. Four additional polypeptides copurified with G_2 in the immunoisolated fraction (Fig. 6, arrows on the left of the gel). These proteins had M_r of 120, 92, 87 and 75 Kd, respectively. Analysis by 2-dimensional gel electrophoresis showed that several spots present in the sucrose gradient fractions before immunoisolation (Fig. 7A) were absent or substantially reduced in the immunoisolated fraction (Fig. 7B), while the 4 spots (arrowheads 1-4) corresponding in M_r to the polypeptides indicated in Fig. 6A were recovered after immunoisolation together with G_2 protein and were well separated from each other.

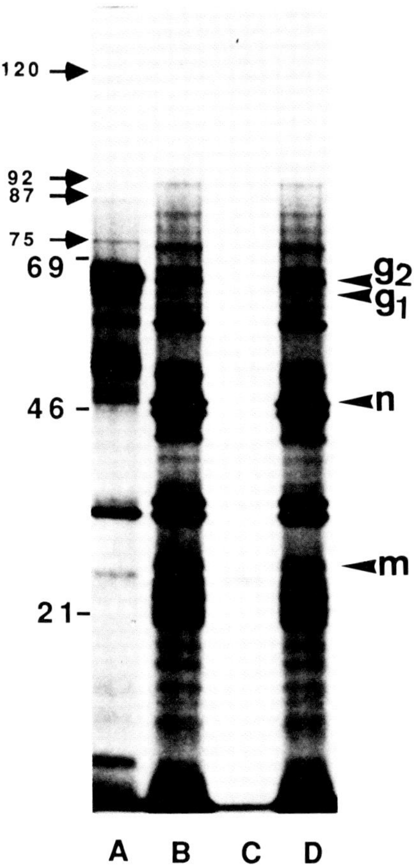

Figure 6. One dimensional gel electrophoresis. BHK cells were metabolically labeled with ^{35}S-methionine (50 μCi/ml of medium) for 18 h before starting infection with VSV ts O45. Infection was carried out in the presence of ^{35}S-methionine. The 20°C incubation was carried out in presence of cycloheximide and excess cold methionine. The cells were subsequently homogenized, fractionated on a sucrose density gradient, and the TGN fraction was prepared as described in Fig.4. Equal volumes of specific fibers (TGN fraction, lane A), control fibers (C) and the respective washes after immunoisolation (B, D) were analyzed by SDS-PAGE and autoradiography. The positions of the viral proteins are indicated on the right , the standards M_r (large numbers) and the M_r of the bands enriched in the TGN fraction (small numbers) are on the left.

One of the advantages of using cells infected with VSV is that the synthesis of cellular proteins is shut off soon after infection with the virus. Under these conditions the G protein is the only protein in transit along the exocytic pathway, and therefore the metabolically labeled polypeptides enriched in the TGN fraction probably represent resident markers of the TGN. One of our future

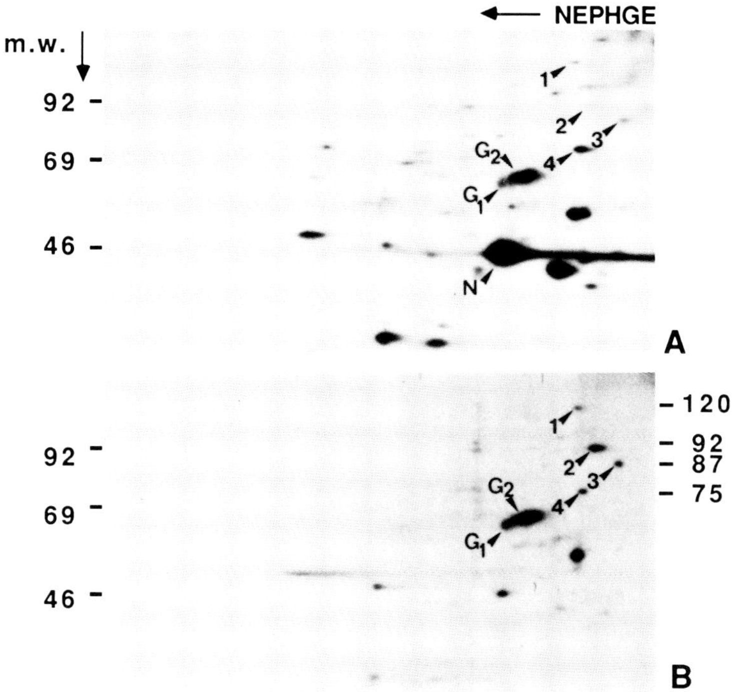

Figure 7. Two-dimensional gel electrophoresis. Immunoisolation of samples from BHK cells, metabolically labeled with ^{35}S-methionine and infected with VSV ts O45 was performed as described in Fig.6. Samples (A) and (B) represent the starting fraction (pooled density gradient fractions) and the immunoisolated fraction from the same experiment, respectively. After running the samples in the first dimension (NEPHGE, Ferro-Luzzi Ames et al., 1976) the gels were put on 10% acrylamide slab gels and run in SDS-PAGE. Spots 1-4 correspond to the polypeptides having the M_r indicated on the right of the gels. The M_r of the standards are indicated on the left. Approximately 8 times more ^{35}S-TCA precipitable cpm were loaded in (A) as compared to (B).

aims is to raise antibodies against these polypeptides which will serve as specific markers for the compartment. Furthermore, the purification of a TGN-enriched fraction is an important prerequisite for constructing an "in vitro" assay to study the sorting and the budding of vesicles from the TGN destined for the plasma membrane of the cell. The membranes of the TGN isolated by the procedures described here are packed with potentially transport competent G proteins.

REFERENCES

Bordier C (1981). Phase separation of integral membrane proteins in Triton X-114 solution. J Biol Chem 256: 1604-1607.

Bradford M (1976). A rapid and sensitive method for the quantitation of microgram quantities of protein utilizing the principle of protein-dye binding. Anal Biochem 72: 248-254.

Burke B, Griffiths G, Reggio H, Louvard D, Warren G (1982). A monoclonal antibody against a 135 Kd Golgi membrane protein. EMBO J 1: 1621-1628.

Dunphy WG, Rothman JE (1983). Compartmentation of asparagine-linked oligosaccharide processing in the Golgi apparatus. J Cell Biol 97: 270-275.

Ferro-Luzzi Ames G, Kaido KN (1976). Two dimension gel electrophoresis of membrane proteins. Biochemistry 15: 616.

Griffiths G, Pfeiffer S, Simons K, Matlin K (1985). Exit of newly synthesized membrane proteins from the trans cisternae of the Golgi complex to the plasma membrane. J Cell Bio 101: 949-964.

Griffiths G, Simons K (1986). The trans-Golgi network: sorting at the exit site of the Golgi complex. Science 234: 438-443.

Knipe DM, Lodish HF, Baltimore D (1977). Localization of two forms of the VSV glycoprotein. J Virol 21: 1121-1127.

Kreis TE (1986). Microinjected antibodies against the cytoplasmic domain of VSV glycoprotein block its transport to the cell surface. EMBO J 5: 931-941.

Laemmli UK (1970). Cleavage of structural proteins during the assembly of the head of bacteriophage

T4. Nature 227: 680-685.

Luzio JP (1977). Immunological approaches to the study of membrane features in adipocytes. In Reid E (ed): "Methodological Surveys in Biochemistry," Chichester: Ellis Horwood Limited, Vol 6, pp 131-142.

Matlin KS, Simons K (1983). Reduced temperature prevents transfer of a membrane glycoprotein to the cell surface but does not prevent terminal glycosylation. Cell 34: 233-243.

Novikoff AB (1964). GERL, its form and function in neurons of rat spinal ganglia. Biol Bull 127: 358A.

Novikoff AB (1976). The endoplasmic reticulum: a cytochemist´s view. Proc Natl Acad Sci USA 73: 2781-2787.

Pesonen M, Ansorge W, Simons K (1984). Transcytosis of the G protein of VSV after implantation into the apical membrane of MDCK cells. I. Involvement of endosomes and lysosomes. J Cell Biol 99: 796-802.

Roth J, Berger EG (1982). Immunocytochemical localization of galactosyl-transferase in Hela cells: codistribution with TPPase in trans-Golgi cisternae. J Cell Biol 92: 223-229.

Roth J, Taatjes DJ, Lucocq JM, Weinstein J, Paulson JC (1985). Demonstration of an extensive trans tubular network continuous with the Golgi apparatus stack which may function in glycosylation: evidence from sialyl-transferase and sialic acid localization in rat hepatocytes. Cell 43: 287-295.

Cell-Free Analysis of Membrane Traffic, pages 115–127

IMMUNO-ISOLATION OF HEPATOCYTE MEMBRANE COMPARTMENTS USING S. AUREUS CELLS

Ann L. Hubbard, William A. Dunn, Susette C. Mueller and James R. Bartles
Dept. of Cell Biology/Anatomy (A.L.H., W.A.D., J.R.B.), Johns Hopkins Univ. School of Medicine, Baltimore, MD 21205, and Dept. of Anatomy (S.C.M.), Georgetown Univ. Medical Center, Washington, D.C. 20007

INTRODUCTION

We have used the immunoadsorption approach to study several aspects of vesicular transport in the isolated perfused rat liver system. Our goal was to establish a simple, fast, yet quantitative and highly-specific method that would be applicable to relatively crude subcellular preparations. We have now characterized several aspects of the immunoadsorption method, in particular those relating to the optimization and simplification of conditions. With this method we have studied when segregation of internalized ligand and recycling receptor occurs (Mueller and Hubbard, 1986). In addition, we have measured the extent to which compartments housing the intracellular epidermal growth factor (EGF) and asialoglycoprotein (ASGP) receptors overlap.

METHODS

Preparation and Characterization of Antibodies

Antibodies to the following membrane proteins were raised in rabbits using protocols similar to that already published (Mueller and Hubbard, 1986): 1) a detergent form of the ASGP receptor (Mueller and Hubbard, 1986); 2) a synthetic peptide, comprising residues #717-726 in the cytoplasmic tail of the human EGF receptor, coupled to keyhole limpet hemocyanin; 3) a synthetic peptide, comprising residues #1-10 in the cytoplasmic tail of the 42 kD subunit of the ASGP receptor coupled to tuberculin PPD. Affinity-purified antibodies were prepared by passage of anti-sera over the relevant receptor-Sepharose-columns and eluting as described (Mueller and Hubbard, 1986). Antibody concentrations were determined by A_{280} (ex coef=1.4 mg/ml). In each case the antibodies were shown by immunoblot analysis to specifically recognize the relevant receptor.

Immunoadsorption of Vesicles Using S. Aureus

The immunoadsorption method determined to be optimal in our hands is presented diagrammatically in Figure 1 and has been published in abbreviated form (Mueller and Hubbard, 1986).

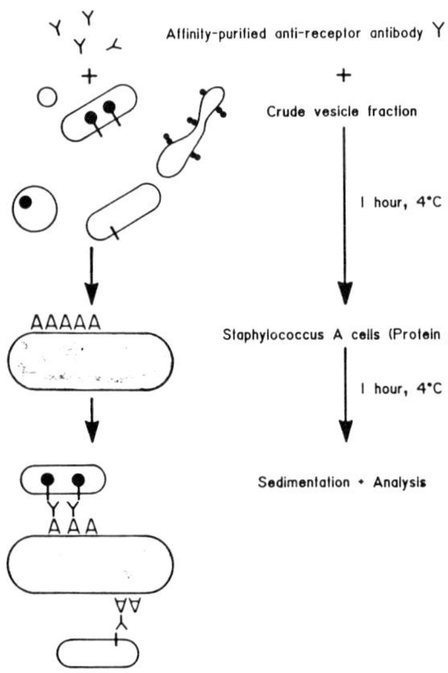

Figure 1. Immunoadsorption with S. aureus cells. Y, antibody molecule: |, receptor recognized by antibody; o, •, ^{125}I-ligand sequestered within membrane-bound vesicle (+/- a receptor).

Membrane vesicles (either crude rat liver microsomes from a 100,000 x g centrifugation step or selected fractions from a continuous sucrose gradient of the microsomes) in 0.25 M sucrose-3 mM imidazole, pH 7.4 (5-300 μg of protein in 50-200 μl) were incubated with affinity-purified antibody (3-6 μg) for 1 h at 4°C with gentle agitation in 15 ml conical centrifuge tubes (Sarstedt). The final volume was adjusted to 1 ml with 1% BSA-0.15 M NaCl-20 mM NaPi, pH 7.4. S. aureus cells (100-125 μl of a 10% fixed suspension (Kessler, 1975)) in the same buffer were added directly to the tube and the incubation continued for an additional hour. The reaction was terminated by addition of 9 ml of 0.25 M sucrose-3 mM imidazole, pH 7.4, and centrifugation at 1500 x g for 25 min (Beckman, TJ-6R). The pellet, containing the S. aureus cells and adsorbed antibody-vesicle complexes, was separated from the supernate (unadsorbed vesicles) and both fractions were analysed.

Analysis

The distributions of sequestered ^{125}I-ligands were determined as described (Mueller and Hubbard, 1986). When enzyme assays or immunoblot analyses were performed, 8 ml of the incubation supernate (unadsorbed vesicles) were centrifuged at 100,000 x g for 60 min (Beckman, Ti 50). The resulting high speed pellet and the initial low-speed pellet (containing the adsorbed vesicles and S. aureus cells) were solubilized for SDS-PAGE or enzyme assay. The immunoblot analysis was performed as described (Wall and Hubbard, 1985). An endoplasmic reticulum (ER) marker, NADH-ferricyanide reductase, was assayed as described (Mueller and Hubbard, 1986; Rogers and Strittmatter, 1973).

Other Methods

Protein A (Pharmacia, Freehold, NJ) was coupled to succinylated aminoethyl-polyacrylamide (Affi-Gel 702) beads according to the manufacturer's instructions (Bio Rad, Richmond, CA). The efficiency of coupling was monitored by inclusion of trace amounts of ^{125}I-protein A. The capacity of the protein A-Affi-Gel beads was compared to that of the S. aureus cells by first saturating the two immunoadsorbents with rabbit anti-mouse IgG (Cooper Biomedical, Malvern, PA), rinsing and then monitoring the binding of ^{125}I-labeled HA 4 mouse monoclonal antibody as described in Figure 5.

RESULTS

Our efforts were directed toward development of a quantitative method with high specificity that could be applied to relatively crude subcellular fractions and utilized standard equipment (e.g. shakers, centrifuges) so that multiple samples could be run simultaneously. The protocol described in some detail above is the one that we have successfully employed to answer several interesting biological questions. We will briefly summarize the validation of this protocol, which has been published (Mueller and Hubbard, 1986), then present recent results using the method to answer a question about a low affinity EGF receptor that we discovered in hepatocytes, and finally, present results of several parameters of the method that we investigated during its development.

Validation of the Protocol

We first wanted to establish that the vesicles containing the membrane antigen of interest were quantitatively immunoadsorbed under our conditions when specific antibody but not when non-immune IgG was used. We chose the ASGP receptor and an antibody to the detergent-solubilized form of this molecule, since the question we were asking was in which endosomal compartment this ligand and its recycling receptor segregated. As shown in Figure 2, all of the ASGP receptor in a low density vesicle fraction from a microsomal gradient was associated with vesicles adsorbed to the _S. aureus_ pellet when specific antibody was used. Substitution with preimmune IgG resulted in no ASGP receptor-positive vesicles in the pellet.

We next applied the method to a cruder fraction, the microsomes, since they contained 60-80% of the total sequestered ^{125}I-ligand, and therefore constituted a more representative population of endosomes for our studies. Since detection of microsomal ASGP receptor by immunoblots was less reliable, our criterion for complete immunoadsorption of receptor-positive vesicles became quantitative immunoadsorption of ^{125}I-ASOR contained in a microsomal fraction from livers that had internalized ^{125}I-ASOR for only 2.5 min. This corresponds to a time when all internalized ligand was likely to be in receptor-positive

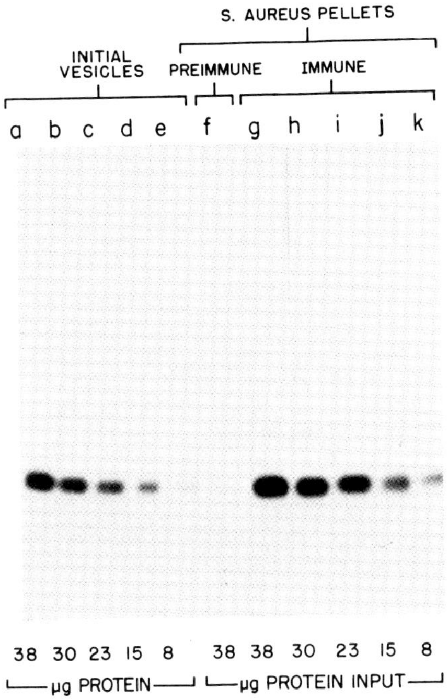

Figure 2. Immunoblot analysis of initial and immunoadsorbed vesicles. Lanes a-e, 38-, 30-, 23-, 15-, and 8-μg protein of an endosome-enriched fraction were separated by SDS-PAGE, transferred to nitrocellulose, and labeled with affinity-purified ^{125}I-anti-ASGP receptor. Lanes f-k, sucrose gradient endosomes at the indicated protein inputs were immunoadsorbed to S. aureus cells with either 2-3 μg preimmune IgG (lane f) or 2-3 μg anti-ASGP receptor (lanes g-k) and the polypeptides present in the final S. aureus pellets labeled with ^{125}I-anti-ASGP receptor. From Mueller and Hubbard, 1986.

endosomes. We found that 98% of the sequestered ^{125}I-ASOR was present in the S. aureus pellet when immune antibody was used in the immunoadsorption reaction, but only 4% in the presence of preimmune IgG (Table 1). Furthermore, only 9% of total microsomal protein ((^{35}S)methionine-labeled) was immunoadsorbed with anti-ASGP receptor versus 2% using preimmune IgG.

To further test the specificity of immunoadsorption from crude microsomes, we assayed both the starting and final vesicles for the presence of membranes expected to contain little or no ASGP receptor (Hubbard et al., 1985). Specifically, we determined whether vesicles derived from bile canalicular membrane or from the ER were adsorbed, using antibody to a specific marker of the apical domain of the hepatocyte, HA 4 (Hubbard et al., 1985), and an assay for the ER enzyme, NADH-ferricyanide reductase. Very little of either marker was found in vesicles incubated

Table 1. Validation of the Immunoadsorption Protocol for Microsomal Vesicles[a]

	Immunoadsorbed %	
	Imm.	Pre-Imm.
Sequestered ^{125}I-ASOR (2.5 min)	98 +/- 9[b]	5 +/- 3
Protein (^{35}S-methionine-labeled)[c]	9	2
Apical membrane protein (HA 4)	5	0
ER enzyme (NADH-ferriCN Red'ase)	8	0

a From livers exposed to ^{125}I-ASOR for 2.5 min at 37°C.
b Represents 70% of total sedimentable ^{125}I-ASOR in homogenate.
c See Mueller and Hubbard, 1986, for details of metabolic labeling.

with either anti-ASGP receptor or preimmune IgG (Table 1). Thus, the membrane vesicles immunoadsorbed using anti-ASGP receptor antibody were not significantly contaminated by these other cellular membranes.

Both the efficiency (% of total antigen adsorbed) and specificity (immune/non-immune adsorption) of our method were quite good when compared to those in other studies. For example, our specificity was among the highest, ranging from 11/1 to 20/1. Others have reported specificities of 3-6/1 (coated vesicle subpopulations by Pfeffer and Kelly, 1985), 5/1 (coated vesicles by Hanover et al., 1984) and 8 or 18/1 (Golgi or ER adsorption by Ito and Palade, 1978). In addition, the efficiency (90-99%) compared favorably to that of others (65%, Ito and Palade, 1978; 70-80%, Roman and Hubbard, 1984; and 95%, Pfeffer and Kelly, 1985).

Use of Immunoadsorption to Study Receptor Distributions

Two ways in which we have used immunoadsorption will be briefly described here. The first study is published and used a combination of approaches to examine when, where and in what compartment segregation of endocytosed ligand from the recycling ASGP receptor occurred (Mueller and Hubbard, 1986). We used immunoadsorption to determine the kinetics and temperature-dependence of ligand movement from receptor-positive (R+) to receptor-negative (R-) endosomes. At early times (2.5 min), 98% of the internalized ligand could be immunoadsorbed and thus was in R+ vesicles (Figure 3). Over the next 12 min of continuous 37°C perfusion with ^{125}I-ASOR, an increasing fraction of the ligand (up to a maximum of 30% after 14 min) was not immunoadsorbed and therefore was present in R-vesicles on its way to lysosomes. Degradation was first detected at 18 min. When livers were maintained at 16°C, ligand was internalized but remained in R+ vesicles. Furthermore, ligand accumulating in R- vesicles at 37°C remained there when livers were cooled to 16°C. Thus, both movement between R+ and R- endosomes and between R- endosomes and lysosomes was inhibited at 16°C.

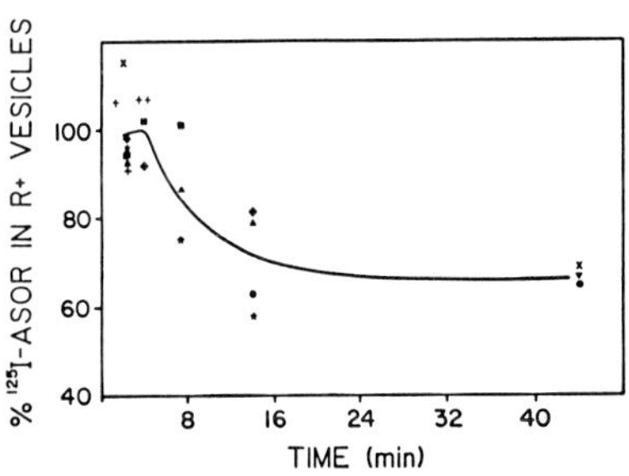

Figure 3. Time course of ^{125}I-ASOR transfer from R+ to R- vesicles during continuous uptake at 37°C. Biopsies were taken from perfused livers at the indicated times, microsomes prepared, vesicles immunoadsorbed and ^{125}I-ASOR in the S. aureus pellet and supernate determined. Symbols are separate experiments. From Mueller and Hubbard, 1986.

Recently, we have used immunoadsorption to measure the extent to which the intracellular pools of EGF and ASGP receptors overlap. Low density vesicles obtained from a sucrose gradient (representing > 90% and 70% of the total intracellular ASGP and EGF receptors) were incubated with affinity-purified anti-receptor antibodies (specific to the cytoplasmic tails of each) and then with fixed S. aureus cells. Anti-EGF receptor antibodies efficiently adsorbed the EGF-receptor-positive vesicles, but only 60% of the ASGP receptor and 20% of the sialyl-transferase (a trans-Golgi marker) were in these vesicles (Figure 4). The remainder of each was in unadsorbed vesicles. Anti-ASGP receptor immunoadsorbed all the vesicles containing the ASGP receptor, and they contained 100% of the low density EGF receptor but, again, only 20% of the sialyl-transferase. These results suggest that intracellular ASGP and EGF receptors reside in the same compartments in hepatocytes and that a majority of the sialyl-transferase is not in this compartment.

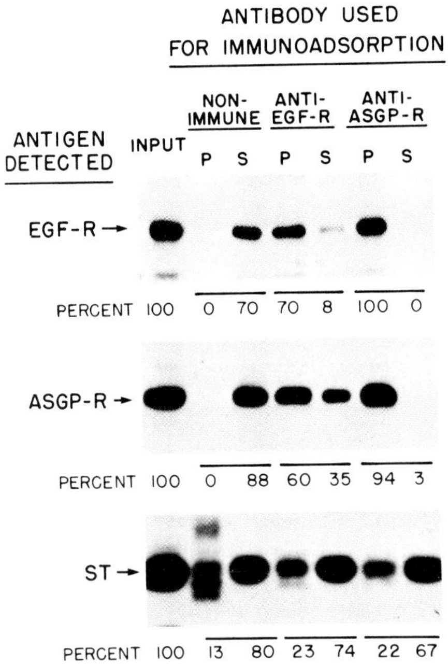

Figure 4. Immunoadsorption of receptor-positive vesicles and quantitation of membrane antigens recovered in the adsorbed and unadsorbed vesicles. Vesicles from a low density fraction of a microsomal sucrose gradient were immunoadsorbed to S. aureus cells with either non-immune IgG (8 μg) or affinity-purified anti-receptor (cytoplasmic domain-specific) antibodies (4-8 μg). The adsorbed vesicles were sedimented (P) and EGF receptor, ASGP receptor, and sialyltransferase (ST) quantified on immunoblots. Unadsorbed vesicles (S) were recovered from the supernate and also analyzed. The values are expressed as the percentage of input and represent an average of three determinations.

Comparison of S. aureus and Protein A-Polyacrylamide as Immunoadsorbents

In early experiments, we examined a protein A-polyacrylamide bead conjugate as an immunoadsorbent. We found that S. aureus cells had a much greater capacity for binding a mouse monoclonal antibody, HA 4, through previously-bound rabbit anti-mouse IgG than did the protein A-Affi-Gel 702 beads (Figure 5).

In Table 2, we have compared these two immunoadsorbents, which are similar in size. Although there is a 17-fold difference in the concentration of protein A per bead or bacterium, the measured IgG binding is 1000-fold higher to the S. aureus cells than to the protein A-bead.

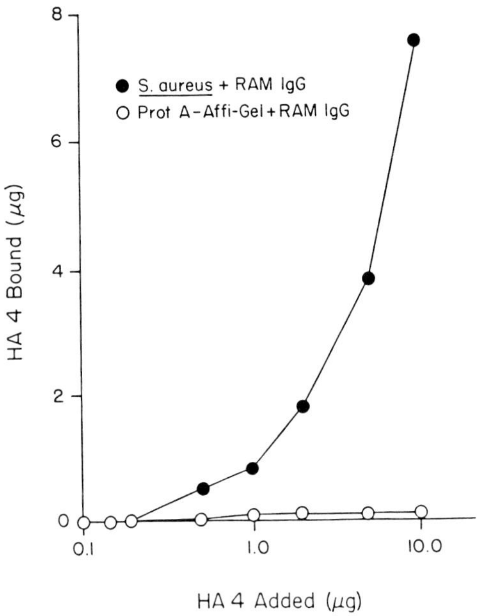

Figure 5. Comparison of S. aureus and protein A-Affi-Gel 702. Fixed cells or protein A-Affi-Gel 702 beads (Prot A-Affi-Gel; ~1 mg of protein A/ml of packed beads) were saturated with rabbit anti-mouse IgG fraction (RAM IgG) by incubation for 2 h at 4°C in 1% BSA-PBS and then rinsed three times by centrifugation. 100 μl aliquots of 10% (w/v) suspensions of the RAM IgG-loaded immunoadsorbents were incubated for 2 h at 4°C with the indicated amount of ^{125}I-HA 4 mouse monoclonal antibody, rinsed twice by centrifugation and the amount of bound HA 4 was determined in a gamma counter.

This may be due to a random orientation of the protein A molecule when coupled to the bead, such that the active site is not available to the F_c portion of the IgG, whereas the correct orientation is presumably presented on the S. aureus cell surface.

Table 2. Comparison of S. Aureus Cells to Protein A-Polyacrylamide Beads

	S. Aureus	Protein A-Affi-Gel 702
Particle size (diam.)	0.5–2 um	1–3 um
Particle concentration of slurry	5% (v/v)	10% (v/v)
	10% (w/v)	13% (w/v)
	per ml of slurry	
Protein A concentration	1.7 mg[a]	0.1 mg
"Estimated" IgG binding capacity	12 mg[b]	0.7 mg
Measured IgG binding capacity[c]	1 mg	1 ug

[a] Surolia A. et al., 1982. Calculation based on bacterium being ∿ 1.7% protein A by weight.
[b] Calculated based upon 2 IgG bound/protein A.
[c] See Figure 5.

Ito and Palade (1978) first reported the use of polyacrylamide as a vesicle immunoadsorbent. They coupled goat anti-rabbit IgG to the beads instead of protein A and then saturated the beads with affinity-purified antibody to an ER membrane enzyme, NADH cytochrome c reductase. It is difficult to compare our results with theirs, since no levels of antigen or antibody were reported. However, they successfully immunoadsorbed a subpopulation of bona fide Golgi elements from 250 μg of Golgi fractions that contained this ER marker. The efficiency was 60% and specificity 8-fold when immune was compared to non-immune antibody. Miljanich et al. (1982) also reported the successful use of polyacrylamide. It is possible that IgG, being larger and more flexible than protein A, is a better molecule to couple to the beads. However, our experience with IgG was no more successful than that with protein A.

Variations of the Immunoadsorption Method

In our effort to achieve the optimal method of immunoadsorption using _S. aureus_ cells, we found three parameters to be important: 1) the order of addition of the reagents; 2) the type of antibody preparation used; and 3) the method of rinsing. We will discuss each of these variations below.

1) _Order of reagents_. When _S. aureus_ cells were first coated with specific antibody (in these early experiments whole anti-serum was used) and then the vesicle fraction added (method #1), as was done by Merisko et al. (1982) and this laboratory earlier (Roman and Hubbard, 1984), we found that the extent of immunoadsorption was less than when vesicles were first coated with antibody (method #2). We speculate that steric hindrance may prohibit interactions between the IgG bound to protein A and the antigen in the membrane. In method #2, there is perhaps a greater probability for several antibodies complexed to antigens on the vesicle membrane to interact with protein A on the _S. aureus_ cells. Another difference that depends on the order of addition of reagents is the amount of antibody required. This is not surprising, since in method #1, antibody must saturate all of the protein A on the _S. aureus_ cells in order to bring multiple antibody molecules into close enough proximity for multivalent interactions with a nearby membrane vesicle. In method #2,

one need only saturate the antigen on the membrane. Thus, when affinity-purified antibody is used, method #1 can be quite costly. For example, we used method #1 to isolate the bile canalicular domain of hepatocyte cell surface and needed 150-250 μg of affinity-purified anti-leucine aminopeptidase. Using method #2, we have needed only 3-6 μg of affinity-purified antibody to any one of a number of antigens (e.g., the ASGP and EGF receptors). Most other studies using S. aureus as the immunoadsorbent have chosen method #1, often with the comment (undocumented) that method #2 led to aggregation of membrane vesicles and increased non-specific adsorption. Although we found small aggregates adsorbed to the S. aureus surface (see Figure 4 of Mueller and Hubbard, 1986), our non-specific binding was quite low. Perhaps this is due to the extensive dilution that terminates our reaction and could disperse aggregates. In their study of brain coated vesicles, Pfeffer and Kelly (1985) also used immunoadsorption method #2 without noticeable aggregation.

2) Type of antibody preparation. When we compared the use of affinity-purified antibody to that of whole anti-serum (on an equivalent specific antibody basis), the amount of specific adsorption was similar, but non-specific binding was much higher. That is, the preimmune controls could be as high as 50% of the immune when whole anti-serum was used, but were routinely 5-15% with IgG. This difference is most likely due to the much lower amounts of total IgG protein present in the affinity-purified preparations as well as the absence of non-IgG components in the serum that might increase aggregation of vesicles. (Non-specific binding also increased when > 15 μg of affinity-purified IgG was used.) The difference was less dramatic using the method #1 order of reagents discussed above (see Table 1 of Roman and Hubbard, 1984) and extensive rinsing; however, more total IgG had to be used (see above). Most investigators have employed affinity-purified antibody for immunoadsorption (but see Merisko et al., 1982).

3) Method of rinsing. In our adopted immunoadsorption protocol, we terminate the reaction by a 10-fold dilution with 0.25 M sucrose-3 mM imidazole, pH 7.4 and one low-speed centrifugation that sediments the S. aureus cells with adsorbed vesicles. We have varied this simple procedure, by either sedimenting the cells once without dilution or by repeatedly resuspending and recentrifuging

them, as most others using immunoadsorption have done. This last method led to loss of both the S. aureus cells and vesicles as well as aggregation of the cells, since the pellets were difficult to resuspend. No dilution prior to one sedimentation led to poor washing of the cells and high backgrounds. Thus, the dilution-sedimentation we currently use is fast, simple, avoids losses, and maximizes specificity (immune vs. nonimmune adsorption).

Concluding Remarks

There are several aspects of the immunoadsorption protocol that we are currently trying to improve. For example, the method is still analytical in scale, yet preparative immunoadsorption would be useful for the purification of receptor-containing endosomes. The desorption of intact vesicles has not yet been achieved, so that assays requiring sealed membranes would have to be performed in the presence of the S. aureus cells. Finally, S. aureus proteins that are solubilized in SDS can interfere with SDS-PAGE and immunoblotting, but repeated extractions in solubilization buffer does decrease this problem. Nonetheless, the immunoadsorption protocol we have described in this article has been very useful in kinetic studies of endocytosis, in the co-localization of endocytic receptors and in membrane biogenesis studies currently underway.

REFERENCES

Hanover JA, Willingham MC, Pastan I (1984). Kinetics of transit of transferrin and epidermal growth factor through clathrin-coated membranes. Cell 39:283-293.

Hubbard AL, Bartles JR, Braiterman LT (1985). Identification of rat hepatocyte plasma membrane proteins using monoclonal antibodies. J Cell Biol 100:1115-1125.

Ito A, Palade GE (1978). Presence of NADPH-cytochrome P-450 reductase in rat liver Golgi membranes. Evidence obtained by immunoadsorption method. J Cell Biol 79:590-597.

Kessler SW (1975). Rapid isolation of antigens from cells with Staphylococcal protein A-antibody absorbent: parameters of interaction of antibody-antigen complexes with protein A. J Immunol 115:1617-1624.

Merisko EM, Farquhar MG, Palade GE (1982). Coated vesicle isolation by immunoadsorption on Staphylococcus aureus cells. J Cell Biol 92:846-857.
Miljanich GP, Brasier AR, Kelly KB (1982). Partial purification of presynaptic plasma membrane by immunoadsorption. J Cell Biol 94:88-96.
Mueller SC, Hubbard AL (1986). Receptor-mediated endocytosis of asialoglycoproteins by rat hepatocytes: receptor-positive and receptor-negative endosomes. J Cell Biol 102:932-942.
Pfeffer SR, Kelly RB (1985). The subpopulation of brain coated vesicles that carries synaptic vesicle proteins contains two unique polypeptides. Cell 40:949-957.
Rogers MJ, Strittmatter P (1973). Lipid-protein interactions in the reconstitution of the microsomal reduced nicotinamide adenine dinucleotide-cytochrome b_5 reductase system. J Biol Chem 248:800-806.
Roman LM, Hubbard AL (1984). A domain-specific marker for the hepatocyte plasma membrane. III. Isolation of bile canalicular membrane by immunoadsorption. J Cell Biol 98:1497-1504.
Surolia A, Pain D, Islam Khan M (1982). Protein A: nature's universal anti-antibody. Trans Biochem Sci 7:74-76.
Wall DA, Hubbard AL (1985). Receptor-mediated endocytosis of asialoglycoproteins by rat liver hepatocytes: biochemical characterization of the endosomal compartments. J Cell Biol 101:2104-2112.

Supported by NIH GM29133.

Cell-Free Analysis of Membrane Traffic, pages 129–139

ORGANELLE-SPECIFIC ANTIBODIES: PRODUCTION OF ANTIBODIES TO GOLGI SUBCOMPARTMENTS

Jaakko Saraste, Mary Bronson, George E. Palade and Marilyn Gist Farquhar

Department of Cell Biology, Yale University School of Medicine, New Haven, Connecticut 06510

INTRODUCTION

After some years of controversy and confusion we now have a general idea about how the Golgi complex is organized and the pathway taken by biosynthetic products in transit through this organelle. Our current concepts of Golgi organization and the flow diagram for products is embodied in the so-called stationary cisternae model (Fig. 1) according to which this organelle is seen to consist of a series of subcompartments arranged in polarized series through which products move sequentially and are progressively modified in transit. At present this model or some variant of this model is widely accepted (Reviewed by Farquhar, 1985).

While this overall framework of Golgi organization is logical and reasonable, detailed information on the localization of Golgi proteins and on the composition and even the exact number of Golgi subcompartments is limited to a few glycosylation processing enzymes -- galactosyltransferase (Roth and Berger, 1982), sialyltransferase (Roth et al., 1985), N-acetylglucosaminyltransferase I (Dunphy et al., 1985), α-mannosidase II (Novikoff et al., 1983) plus a few cytochemical enzyme markers (TPPase, CMPase, and NADPase) as shown in Figure 1.

Recently we set out to attempt to generate markers for Golgi subcompartments using Golgi subfractions enriched in cis, middle and trans Golgi elements as antigens in order to produce antibodies that specifically recognize a single, morphologically distinct Golgi subcompartment.

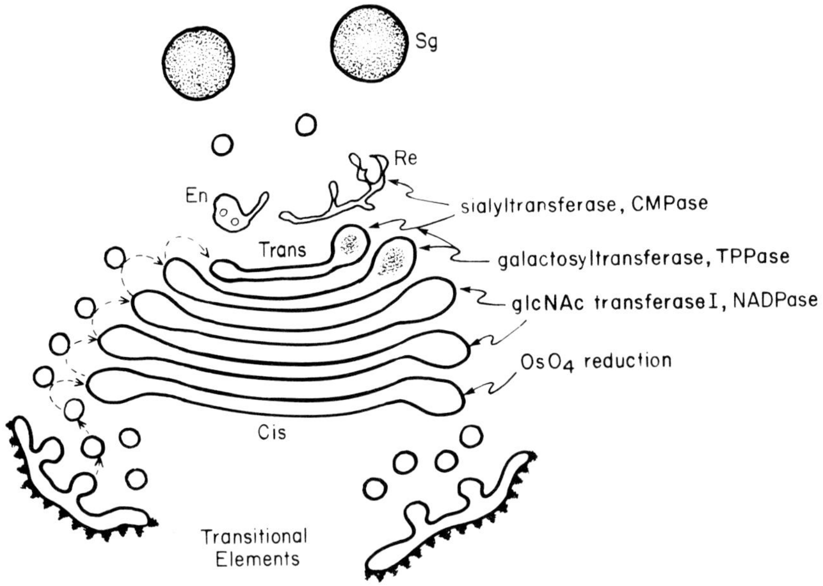

Figure 1. Stationary cisternae model of the Golgi complex. The membrane components that have been localized _in situ_, together with their most frequent localization in either cis, middle, or trans cisternae, are indicated on the right. Note that exceptions and functional modulations in these distributions have also been reported. The flow of biosynthetic products through the Golgi complex is diagrammed on the left. The main features of this model are that (1) each cisterna represents a separate subcompartment with a distinctive membrane composition and internal milieu; (2) products move vectorially from rough ER to transitional elements (part rough, part smooth) located on the cis side of the Golgi complex, and then unidirectionally across the stacks (cis to trans), traversing the cisternae one by one; (3) transport along the route is effected by vesicular carriers; and (4) the main flow of traffic is to the dilated rims of the cisternae. Sg-secretory granules; En-endosome; Re-reticular element or trans Golgi reticulum. (From Farquhar, 1985.)

PREVIOUS ATTEMPTS TO PRODUCE ANTI-GOLGI ANTIBODIES

There have been a few previous attempts to generate antibodies against Golgi components. The first successful attempt was by Louvard, Reggio and Warren (1982) whose goal it was to produce antibodies that distinguish the Golgi complex from other cell organelles. The strategy they used was to immunize rabbits with Golgi fractions prepared from rat liver and to absorb out unwanted antibodies that stained other organelles (e.g., plasmalemma, ER) by absorbtion with the appropriate membrane fractions. With this approach they succeeded in preparing an antibody that specifically recognizes a 135 kD membrane glycoprotein that stains the Golgi complex by immunofluorescence and is broadly distributed across the Golgi stack when localized at the electron microscope level. This antibody has provided a valuable Golgi marker that has been used by many of us to distinguish the Golgi from other cell components. However, due to its broad distribution it is not useful as a marker for specific Golgi subcompartments.

More recently, Chicheportiche, Vassalli and Tartakoff (1984) generated a monoclonal antibody using a Golgi fraction derived from rat myeloma cells that recognizes a 54 kD band by immunoprecipitation, appears to be a peripheral membrane protein (since it is extracted by KCl treatment), and stains primarily middle Golgi cisternae when localized at the electron microscope level. Two other monoclonals have been generated more or less by accident. Lin and Queeley (1982) obtained a monoclonal antibody against a 110 kD heat shock protein that proved to stain both the nucleus and the Golgi complex. Smith et al. (1984) generated another monoclonal antibody that recognizes as 103-108 kD protein present in trans Golgi elements. Both the latter are IgM's and therefore of limited use for immunocytochemical studies.

Thus none of the anti-Golgi antibodies generated so far has proved to be useful as a subcompartment-specific Golgi marker. As stated above, to date only antibodies raised against several purified glycosyltransferases (galactosyl- and sialyltransferase, which are localize in trans Golgi cisternae, and N-acetylglucosaminyltransferase I and α-mannosidase II, localized in middle Golgi cisternae) have been used for this purpose. It is noteworthy that there is currently no known marker for <u>cis</u> Golgi elements.

STRATEGY FOR PRODUCING GOLGI SUBCOMPARTMENT-SPECIFIC ANTIBODIES

The general approach we decided to adopt (Saraste et al., 1987) consisted of 5 main steps: Step 1 was to isolate Golgi subfractions enriched in cis, middle or trans Golgi elements. Step 2 was to separate the hydrophobic membrane proteins from the hydrophilic secretory and membrane proteins of a given fraction using the Triton X-114-phase separation procedure of Bordier (1981). Step 3 was to immunize rabbits with the proteins that partition into the detergent phase (enriched in hydrophobic membrane proteins) using the lymph node injection procedure and protocol of Louvard et al. (1982) which favors production of antibodies to the major antigens present in the mixture. Step 4 was to screen the antibodies on various cell lines (whose tissue and species of origin differed from that of the original antigen) in order to search for widely distributed Golgi antigens. Step 5 was to characterize any antigens identified by immunoblotting and immunoprecipitation and to determine the distribution of the antigen within the Golgi complex by immunoelectron microscopy.

PREPARATION OF GOLGI SUBFRACTIONS FROM RAT PANCREAS

The first decision to be made was the choice of the source and procedure for producing Golgi subfractions. It had been shown by Dunphy et al. (1981) that in CHO cells early and late acting Golgi enzymes separate into heavy and light fractions by fractionation of a total microsomal fraction on sucrose gradients. Subsequently, using the same procedures, Goldberg and Kornfeld (1983) showed that in a mouse lymphoma cell line, the order of fractionation (heavy to light) of a number of enzymes involved in glycoprotein processing is identical to the sequence which they act *in vivo*. The ability to separate Golgi enzyme activities has been quite valuable as an analytical tool in assigning a particular function to a particular region of the Golgi complex. However, the resolution between enzyme activities achieved in cultured cells by us (Brown and Farquhar, 1985,1987) and others was not sufficient for preparative isolation of Golgi subfractions because there was always considerable overlap between marker enzymes.

To get around this problem, we decided to try to prepare Golgi subfractions from a well-organized or compartmentalized cell type *in situ*. For this purpose we chose the pancreatic exocrine cell -- the cell type that has served as the model for studies on the secretory pathway (Palade, 1975). Although techniques for the preparation of reliable cell fractions from pancreatic homogenates had been well worked out, there had been no previous attempt to make Golgi subfractions from pancreas.

To produce Golgi subfractions from rat pancreas we utilized flotation of a total microsome (TM) fraction in sucrose gradients as had been done on cultured cells. Following such a procedure we obtained 5 visible bands, B1-B5 (Saraste et al., 1986). B1, the heaviest band (d=1.20), contained 85% of the total protein and 95% of the RNA (Fig. 2A) and by electron microscopy was seen to consist of rough microsomes. B4 had the highest specific activity for galactosyltransferase (Fig. 2B), a trans Golgi marker. By following the movement of a cohort of radiolabelled secretory proteins, we established that secretory proteins are transported to B2 at 16° and to B3 at 20° (Figs. 3A and B). Based on the enzyme and kinetic marker data we concluded that B2, B3 and B4 represent cis, middle and trans Golgi elements, respectively. An important finding was that the separation achieved between Golgi subfractions obtained from the rat pancreas was much better than had been obtained previously by anyone on cultured cells. This suggested that the subfractions might represent suitable starting material for the generation of antibodies against specific Golgi subcompartments.

GENERATION OF POLYCLONAL ANTIBODIES AGAINST HEAVY (CIS) GOLGI SUBFRACTIONS

We decided to test the strategy outlined above for producing antibodies that specifically recognize only a single, morphologically-distinct Golgi element using membrane proteins derived from B2 (which is enriched in cis Golgi elements) as an immunogen (Saraste et al., 1987). B2 was chosen for our initial trial because there is no known marker to date for the cis Golgi subcompartment, and no cis Golgi resident protein has yet been purified and localized. Toward this end we collected B2 fractions, carried out Triton X-114-phase separation (Bordier, 1981), collected the

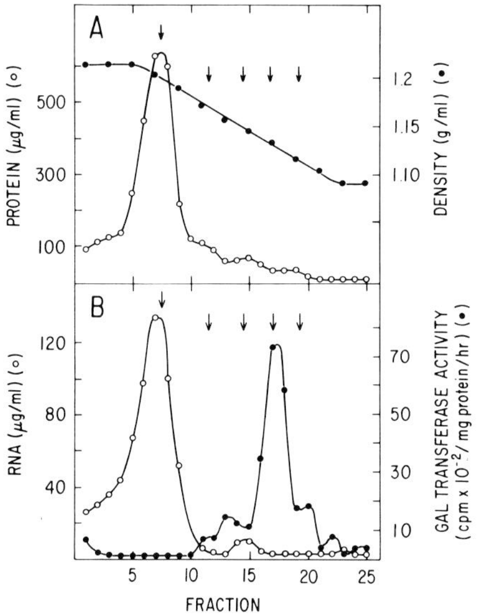

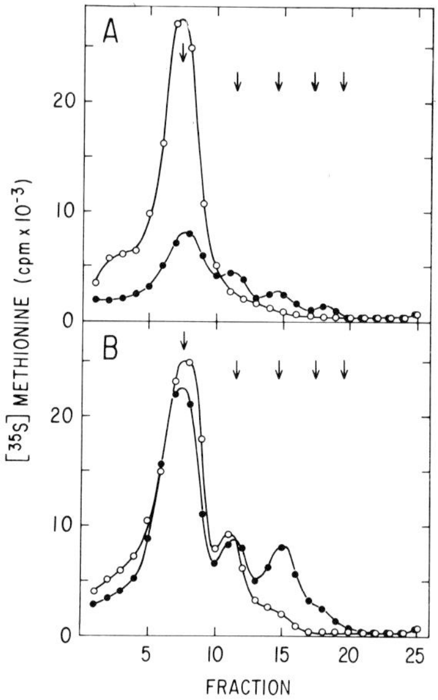

Figure 2 (left). Subfractionation of total pancreatic microsomes (TM) prepared from homogenized lobules by flotation in 20-50% sucrose gradients. Arrows indicate locations of the five bands detected after centrifugation (d=1.19-1.2 (B1), 1.17 (B2), 1.15 (B3), 1.13-1.14 (B4), and 1.12 (B5) g/ml). 25 gradient fractions were collected and their density (A), protein (A) and RNA content (B), and galactosyltransferase activity (B) were determined. The major protein and RNA peaks mark the location of rough microsomes (B1). Galtransferase colocalized with B4.

Figure 3 (right). Analysis of secretory protein transport at reduced temperatures. Pancreatic lobules were labeled for 2 min with [^{35}S]methionine and chased for 60 min at 37, 20, or 16°C. TM were prepared, subfractionated and the fractions assayed for protein radioactivity. Distribution of radioactivity (A) after 60-min chase at 37°C (o) or (B) at the end of a 2-min pulse at 37°C followed by 60 min chase at 16° (o) or 20°C (•). Arrows indicate the locations of the 5 visible bands (B1-B5). At the end of the pulse the incorporated radioactivity was associated with rough microsomes (B1), whereas after chase at 16, 20, and 37°C it appeared in one (B2), two (B2 and B3), and three (B2-B4) smooth membrane subfractions, respectively. (From Saraste et al., 1986.)

proteins that fractionated with the detergent phase and immunized two rabbits by the procedure of Louvard et al. (1982). Since we were interested in obtaining a general Golgi marker for Golgi components, we initially screened by immunofluorescence for Golgi staining on NRK cells (Fig. 4), and mouse RPC 5.4 myeloma cells. One of the rabbits, PC2, produced an antibody that gave a strong Golgi signal by immunofluorescence on both cell types whereas the other, PC1, gave a more punctate cytoplasmic staining. PC2 also stained the Golgi region of pancreatic exocrine cells by immunofluorescence. The results of the initial screening suggested that PC2 might be useful for our purposes and was worthy of further characterization.

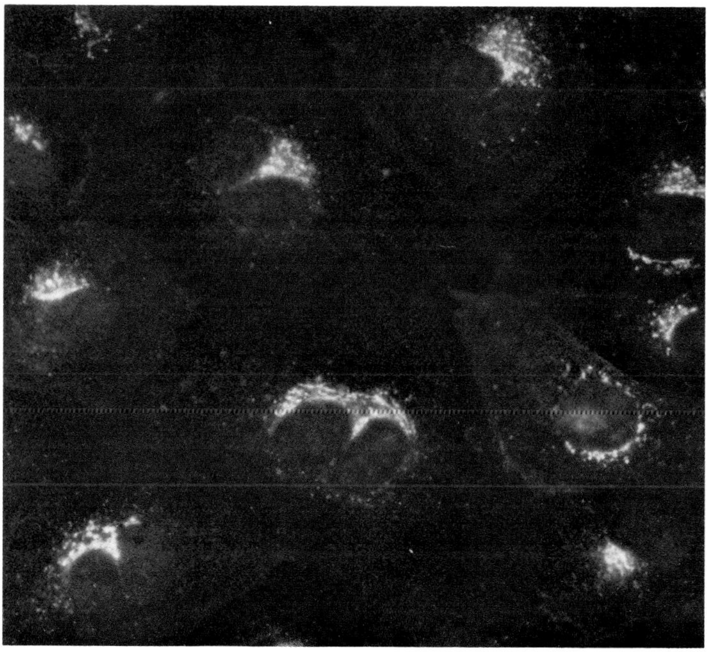

Figure 4. Immunofluorescence preparation of PC2 on aldehyde-fixed NRK cells, demonstrating localization of the antigen in the Golgi region located at one pole of the nucleus. X 1,000.

IDENTIFICATION OF A 58 KD CIS GOLGI MARKER PROTEIN

When immunoblotting was carried out on solubilized B2 membranes with PC2, the antibodies proved to recognize several membrane proteins, including a 58 kD band. In NRK cells, however, PC2 recognized mainly the 58 kD protein plus additional minor bands, and in myeloma cells they recognized exclusively the 58 kD protein by immunoblotting and immunoprecipitation, suggesting that it is the 58 kD antigen that is responsible for the Golgi staining in all these cells. To obtain confirmation of this point, anti-58 kD antibodies were affinity purified by absorption onto the 58 kD band transferred to nitrocellulose. The eluted, affinity-purified antibodies recognized only the 58 kD band by immunoblotting on pancreatic B2 membranes and myeloma cell extracts (Fig. 5), and gave a Golgi staining signal by immunofluorescence. When the antibodies were used to localize the 58 kD protein within Golgi elements of myeloma cells by immunoelectron microscopy, the 58 kD antigen was found to be restricted in its distribution to cis Golgi elements (Fig. 6). Similar findings were obtained on anterior pituitary cells. These results suggest that the 58 kD protein is a resident membrane protein restricted primarily to cis Golgi elements.

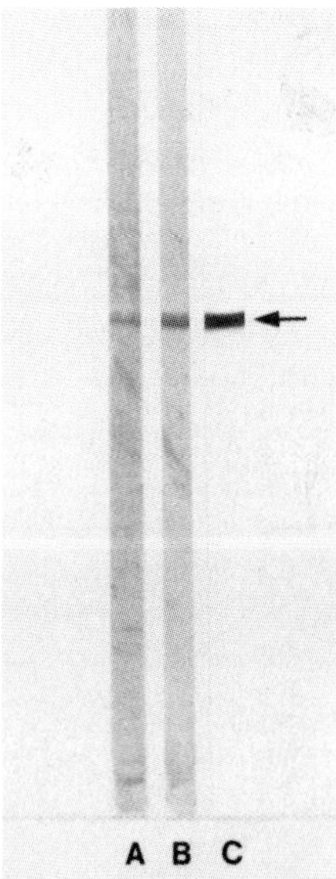

Figure 5. Immunoblots of pancreas B2 proteins (lanes A and C) or myeloma TM (B). Myeloma TM proteins and pancreas B2 proteins were separated by SDS-PAGE (100 μg protein/lane), the proteins were transferred to nitrocellulose, and strips of nitrocellulose were reacted with PC2 IgG (lanes A and B) or affinity purified, anti-58 kD IgG (C). In myeloma cells (B) the whole IgG detects only the 58 kD protein. In pancreas B2 membranes the affinity purified anti-58 kD IgG recognizes only the 58 kD protein (C), whereas the whole IgG reacts with several additional minor bands (A).

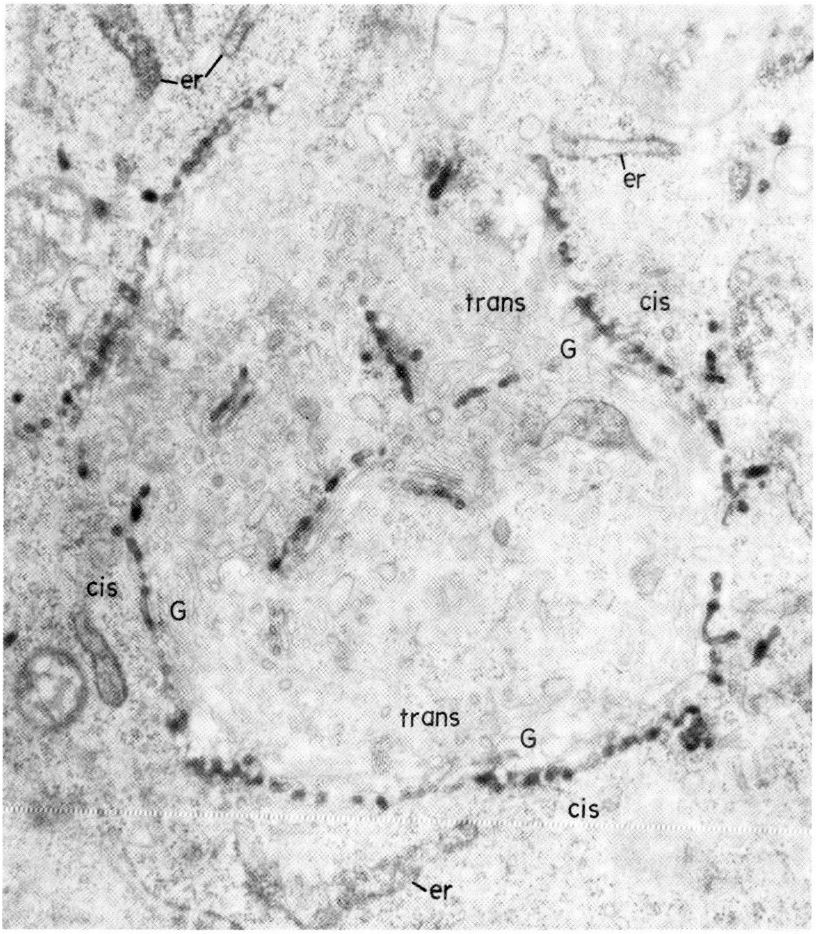

Figure 6. Immunoperoxidase preparation of PC2 IgG in mouse myeloma (RPC 5.4) cells, demonstrating the localization of the 58 kD protein in a single cis Golgi cisterna and in vesicles (ve) located in close proximity to the cis cisterna. The protein is not detected in middle or trans cisternae of the Golgi stack (G). Weak peroxidase reaction is also seen in some of the rough ER cisternae (er).

SUMMARY AND CONCLUSIONS

We have devised a strategy for producing Golgi subcompartment-specific antibodies that involves utilizing Triton X-114-phase separated membrane proteins derived from rat pancreatic Golgi subfractions as immunogens. When we tested the approach by immunizing rabbits with membrane proteins derived from heavy Golgi subfractions that are known to be enriched in cis Golgi elements, we succeeded in generating an antibody that recognized a 58 kD protein that was restricted in its distribution to cis Golgi cisternae in several cell types. Thus we have demonstrated the feasibility of the approach we devised for generation of Golgi subcompartment-specific antibodies, and we have also succeeded in identifying a heretofore unknown cis Golgi marker protein.

REFERENCES

Bordier C (1981). Phase separation of integral membrane proteins in Triton X-114 solution. J Biol Chem 256:1604-1607.

Brown WJ, Farquhar MG (1985). Mannose-6-phosphate (Man6P) receptors are concentrated in heavy (cis) Golgi fractions in Clone 9 hepatocytes. J Cell Biol 101:138a.

Brown, WJ, Farquhar MG (1987). The distribution of 215 kD mannose-6-phosphate receptors within cis (heavy) and trans (light) Golgi subfractions varies in different cell types. Proc Natl Acad Sci USA "in press".

Chicheportiche Y, Vassalli P, Tartakoff AM (1984). Characterization of cytoplasmically oriented Golgi protein with a monoclonal antibody. J Cell Biol 99:2200-2210.

Dunphy WG, Fries E, Urbani LJ, Rothman JE (1981). Early and late function associated with the Golgi apparatus reside in distinct compartments. Proc Natl Acad Sci USA 78:7453-7457.

Dunphy WG, Brands R, Rothman JE (1985). Attachment of terminal N-acetylglucosamine to asparagine-linked oligosaccharides occurs in central cisternae of the Golgi stack. Cell 40:463-472.

Farquhar MG (1985). Progress in unraveling pathways of Golgi traffic. Ann Rev Cell Biol 1:447-488.

Goldberg DE, Kornfeld S (1983). Evidence for extensive subcellular oganization of asparagine-linked oligosaccharide processing and lysosomal enzyme phosphorylation. J Biol Chem 258:3159-3165.

Lin JJ-C, Queally SA (1982). A monoclonal antibody that recognizes Golgi-associated protein of cultured fibroblast cells. J Cell Biol 92:108-112.

Louvard D, Reggio H, Warren G (1982). Antibodies to the Golgi complex and the rough endoplasmic reticulum. J Cell Biol 92:92-107.

Novikoff PM, Tulsiani DRP, Touster O, Yam A, Novikoff AB (1983). Immunocytochemical localization of α-D-mannosidase II in the Golgi apparatus of rat liver. Proc Natl Acad Sci USA 80:4364-4368.

Palade GE (1975). Intracellular aspects of the process of protein secretion. Science 1819:3477-358.

Roth J, Berger EG (1982). Immunocytochemical localization of galactosyltransferase in HeLa cells: Co-distribution with thiamine pyrophosphatase in trans-Golgi cisternae. J Cell Biol 92:223-229.

Roth J, Taatjes DJ, Lucocq JM, Weinstein J, Paulson JC (1985). Demonstration of an extensive trans-tubular network continuous with the Golgi apparatus stack that may function in glycosylation. Cell 43:287-295.

Saraste J, Palade GE, Farquhar MG (1986). Temperature-sensitive steps in the transport of secretory proteins through the Golgi complex in exocrine pancreatic cells. Proc Natl Acad Sci USA 83:6425-66429.

Saraste J, Palade GE, Farquhar MG (1987). Antibodies to Golgi subfractions: Identification of a 58 kD cis Golgi protein. J Cell Biol "in press".

Smith ZDJ, D'Eugenio-Gumkowski F, Yanagisawa K, Jamieson JD (1984). Endogenous and monoclonal antibodies to the rat pancreatic acinar cell Golgi complex. J Cell Biol 98:2035-2046.

ACKNOWLEDGEMENT

This research was supported by National Institutes of Health Grants DK17780 (to MGF) and GM27303 (to GEP). Dr. Saraste was supported by a long term EMBO Fellowship, ALTF 270-1983.

Cell-Free Analysis of Membrane Traffic, pages 141–152

ANALYSIS OF COATED VESICLE STRUCTURE AND FUNCTION USING ANTI-CLATHRIN MONOCLONAL ANTIBODIES

Frances M. Brodsky

Becton Dickinson Immunocytometry Systems, Mountain View, CA 94043. Present address: Dept. of Pharmacy, UCSF, San Francisco, CA 94143

INTRODUCTION

A clathrin-coated vesicle (CCV) is one of the most characterized organelles which has been implicated in mediating intracellular traffic (Pearse and Bretscher, 1981). Although its structural properties are fairly well documented, its precise function in membrane traffic is less defined. Morphologic and biochemical evidence indicate that CCV are responsible for receptor-mediated endocytosis (Goldstein et al., 1985). However, their role in protein export has recently been questioned because of experiments which demonstrate that clathrin-minus yeast mutants are able to transport enzymes to vacuoles as well as secrete them (Payne and Schekman, 1985). Furthermore, electron microscopic studies have indicated that clathrin does not co-localize with newly synthesized viral glycoproteins in infected mammalian cells (Griffiths and Simons, 1986). However, there is strong evidence that, in mammalian cells, CCV contain lysosomal enzymes destined for lysosomes and proteins destined for secretory granules (Brown and Farquhar, 1984; Schulze-Lohoff et al., 1985; Helmy et al., 1986; Tooze and Tooze, 1986). One interpretation of this data is that CCV do not transport newly synthesized proteins which are constitutively expressed but only transport those proteins which must be targeted to organelles following biosynthesis. With lysosomal enzymes, their transport is receptor-mediated and they seem to be sorted in an "endosome-like" compartment before reaching the lysosome (Sly and Fischer, 1982; Griffiths and Simons, 1986). It is possible that proteins destined for secretory granules are similarly directed. Thus, it seems that CCV may play the same role during protein export as they play during endocytosis. That is, they transport proteins which are bound by receptors, to a sorting compartment which allows segregation and targeting of the transported protein and recycling of the receptor. Additional evidence for this more focused definition of clathrin-mediated transport has been provided by the observation of a second type of coated vesicle which seems to contain newly synthesized, constitutively expressed viral glycoproteins in infected mammalian cells (Orci et al., 1986).

The fact that clathrin may be purified in large amounts, has facilitated production of a panel of anti-clathrin monoclonal antibodies (MAbs) (Brodsky, 1985a). Such reagents are readily applicable to addressing questions about the function and assembly of CCV, since they react with clathrin in its native form. They have also provided further structural information about the organelle, which in turn has indicated how CCV may participate in tissue-specific activities. In this respect, CCV provide a model system for the use of MAbs in studying organelles involved in intracellular traffic. The CCV has the relatively simple advantage that the outside is completely coated with a protein composed of only three polypeptides (the 180KD heavy chain and the 30-40KD light chains, LC_a and LC_b) (Pearse, 1978; Kirchhausen and Harrison, 1981; Ungewickell and Branton, 1981; Brodsky and Parham, 1983). However, it is only a matter of time before marker proteins are discovered for other organelles. This discovery will undoubtedly parallel the definition of lymphocyte markers which revolutionized the field of cellular immunology several years ago (Ledbetter et al., 1981). Subsets of lymphocytes share many surface proteins (as do organelles) so it was extremely difficult to produce subset-specific antisera. However, production of MAbs allowed identification of molecules or combinations of molecules unique to particular subsets (Lanier and Phillips, 1986). Even MAbs to common antigens were useful for fractionation of material used for immunization (Radka et al., 1982). The present efforts to make MAbs to organelles may first reveal the common markers but once these are defined, it should be easier to select MAbs to the unique markers. Eventually, it should be possible to study the structure, assembly and function of other membrane organelles using the approaches applied to CCV described below.

PRODUCTION OF ANTI-CLATHRIN MAbs

It has been possible to obtain a panel of MAbs (about 30 total) reacting with a number of different determinants on the clathrin molecule by varying the form of the antigen, the immunization schedule and the strain or species immunized (Kirchhausen et al., 1983; Brodsky, 1985a; Blank and Brodsky, 1986; Brodsky et al., 1987). Clathrin purified from both bovine brain and human brain has been used to immunize mice and rats. Starting with material from evolutionarily distant species may have advantages over immunization with material from closely related species. Some antibodies obtained are species specific. Others are broadly cross-reactive and some even reacted with clathrin from the species immunized. Broad species cross-reactivity is not necessarily expected but can be selected for during screening. Crude protein extracts, purified clathrin triskelions, separated clathrin subunits and large peptides (~6KD) produced by proteolytic digestion have been used as immunogen. The specificity of the antibody obtained is more predictable, the more highly fractionated the antigen. However, all immunizations, except those with peptide antigen, have elicited some antibodies to highly immunogenic minor contaminants.

Anti-clathrin MAbs have only been obtained when methods of hyperimmunization have been used. Antigen for the first immunization was combined 1:1 with Complete Freund's Adjuvant and antigen for the second was combined with Incomplete Freund's Adjuvant, regardless of the protocol used. Concentrations of antigen have varied from 50-500μg/ml. Two protocols for hyperimmunization are as follows:

Protocol I. (used with mice)
Day 1: 200μl antigen intraperitoneal
Day 30: 200μl antigen intraperitoneal
Day 60: 100μl antigen intravenous (tail)
Day 63: 100μl antigen intravenous
Day 64: 100μl antigen intravenous
Day 65: 100μl antigen intravenous
Day 66: 100μl antigen intravenous
Day 67: remove spleen for fusion of splenocytes with mouse myeloma cells

Protocol II. (used with mice and rats)
Day 1: 20-100μl antigen subcutaneous in the footpad
Days 4,7,10,13,17: as above

Day 18: remove popliteal lymph nodes for fusion of lymphocytes with myeloma cells

Since most animal care committees discourage footpad immunizations, it is worth attempting local subcutaneous injections in the vicinity of the popliteal lymph node as a substitute for Protocol II. A test of the immunized animal's serum will indicate whether an antibody response has been produced.

Different mouse strains have produced different antibody responses to the clathrin subunits. The immunodeficient SJL strain was particularly productive (Wanebo et al., 1966). Another type of response was obtained in rats. Lymphocytes isolated from immunized mice were fused with the mouse myeloma cell line SP2/0 (Galfrè et al, 1977; Shulman et al., 1978). When strains other than BALB/c were used, the resulting hybridoma cell lines were grown in ascites fluid induced in irradiated (300 rad) BALB/c mice. Lymphocytes isolated from immunized rats were fused with the rat myeloma cell line, YB2/0, and hybridomas grown in rat ascites fluid (Kilmartin et al., 1982).

STRUCTURAL FEATURES RECOGNIZED BY ANTI-CLATHRIN MAbs

A large number of methods have been used to determine the nature and location of the binding sites recognized by anti-clathrin MAbs. As many of these methods are applicable to other systems, they will be described briefly. The most useful antibodies have been ones which react with native antigenic determinants which remain intact even when the protein is denatured or proteolyzed. Thus far, such antibodies have been generated by immunization with intact protein and peptides rather than by starting with denatured material. The easiest way to screen for reactivity with denatured material is by Western blotting. As might be expected, only one out of ten anti-heavy chain MAbs reacts in Western blots (Brodsky, 1985a) but fifteen out of eighteen anti-light chain MAbs react in Western blots (Kirchhausen et al., 1982; Brodsky et al., 1987). Western blot reactivity can also be used to identify the protein subunit with which an antibody reacts and in some cases, large peptides react in Western blots (Blank and Brodsky, 1987).

MAbs which only react with proteins in native form are still extremely useful in studying organelles *in situ*. For example, they are applicable to studying organelle biosynthesis and determination of structure *in situ*. In the case of clathrin, MAbs have been used to immunoprecipitate biosynthetically labeled clathrin from fractionated tissue culture cell homogenate and thereby determine the *in vivo* form of clathrin (Brodsky, 1985b). It was found that the triskelion form of the molecule, identified *in vitro* (Ungewickell and Branton, 1981; Kirchhausen and Harrison, 1981; Crowther and Pearse, 1981), also exists *in vivo*. Furthermore, the triskelion form, rather than free subunits, appears to be the functional unit which undergoes polymerization and depolymerization *in vivo*.

It is possible to localize the binding site of antibodies which do not react with denatured protein using two approaches. For proteins which are visible by electron microscopy following rotary shadowing, immunoelectron microscopy can be effective. The clathrin triskelion is sufficiently elongated, such that a bound antibody is visible as a bump along the triskelion arm, when triskelion and bound antibody are shadowed together. This technique was used by Kirchhausen et al. (1983) to demonstrate that an antigenic determinant on the clathrin light chains is located at the vertex of the triskelion. We have recently taken a second approach to mapping "native" antigenic sites (Blank and Brodsky, 1987). It has been demonstrated that large fragments of clathrin triskelions can be generated by mild proteolysis of clathrin baskets (Schmid et al., 1982; Kirchhausen and Harrison, 1984). Limited proteolysis cleaves off a globular "terminal domain" and generates a "core" which retains the triskelion conformation, lacking the clathrin light chains. Using a combination of HPLC and analysis by gel electrophoresis, we have refined these methods to produce a series of increasingly smaller, truncated triskelions. By testing antibody binding to these truncated triskelions, using an indirect solid phase radioimmunoassay (Brodsky, 1985a), it has been possible to localize binding sites to regions which are progressively cleaved off the triskelion. Specifically, two antigenic sites on the clathrin heavy chain, which have been shown to be involved in clathrin-clathrin interaction during assembly, have been mapped to two 6-8KD regions located at the triskelion elbow using this approach (Blank and Brodsky, 1987). Neither of the reactive antibodies bind to clathrin heavy chain in Western blots. However, the truncated triskelions allow localization of their binding sites by progressive loss of reactivity. By this method, it has been possible to demonstrate that all of the anti-heavy chain MAbs generated so far react with the "core" of the triskelion form and none bind to the terminal domain.

Localization of binding sites for antibodies which do react with denatured protein can be accomplished by identification of reactive peptides. Mapping the binding sites of MAbs against clathrin light chains was facilitated by the efforts of Parham and Drickamer in purifying and sequencing light chain peptides, as an approach to cloning the genes (Jackson et al., 1987). Peptides were generated from clathrin light chains using CNBr, clostripain and elastase cleavage. Peptides were isolated by reverse-phase HPLC and their protein sequence determined. Anti-light chain MAbs were tested for peptide binding by indirect solid-phase radioimmunoassay. This involves drying peptides from fractions collected following HPLC (3µl at 100µg/ml) onto the wells of polyvinyl chloride plates, rehydrating in phosphate buffered saline and blocking with excess bovine serum albumin. MAb binding is then assayed with radioactive anti-immunoglobulin. Using this approach, it has been possible to localize some antibody binding sites to within 12 amino acids. Now that cDNA clones of the light genes have been isolated and sequenced, the MAb-reactive peptides have been localized within the total protein sequence. Thus far, monoclonal antibodies reacting with sequences in the carboxy-terminal two thirds of both light chains have been identified. The amino-terminal third does not seem to have been immunogenic in the protocols used.

The regions of the clathrin heavy and light chains which have not yet elicited reactive antibodies may be highly conserved, which implies they have significant functions. Generation of MAbs to these regions will be achieved by immunizing with synthetically produced peptides based on protein sequence in the case of the light chains, and by immunizing with isolated terminal domain fragments in the case of the heavy chain. We have preliminary results that this latter approach will be successful. Once MAbs to these regions are available, the significance of these sites in clathrin function can be confirmed. It has been possible to carry out extensive structure-function correlation for MAbs which are already available and recognize more immunogenic sites. This has been achieved by the series of functional assays described in the next section.

FUNCTIONAL FEATURES RECOGNIZED BY ANTI-CLATHRIN MAbS

To associate functional properties with the epitopes recognized by MAbs, we have developed three approaches. The first involves inhibition of protein-protein interactions, using MAbs to identify sites which participate in these interactions. A number of laboratories have demonstrated that purified clathrin assembles into coated vesicle-like structures (baskets) in response to appropriate changes in pH and ionic conditions (Woodward and Roth, 1978; Schook et al., 1979; Keen et al., 1979; Nandi et al., 1980). This assembly can be monitored spectrophotometrically by measuring absorbance at 320 nm (Crowther and Pearse, 1981). Reassembled clathrin baskets scatter light at this wavelength, whereas, disassembled clathrin has no effect. It has been possible to identify sites on the clathrin heavy chain which interact during assembly by assembling clathrin in the presence of anti-heavy chain MAbs (Blank and Brodsky, 1986). It was necessary to use Fab fragments of MAbs to interpret these studies clearly, since cross-linking by IgG caused a large rise in light scattering even when inhibition of proper assembly was occurring. Fab fragments of three MAbs caused a reduction in light scatter generated during the assembly reaction. Examination of the samples by electron microscopy indicated that the Fabs were preventing normal assembly and causing formation of abnormal structures. Further analysis of MAb binding indicated that these three Fab fragments recognized two different sites on the heavy chain, critical for heavy chain interactions during assembly. Correlation of Fab fragment inhibition of assembly function with the proteolysis studies mapping these Fab fragment epitopes to the triskelion elbow, it was possible to develop a model of how triskelion interaction occurs during assembly (Blank and Brodsky, 1986; 1987). This analysis indicates that the triskelion elbow is a critical structural feature which determines correct assembly.

We have also used Fab fragments to probe the interaction between clathrin light chains and the clathrin heavy chain (Brodsky et al., 1987). An assay was established to measure binding of purified, iodinated light chain to heavy chains from which the light chains were removed by mild proteolysis. This reassociation was inhibited by unlabeled, purified light chains and by Fab fragments derived from four MAbs. Of these MAbs, one was LC_b-specific, one was LC_a-specific and two reacted with both light chains. All four MAbs reacted with peptides which localized to corresponding regions on each light chain, strengthening the evidence that this region is involved in contacting the heavy chain when the subunits are associated. This region contains sequences which are homologous to those which form alpha helices in intermediate filaments (Jackson et al., 1987). These results imply that the light chain binding site on the heavy chain may have an alpha helix binding structure.

A second test of MAb function which has proven informative when correlated with structural information has been to assay for organelle binding. MAbs which react with organelle subunits, such as the clathrin heavy and light chains, do not necessarily react with the intact organelle. Characterization of exposed sites on an organelle indicates the regions which are in contact with the cytoplasm and provides target binding sites for staining organelles with antibodies. Exposed sites on CCV have been identified by testing for co-sedimentation of antibodies with CCV and by staining CCV with fluorescent antibodies and analyzing them on a fluorescence-activated cell sorter (FACS). Co-sedimentation can be used to screen a panel of antibodies since iodinated anti-immunoglobulin is used to detect MAb binding. CCV can be sedimented and washed between antibody binding steps in a microfuge (Brodsky et al., 1987). FACS analysis must be done using directly conjugated MAbs since washing causes aggregation of CCV. The fact that CCV stained with anti-clathrin MAbs conjugated to fluorescein or phycoerythrin can be detected using the FACS, suggests that MAbs binding to exposed sites on CCV might be used to detect the presence of CCV subpopulations (Androlewicz et al., 1986). Using both these methods, epitopes on the clathrin light chains which are exposed on CCV have been identified. For both the LC_a and LC_b light chains, these epitopes are adjacent to the insertion sequences which are characteristic of the clathrin light chains found in brain. The fact that this region exhibits tissue polymorphism and is exposed to the cytoplasm suggests that it may be a recognition site for cytoplasmic proteins which bind CCV in a tissue specific fashion, allowing CCV to mediate tissue-specific functions.

The experiments described above use MAbs to identify functional regions of the clathrin molecule. To study the general function of the clathrin molecule itself, it has been possible to examine the effects of functionally active MAbs on intracellular processes by delivering anti-clathrin MAbs and Fab fragments into the cytoplasm of cells. This cytoplasmic delivery is achieved by loading erythrocyte ghosts with antibody and fusing them to target cells via a fusion protein (hemagglutinin) expressed on the surface of the target cell. This technique was developed by Doxsey et al. (1985) and the experiments described below were carried out in collaboration with Doxsey, Blank and Helenius (1987).

Anti-clathrin antibodies in both IgG and Fab fragment form were delivered into the cytoplasm of CV1 cells. Endocytosis was assessed by measuring cellular uptake and degradation of Semliki forest virus (SFV). This process is known to be receptor-mediated and uptake occurs in clathrin-coated pits. In the presence of several anti-clathrin MAbs and Fab fragments, SFV uptake was reduced to half its normal value. Reduction in virus uptake was also measured by a reduction in viral RNA replicated. In addition, the number of clathrin-coated pits on the surface was below normal and the pits had an altered morphology. These results all suggest that the presence of anti-clathrin MAbs in the cytoplasm inhibited receptor-mediated endocytosis and thereby confirm that this is a clathrin-dependent process. Interestingly, the biosynthesis and expression of hemagglutinin on the cell surface was not affected by cytoplasmic delivery of anti-clathrin MAbs. This further supports the evidence which suggests that export of viral glycoproteins to the cell surface is not clathrin-mediated.

CONCLUSIONS

This chapter has reviewed methods which have been used in applying monoclonal antibodies to the study of clathrin-coated vesicle structure and function. As the details of these methods are available in the listed primary references, this summary has been confined to generalized comments about the approaches used. This discussion should be applicable to using monoclonal antibodies to study other organelles. We have demonstrated that monoclonal antibodies provide an effective link between organelle structure and function. In the case of clathrin coated vesicles, studies with monoclonal antibodies have identified molecular sites involved in organelle assembly and analyzed organelle assembly *in situ*. They have identified structural features on the light chains which are exposed to the cytoplasm, indicating a recognition role in CCV function. Finally, monoclonal antibodies introduced cytoplasmically have confirmed the role of CCV in receptor-mediated endocytosis by inhibition of this intracellular process.

ACKNOWLEDGMENTS

The work cited in this chapter was carried out in collaboration with M Androlewicz, G Blank, S Doxsey, K Drickamer, C Galloway, A Helenius, T Kirchhausen and P Parham, with technical assistance from B Abrams and L Kelley.

REFERENCES

Androlewicz MA, Loken MR, Keij J, Callas E, Galloway CJ, Brodsky FM (1986). Analysis and Sorting of Clathrin-Coated Vesicles by Flow Cytometry. J Cell Biol 103:55a.

Blank GS, Brodsky FM (1986). Site-specific disruption of clathrin assembly produces novel structures. The EMBO J 5:2087-2095.

Blank GS, Brodsky FM (1987). The elbow of the clathrin triskelion mediates clathrin assembly and light-chain binding. In preparation.

Brodsky FM (1985a). Clathrin structure characterized with monoclonal antibodies. I. Analysis of multiple antigenic sites. J Cell Biol 101:2047-2054.

Brodsky FM (1985b). Clathrin structure characterized with monoclonal antibodies. II. Identification of in vivo forms of clathrin. J Cell Biol 101:2055-2062.

Brodsky FM, Galloway CJ, Blank GS, Jackson AP, Seow H-F, Drickamer K, Parham P (1987). Topographical segregation of clathrin light chain sequences mediating heavy chain binding and coated vesicle diversity. Nature. In press.

Brodsky FM, Parham P (1983). Polymorphism in clathrin light chains from different tissues. J Mol Biol 167:197-204.

Brown WJ, Farquhar M (1984). Accumulation of coated vesicles bearing mannose-6-phosphate receptors for lysosomal enzymes in the Golgi region of I-cell fibroblasts. Proc Natl Acad Sci USA 81:5135.

Crowther RA, Pearse BMF (1981). Assembly and packing of clathrin into coats. J Cell Biol 91:790-797.

Doxsey S, Helenius A, Blank GS, Brodsky FM (1987). Inhibition of endocytosis by anti-clathrin antibodies. In preparation.

Doxsey SJ, Sambrook J, Helenius A and White J (1985). An efficient method for introducing macromolecules into living cells. J Cell Biol 101:19-27.

Galfrè G, Howe SC, Milstein C, Butcher GW, Howard JC (1977). Antibodies to major histocompatibility antigens produced by hybrid cell lines. Nature 266:550-552.

Goldstein JL, Brown MS, Anderson RGW, Russel DW and Schneider WJ (1985). Receptor-Mediated Endocytosis. In Palade GE, Alberts BM, Spudich JA (eds): "Annual Review Cell Biology," Palo Alto: Annual Reviews Inc., pp 1-39.

Griffiths G, Simons K (1986). The trans Golgi network: sorting at the exit site of the Golgi complex. Science 234: 438-443.

Helmy S, Porter-Jordan K, Dawidowicz EA, Pilch P, Schwartz AL, Fine RE (1986). Separation of endocytic from exocytic coated vesicles using a novel cholinesterase mediated density shift technique. Cell 44:497-506.

Jackson AP, Seow H-F, Holmes NJ, Drickamer K, Parham P (1987). Clathrin light chains contain brain-specific insertion sequences and a region of homology with intermediate filaments. Nature. In press.

Keen JH, Willingham MC, Pastan IH (1979). Clathrin-coated vesicles: Isolation, dissociation and factor-dependent reassociation of clathrin baskets. Cell 16:303-312.

Kilmartin JV, Wright B, Milsten C (1982). Rat monoclonal antitubulin antibodies derived by using a new nonsecreting rat cell line. J Cell Biol 93:576-582.

Kirchhausen T, Harrison SC (1981). Protein organization in clathrin trimers. Cell 23:755-761.

Kirchhausen T, Harrison SC (1984). Structural domains of clathrin heavy chains. J Cell Biol 99:1725-1734.

Kirchhausen T, Harrison SC, Parham P, Brodsky FM (1983). Location and distribution of the light chains in clathrin trimers. Proc Natl Acad Sci USA 80:2481-2485.

Lanier LL, Phillips JH (1986). Evidence for three types of cytotoxic lymphocyte. Immunol Today 7:132.

Ledbetter JA, Evans RL, Lipinski M, Cunningham-Rundles C, Good RA, Herzenberg LA (1981). Evolutionary conservation of surface molecules that distinguish T lymphocyte helper/inducer and cytotoxic/suppressor subpopulations in mouse and man. J Exp Med 153:310-323.

Nandi PK, Pretorius HT, Lippoldt RE, Johnson ML, Edelhoch H (1980). Molecular properties of the reassembled coat protein of coated vesicles. Biochemistry 19:5917-5921.

Orci L, Glick BS, Rothman JE (1986). A new type of coated vesicular carrier that appears not to contain clathrin: Its possible role in protein transport within the Golgi stack. Cell 46:171-184.

Payne GS, Schekman R (1985) A test of clathrin function in protein secretion and cell growth. Science 230:1009-1014.

Pearse BMF (1978). On the structural and functional components of coated vesicles. J Mol Biol 126:803-812.

Pearse BMF, Bretscher MS (1981). Membrane recycling by coated vesicles. Ann Rev Biochem 50:85-101.
Radka SF, Kostyu, D Amos, DB (1982). A monoclonal antibody directed against the HLA-Bwb epitope. J Immunol 128:2804-2806.
Schmid SL, Matsumoto AK, Rothman JE (1982). A domain of clathrin that forms coats. Proc Natl Acad Sci 79:91-95.
Schook W, Puszkin S, William B, Ores C, Kochwa S (1979). Mechanochemical properties of brain clathrin: Interactions with actin and α-actin and polymerization into basketlike structures or filaments. Proc Natl Acad Sci USA 76:116-120.
Schulman M, Wilde CD, Kohler G (1978). A better cell line for making hybridomas secreting specific antibodies. Nature 276:269-270.
Schulze-Lohoff E, Haslik A, von Figura K (1985). Cathepsin D precursors in clathrin-coated organelles from human fibroblasts. J Cell Biol 101:824-829.
Sly WS, Fischer HD (1982). The phosphomannosyl recognition system for intracellular and intercellular transport of lyosomal enzymes. J Cell Biochem 18:67.
Tooze J, Tooze SA (1986). Clathrin-coated vesicular transport of secretory proteins during the formation of ACTH-containing secretory granules in AtT20 cells. J Cell Biol 103:839-850.
Ungewickell E, Branton D (1981). Assembly units of clathrin coats. Nature 289:420-422.
Wanebo HJ, Gallmeier WM, Boyse EA, Old LJ (1966). Paraproteinemia and reticulum cell sarcoma in an inbred mouse strain. Science 154:901-903.
Woodward MP, Roth TF (1978). Coated vesicles: characterization, selective dissociation and reassembly. Proc Natl Acad Sci USA 75:4394-4398.

C. DENSITY MODIFICATION, PHASE PARTITION, AND OTHER COMPLEMENTARY TECHNIQUES

Cell-Free Analysis of Membrane Traffic, pages 155–167

HEPATIC ENDOSOME SUBFRACTIONS: PREPARATION, BIOCHEMICAL PROPERTIES AND FUNCTIONAL ROLES

W. Howard Evans and Carlos Enrich

Laboratory of Protein Structure, National Institute for Medical Research, Mill Hill, London NW7 1AA

INTRODUCTION

A comprehensive analysis by Leighton et al. (1966) of the constituents of rat liver homogenates using centrifugation in analytical (Beaufay) zonal rotors indicated that an appreciable quantity of protein could not be assigned to any subcellular location according to the then known enzymic and chemical markers. Substantial progress has since been made in resolving, and biochemically characterising subcellular membranes and organelles. The identification of endosomes as one of the previously unassigned constituents recovered in microsomal fractions in the sucrose density range 1.10-1.14g/ml is a major step forward in the subcellular fractionation of animal cells (Evans, 1987). This chapter describes the isolation and characterisation from rat liver of membrane vesicular subfractions that originate from the endocytic compartment. This compartment has followed the Golgi apparatus in reaching centre stage and is emerging as a multifunctional organelle centrally implicated in controlling the sorting of a wide range of circulating ligands internalised by cells in concert with their respective cell surface receptors. Other functions are also becoming apparent.

Ligands as Subcellular Markers

Studies in which cells were exposed to fluorescently-labelled ligands have long indicated the ease and speed of uptake by cells of macromolecules that combine with cell surface receptors. Vesiculo-tubular structures located subplasmalemmaly and containing tagged ligands were termed receptosomes (Pastan and Willingham, 1981), but such a description at the time was premature and perhaps a more appropriate term would have been ligandosomes. Subsequent studies combining subcellular fractionation and immunolocalisation approaches have indicated that the endocytic structures are pathways for the processing of ligand-receptor complexes. Indeed, not only ligand-receptor complexes but also ion channels (Strong and Evans, 1987) and extracellular matrix components (Enrich and Evans, 1987) are also now known to be present in the endocytic compartment in addition to the cell surface.

Iodinated ligands, e.g. insulin, perfused into livers to tag receptors on the sinusoidal plasma membrane domain of hepatocytes, were rapidly internalised and recovered in microsomal fractions (Carey and Evans, 1977). A major extension of these studies was undertaken using human asialotransferrin Type 3 as an iodinated probe of high specific activity (Regoeczi et al., 1982a). This ligand combined initially with the asialoglycoprotein receptor at the hepatocyte's surface before being transferred within minutes into an intracellular compartment where the ligand was recovered in high yield (> 16%) in a 'light' membrane fraction (Debanne et al., 1982). These experiments provided the basis for the isolation subsequently of endosome fractions of high purity.

Preparation of Hepatic Endosomes

Livers were removed from rats 5-30 min after portal vein perfusion with iodinated asialoglycoproteins and polypeptide hormones and the homogenates were then fractionated in sucrose gradients. These experiments showed that the ligands were located predominantly as a symmetrical peak located within a narrow density range (1.097-1.117g/ml). A similar density position was adopted by these ligands in analytical gradients of microsomal fractions and the membrane components entrapping the ligands were designated 'late' endosomes. Before the ligand-containing components equilibrated at this density, they were first briefly associated in the gradients with a heavier fraction, density 1.12-1.14g/ml, designated 'early' endosomes. A more comprehensive analysis of the location in sucrose gradients of a range of ligands internalised by liver indicated that whereas the

Figure 1a

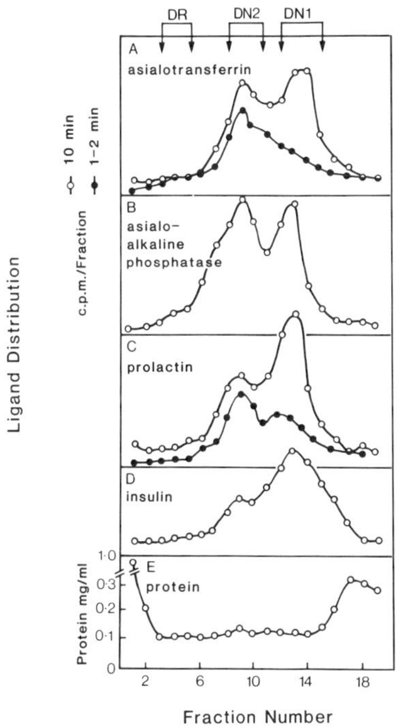

Figure 1b

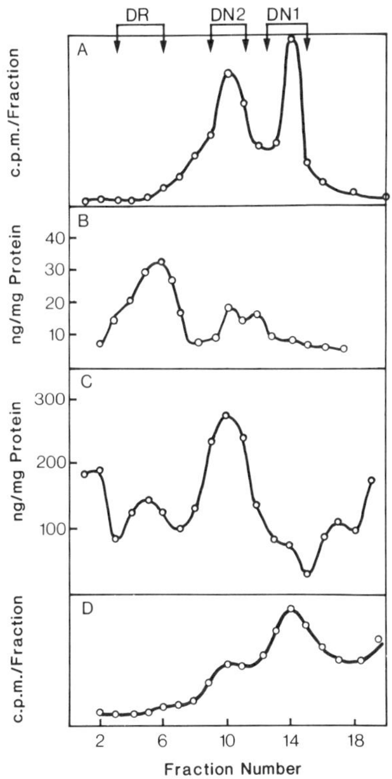

Figure 1. a) Distribution in endosome fractions of various radio-iodinated ligands injected via the portal vein 1-2 min or 10 min before removal and homogenisation of livers. The microsomal fractions were first resolved into ligand containing regions by centrifugation for 3h at 100,000g in sucrose density gradients (Evans, 1985) and then recentrifuged in isoosmotic Nycodenz gradients. For details of fractions DR, DN-2 and DN-1, see text. The origin of these fractions is also discussed in the text and in Figure 2.

b) Comparison of distribution in Nycodenz gradients of (A) endocytosed ^{125}I-asialotransferrin, (B) asialoglycoprotein receptors, (C) insulin receptors, and (D) endocytosed ^{125}I-insulin.

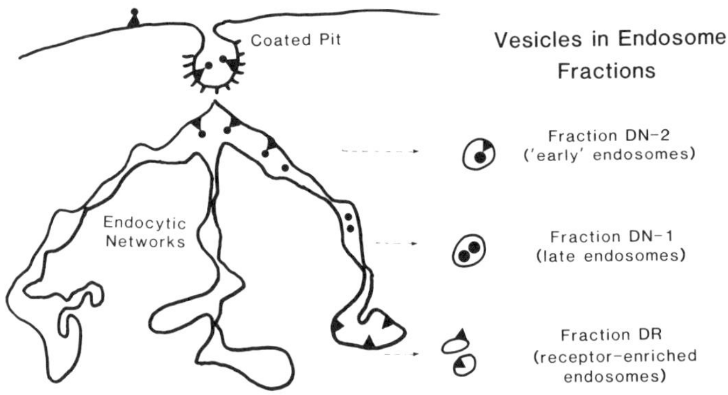

Figure 2. Diagramatic representation of the endocytic compartments depicting the possible origin of the subfractions prepared by fractionation in Nycodenz gradients. After transfer from coated pits on the plasma membrane, receptor-ligand complexes are dissociated and segregated. For further details, see text.

more physiological ligands were located within 5-30 min of uptake in the 'late' endosome fraction, 'foreign' ligands such as apamin, a bee venom polypeptide that combined with Ca^{2+}-activated K-channels (Strong and Evans, 1987), and agonists and antagonists to beta-adrenergic receptors were only associated briefly with 'early' endosomes before becoming associated with a lysosomal position in these gradients (Hadjiivanova et al., 1984). These results suggested that mechanisms exist within the endocytic compartment for rapid transfer of potentially deleterious ligands to lysosomes, whereas ligands that influence cellular metabolism remain for longer periods in the compartment.

A preparative technique to isolate 'early' and 'late' endosome vesicles from the post-mitochondrial supernatants of liver homogenates was devised (Evans, 1985; Evans and Flint, 1985). This technique, involving centrifugation of the microsomal fraction in isoosmotic Nycodenz gradients, and the separation of the subfractions is shown in Fig. 1A. Various ligands were recovered, initially in subfraction

DN-2 and subsequently in DN-1 of the density gradients. Analysis of the distribution of insulin and asialoglycoprotein binding sites indicated that these were found mainly at higher densities in the Nycodenz gradients in subfractions DR and DN-2 (Fig. 1B) (Evans and Flint, 1985). The procedure thus provided vesicles containing ligands (Fraction DN-1), ligands and receptors (Fraction DN-2) and receptors (Fraction DR).

Origin and Biochemical Properties of Endosome Vesicles

The possible origin of the endosome vesicles isolated is shown diagrammatically in Fig. 2. Here, the endocytic compartment is envisaged as comprising large central tubular elements extending towards the nuclear environs with many radiating arms as suggested by three-dimensional analysis of consecutive thin sections of labelled endosomes (Marsh et al., 1986). Receptors and ligands are considered to move along the plane of the membrane. After dissociation of receptor-ligand complexes, receptors and ligands become segregated as suggested by immunocytochemical evidence (Geuze et al., 1983). This model contrasts with other depictions of endocytosis in which separate vesicles corresponding to discrete 'early' and 'late' endosomes exist. The lipid and protein composition of the 'early' and 'late' endosome subfractions are similar (see below) and the major difference is density, probably a reflection of a different lipid to protein ratio of the various regions of a compartment subject to much fusogenic membrane interactions as the sorted receptors and ligands are dispatched towards the plasma membrane/ lysosomes etc.

Membrane vesicles containing receptors (Fraction DR) or ligands (Fraction DN-1) can originate from membrane regions where segregation has already occurred 10 min after uptake, whereas membrane vesicles containing both receptors and ligands originate from regions more proximal to the plasma membrane where the receptor and ligand are associated (Fraction DN-2). A mechanism allowing "self-sorting" of receptors and ligands in such a compartment has been proposed (Rome, 1985).

TABLE 1. Comparison of the biochemical properties of plasma membranes and endosomes of rat liver

	Sinusoidal plasma membranes	Endosomes	
		'Early'	'Late'
Adenylate cyclase[1]			
(i) Isoprenaline stimulated	12.4	0.31	0.39
(ii) NaF stimulated	26.6	0.14	0.32
(iii) Forskolin stimulated	9.8	0.04	0.14
(iv) Glucagon stimulated	7.0	0.11	0.19
5'-Nucleotidase[2]	23.0 (8)	5.3 (2.1)	10.5 (4.1)
Leucylnaphthylamidase[2]	1.51 (5.2)	0.1 (0.5)	-
Alkaline phosphodiesterase[2]	45.3 (22)	6.6 (3.4)	10.4 (5.3)
Na^+ K^+-ATPase[2]	17 (17)	-	5 (5)
Monensin activated Mg^{2+} ATPase[2]	<1	21.7 (37)	49.0 (64)
Clathrin and tubulin	Present	Absent	Absent
Sialic acid[3]	91	69	92
Cholesterol/phospholipid ratio	0.53	0.69	0.80
Sphingomyelin (% phospholipid)	12.8	18.5	21.8
Insulin[4]	+++	++	-
Prolactin[4]	+++	++	-
Asialoglycoprotein[4]	++	++	-
Ca^{2+}-activated K^+ channels[4]	+++	++	+

[1] n.moles cyclic AMP generated/h/mg protein. Note that 'total' plasma membranes were used.

[2] µmole substrate hydrolysed/h/mg protein. Values in brackets show enrichment over homogenate.

[3] n.moles/mg protein.

[4] Receptor distribution is expressed as arbitrary relative value +++ = high, + = low, - = undetected.

Table 1 summarises the biochemical properties of the 'early' and 'late' endosome subfractions prepared by the current methodology. When compared to plasma membranes, there are many common features, e.g. various receptors, sialic acid and lipid content, but clear differences also emerge. The major enzymic difference is the confinement of receptor-activated adenylate cyclase activity to the plasma membrane, implying that those receptors that signal via the intermediation of cyclic AMP must carry out this function at the plasma membrane before the receptor is internalised. On the contrary, a monensin-activated Mg^{2+}-ATPase is highly enriched in endosomes, and it is likely that this is the enzyme that acidifies the lumen of the endocytic compartment. The enzyme is low in activity in plasma membrane fractions, but these membranes when isolated lack the coated regions seen in intact cells, that are recovered independently as coated vesicles that were shown to acidify *in vitro* (Stone et al., 1983). Other enzymes, normally considered as plasma membrane markers, e.g. 5'-nucleotidase and alkaline phosphodiesterase are also present in the endosome subfractions, probably resulting from their migration between the two membrane systems, a property possibly related to their glycoprotein nature. In liver, the low acid phosphatase indicated that little difficulty was encountered in separating endosomes from lysosomes in view of their radically different densities (Debanne et al., 1982).

Endosome Acidification

Despite the prolonged procedures used in the preparation of the 'early' and 'late' endosome fractions, they were shown to undergo acidification in the presence of Mg^{2+}-ATP as demonstrated by aminoacridine fluorescence (Fig. 3). Addition of monensin resulted in a gradual reversal of acidification. Sinusoidal plasma membranes did not acidify, and the proton pumping activity was postulated to be measured as a monensin-activated ATPase (Saermark et al., 1985). Although the degree of acidification *in vitro* was not determined, studies with mouse hepatocytes showed, using fluorescent asialoglycoproteins, that the pH in endosomes was 5.0 - 5.5 (Geisow and Evans, 1984).

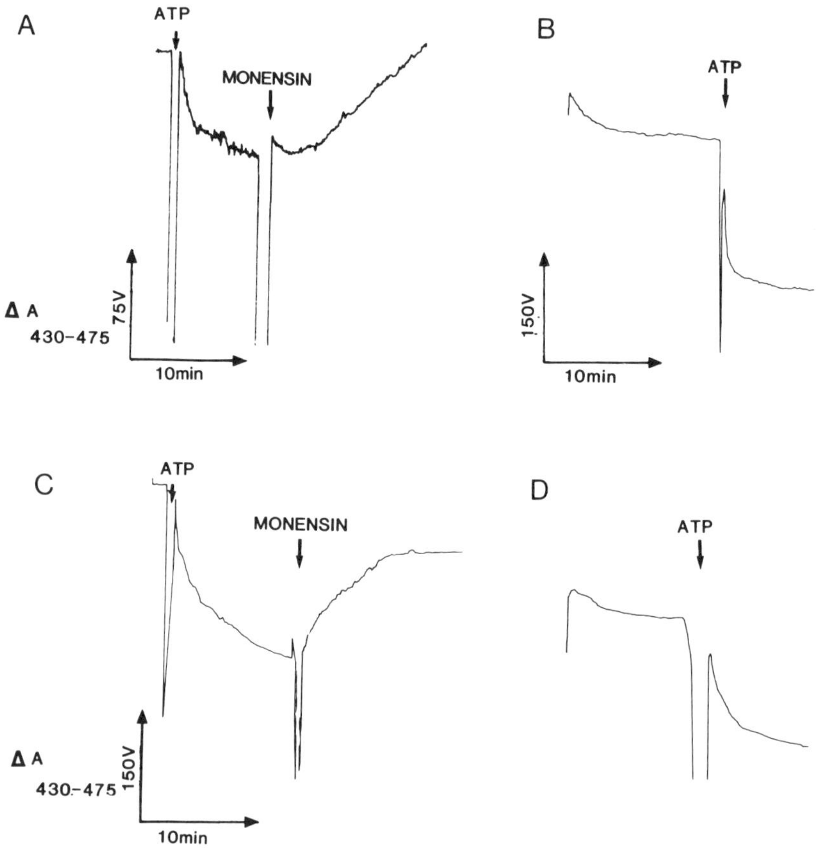

Figure 3. Acidification of rat liver endosomes *in vitro*. A and C, 'early' and 'late' endosomes collected from sucrose gradients; B, and D, purified 'early' and 'late' endosomes collected from Nycodenz gradients, respectively. Endosomes were incubated at 25°C for 30 min in 20mM-Tes, 0.5mM EGTA, 1mM $MgCl_2$, 0.5mM ouabain and 0.25M sucrose, pH 7.0. After equilibration in 1µM 9-aminoacridine, ATP, monensin and oligomycin were added to final concentrations of 0.2mM, 1µM and 10µg/ml respectively, as indicated.

Endosome Proteins

The overall protein and glycoprotein composition of endosomes has been compared by polyacrylamide gel electrophoresis in one dimension (Fig. 4) (Evans and Hardison, 1985), and these studies have now been extended using two-dimensional gel electrophoresis, and by analysis of the hydrophobic and hydrophilic constituents separated in detergent two-phase systems (Enrich and Evans, 1987, in preparation). Figure 5 shows that, by two-dimensional analysis, further differences in protein composition evident in 1D gels begin to emerge between liver plasma membranes and endosomes. No differences were discerned between 'early' and 'late' endosomes. These studies, in combination with a more comprehensive analysis of lysosomal and Golgi membranes, form the basis of a strategy for identifying endosome-specific constituents, and to unravel the routes followed by specific membrane proteins between Golgi, cell surface, endosome and lysosomal membranes.

Functions of the Endocytic Compartment

The endocytic compartment continues to accrue additional functions. Among its varied functions are, 1) it constitutes a major membrane-bound port of entry for a wide range of circulating ligands, 2) it is a locus for uncoupling of pH-sensitive ligands, 3) it has a pivotal role in the sorting and routing of membrane proteins to other intracellular and surface destinations, 4) it has a possible role in signal transduction in view of the presence of epidermal growth factor receptor kinase activity (Cohen and Fava, 1985), 5) it features in the repair of recycling ligands and receptors (Regoeczi et al., 1982b; Snider et al., 1985), 6) it can control, through ion channels and pumps, the ionic conditions in the lumen, thereby governing the conformational and other requirements for entry of ligands across the endosome membrane into the cell, or their transfer to other organelles, e.g. lysosomes, 7) the presence of fibronectin in endosomes indicates a possible functional continuity with the extracellular matrix, suggesting a role in the turnover of fibronectin and its receptor (Enrich and Evans, 1987), 8) proteolytic processing of internalised ligands, e.g. epidermal growth factor is initiated in endosomes (Schaudies et al., 1987).

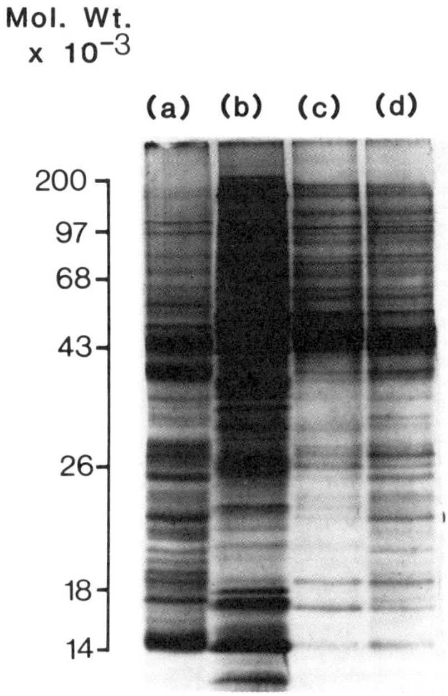

Figure 4. Electrophoretic separation of polypeptides in (a) fraction DR, (b) plasma membranes, (c) fraction DN-1, and (d) fraction DN-2 in 10% SDS-polyacrylamide gels. The endosome subfractions correspond to those shown in Fig. 1. Coomassie-blue staining.

The biochemical analysis of isolated purified hepatic endosomes suggests appreciable compositional commonality with the cell's plasma membrane, indicating that the endocytic compartment may comprise an 'intracellular' domain of the plasma membrane where the external environment is modifiable by the enzymes and ionic channels now identified in the endosome membrane. The preparation of antibodies specific to endosome membrane components should prove instrumental in allowing the fine details of the above functions to be examined, and also to identify new functions.

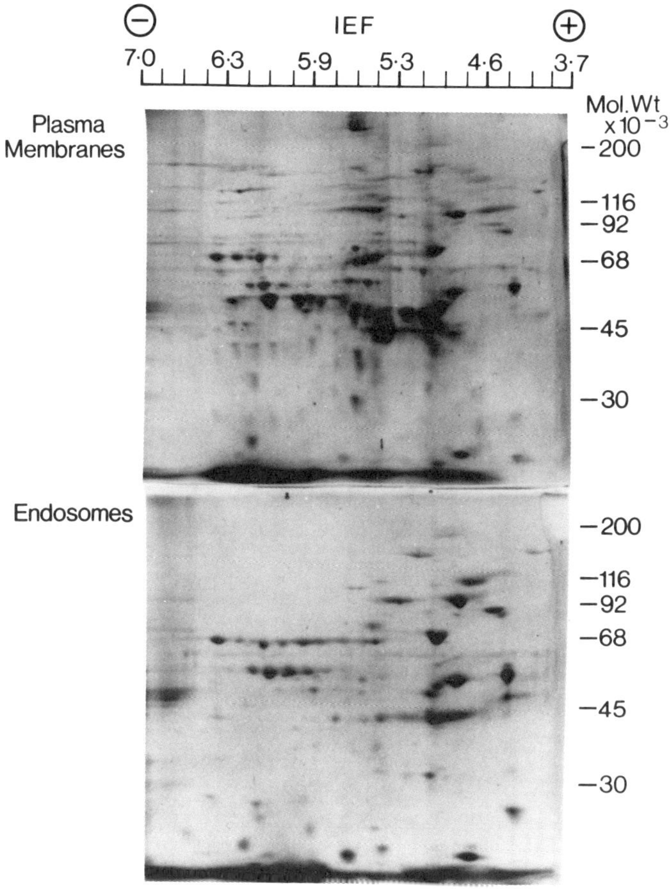

Figure 5. Comparison of liver plasma membranes and endosomes (Fraction DN-1) by two-dimensional electrophoresis. Membrane fractions were first electrofocused (IEF) followed by electrophoresis in SDS-polyacrylamide gels and then stained with Coomassie blue.

Acknowledgements

C. Enrich thanks the Wellcome Trust for support. We thank Mrs Eugénie Pearson for organising the chapter.

REFERENCES

Carey F, Evans WH (1977). Identification of blood-sinusoidal plasma membrane fractions from rat liver homogenates by radio-iodinated ligand binding. Biochem Soc Trans 5:103-104.

Cohen S, Fava RA (1985). Internalisation of functional epidermal growth factor: receptor kinase complexes in A-431 cells. J Biol Chem 260:12351-12358.

Debanne MT, Evans WH, Flint N, Regoeczi, E (1982). Receptor-rich intracellular membrane vesicles transporting asialo-transferrin and insulin in liver. Nature 298:398-400.

Enrich C, Evans WH (1987). Evidence for a role of the hepatic endocytic compartment in the modulation of the extracellular matrix. Exptl Cell Res, in press.

Evans WH (1984). Preparation of low density 'endosome' and 'endosome'-depleted Golgi fractions from rat liver. Methods Enzymol 109:246-257.

Evans WH (1987). Organelles and membranes of animal cells. In: Biological Membranes, a practical approach (ed Findlay JBC, Evans WH) pp 1-35. IRL Press, Oxford.

Evans WH, Flint N (1985). Subfractionation of hepatic endosomes in Nycodenz gradients and by free-flow electrophoresis. Separation of ligand-transporting and receptor-enriched membranes. Biochem J 232:25-32.

Evans WH, Hardison WGM (1985). Phospholipid cholesterol polypeptide and glycoprotein composition of hepatic endosome subfractions. Biochem J 232:33-36.

Geisow MJ, Evans WH (1984). pH in the endosome. Measurements during pinocytosis and receptor-mediated endocytosis. Exptl Cell Res 150:36-46.

Geuze HJ, Slot JW, Strous GJAM, Lodish HF, Schwartz AL (1983). Intracellular site of asialoglycoprotein receptor-ligand uncoupling. Double-label immunoelectron microscopy during receptor-mediated endocytosis. Cell 32:277-287.

Hadjiivanova N, Flint N, Evans WH, Dix C, Cooke BA (1984). Endocytosis of beta-adrenergic ligands by rat liver. Biochem J 222:749-754.

Leighton F, Poole B, Beaufay H, Baudhuin P, Coffey JW, Fowler S, De Duve C (1968). The large scale separation of peroxisomes, mitochondria and lysosomes from the livers of rats injected with Triton WR-1339. J Cell Biol 37:482-513.

Marsh M, Griffiths G, Dean GE, Mellman I, Helenius A (1986). Three-dimensional structure of endosomes in BHK-21 cells. Proc Natl Acad Sci USA 83:2899-2903.

Pastan I, Willingham MC (1981). Journey to the center of the cell. Science 214:504-509.

Regoeczi E, Chindemi PA, Debanne MT, Hatton MWC (1982a). Dual nature of the hepatic lectin pathway for human asialotransferrin Type 3 in the rat. J Biol Chem 257:5431-5436.

Regoeczi E, Chindemi PA, Debanne MT, Charlwood P (1982b). Partial resialylation of human asialotransferrin Type 3 in the rat. Proc Natl Acad Sci USA 79:2226-2230.

Rome LH (1985). Curling receptors. Trends in Biochem Sci 10:151.

Saermark T, Flint N, Evans WH (1983). Hepatic endosome fractions contain an ATP-driven proton pump. Biochem J 225:51-58.

Schaudies RP, Gorman RM, Savage CR, Portez RD (1987). Proteolytic processing of epidermal growth factor within endosomes. Biochem Biophys Res Commun 143:710-715.

Snider MD, Rogers OC (1985). Intracellular movement of cell surface receptors after endocytosis. Resialylation of asialotransferrin receptor in human erythroleukemia cells. J Cell Biol 100:826-834.

Stone DK, Xie XS, Racker E (1983). An ATP-driven proton pump in clathrin-coated vesicles. J Biol Chem 258:4059-4062.

Strong PN, Evans WH (1987). Receptor-mediated endocytosis of apamin by liver cells. Eur J Biochem 163:267-273.

Cell-Free Analysis of Membrane Traffic, pages 169–183

THE DAB-INDUCED DENSITY SHIFT : PRINCIPLE, VALIDITY AND APPLICATIONS TO ENDOSOMES.

Pierre J. Courtoy, Joël Quintart, Jean-Pierre Draye and Pierre Baudhuin.

Laboratoire de Chimie Physiologique, Louvain University Medical School and the International Institute of Cellular and Molecular Pathology, 75 Avenue Hippocrate, B-1200 Brussels, Belgium.

INTRODUCTION

Whereas the general framework of receptor-mediated endocytosis is now well accepted, the nature and properties of the endosomes still remain largely unknown (e.g. Helenius et al., 1983). Direct answers ultimately depend on a good purification of endosomes, a task made difficult by the considerable polydispersity of these structures. Firstly, endosomes are structurally heterogeneous in size, shape and topography, presumably as a reflection of functional differences in maturation, content and destination. Secondly, the composition and properties of endosomes may overlap with those of the other structures they interact with (pericellular membrane, lysosomes, Golgi complex...). Thirdly, because endosomes act as intracellular sorting and delivering organelles, they may turn out to be made of structurally and functionally distinct subcompartments, that we would ultimately have to resolve. Thus, it appears that endosome-specific isolation techniques must be developed.

PRINCIPLE

The DAB shift procedure (Courtoy et al., 1984) was discovered during our attempts to complement the usual fractionation techniques based on size and equilibrium density, by a density shift that would be specific to endosomes or endosomal compartments. It combines the use of ligands conjugated to horseradish peroxidase (HRP), and the application of the 3,3'-diaminobenzidine (DAB) cytochemistry (Graham and Karnovsky, 1966), to unfixed subcellular fractions.

Under correct conditions, ligand-HRP conjugates are faithful substitutes to the native ligands, and they can be demonstrated cytochemically, both in intact cells and in fixed subcellular fractions (Quintart et al., 1983). Thus, an appropriate choice of ligand and time of internalization, together with the use of competitors or inhibitors, allows to introduce HRP into selected, predefined endosomal compartments. The DAB-density shift procedure takes advantage of the finding that peroxidase cytochemistry can also be applied to unfixed subcellular particles, and of the fortuitous observation that the DAB-reaction product is very dense.

Endosomes containing ligand-HRP are purified by a two-step protocol. After ligand internalization, a ligand-enriched subcellular fraction is prepared by differential sedimentation, and submitted to a first isopycnic centrifugation, in order to clear the preparation from the organelles equilibrating at high densities. The low-density organelles, including endosomes, are next incubated in a modified DAB medium, with (test) or without H_2O_2 (control). The enzymic reaction causes the selective accumulation of DAB-reaction product in the HRP-containing endosomes, which results in a considerable increase of their equilibrium density. Thus, a second isopycnic centrifugation will place those particles at their new (heavier) density, in a region which has been cleared after the first gradient. The procedure is outlined in Fig. 1, detailed in Table 1 and exemplified in Fig. 2.

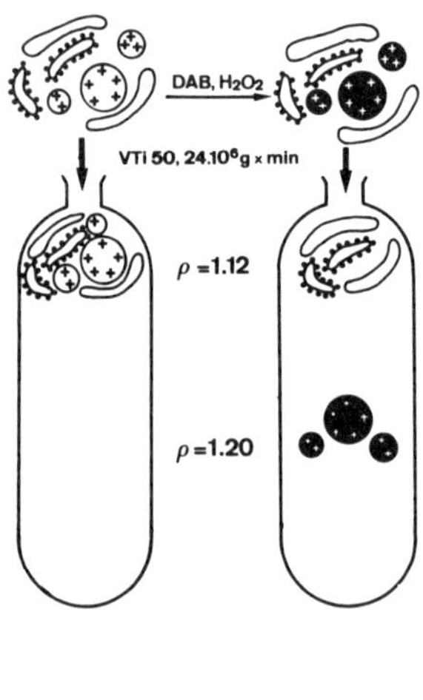

Figure 1. Principle of the density shift. DAB as a monomer diffuses slowly in and out all organelles, but is trapped when polymerized by HRP in the presence of H_2O_2. Accumulation of dense reaction product results in a specific displacement of the equilibrium density (ρ) of HRP-containing particles. From Courtoy et al., 1984.

TABLE 1. Procedure for the DAB-induced density shift.

1. Define the protocol of endocytosis in order to concentrate HRP into the endocytic compartment of interest (conjugation to ligand, time of uptake, competitors, inhibitors, etc...). For uptake by receptor-mediated endocytosis, conjugate the appropriate ligand to HRP of the highest specific activity, and clear the preparation of unconjugated ligand and HRP.
2. After uptake, wash extensively. Whenever possible, remove ligand-HRP adsorbed to the pericellular membrane by treating cells with cold saline containing 1 mM EGTA.
3. Isolate the subcellular fraction most enriched in ligand by differential centrifugation. Resuspend gently to avoid damaging organelles and releasing HRP.
4. Further purify the ligand-containing organelles by a first equilibration in a density gradient (e.g. 10-60 % w/v sucrose buffered with 3 mM imidazole/HCl, metrizamide, Percoll ...). This step also removes free HRP, which should remain in the loading zone. Remember that equilibration can be performed by sedimentation or floatation.
5. Collect the density-purified preparation and incubate it for 15-30 min at 25°C in the dark, in a freshly prepared solution of 5.5 mM 3,3'-diaminobenzidine (e.g. Sigma, Grade II), dissolved in the medium corresponding to the density where organelles were initially recovered, and cleared through 0.45 μm filters. Add 11 mM H_2O_2 for tests. Measure activities of constituents before and after incubation.
6. Equilibrate again in a density gradient starting at the density where organelles were initially recovered. The organelles containing HRP should be shifted to higher densities.
7. To subtract contamination, incubate controls in the absence of H_2O_2.
8. To detect a possible organelle aggregation, it is recommended to incubate in DAB and H_2O_2 a mixture of the fraction purified in ligand-HRP, together with a similar fraction isolated from cells exposed to a ligand not conjugated with HRP, and labeled with a different isotope. The latter material should not be displaced (see Fig 3a). In case of aggregation, try thimerosal (see below).

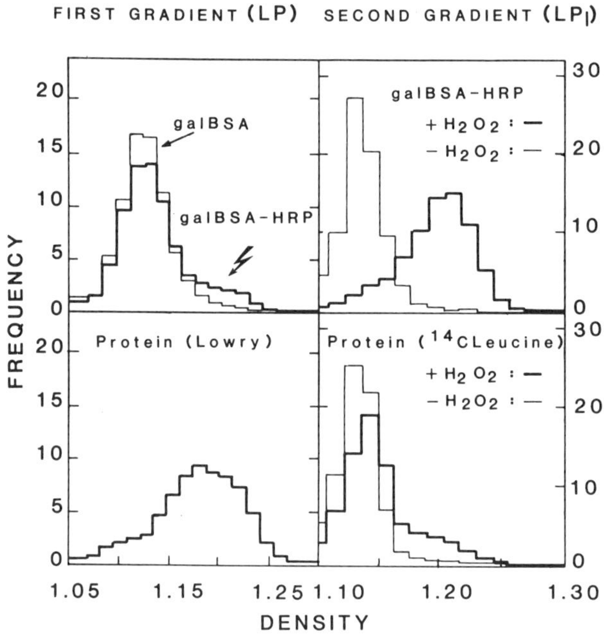

Figure 2. Application to rat liver endosomes. Rats were injected with galBSA or galBSA-HRP. Ten min later, the liver was flushed, homogenized and a LP fraction (organelles sedimenting between 33,000 and 3 x 10^6 g x min) was equilibrated in a first linear sucrose gradient (from 1.05 to 1.30 g/ml in density). Ligand distributions are very similar and already separated from the bulk of liver protein. The organelles recovered between 1.11 and 1.13 g/ml in density (LP_1 pool) were incubated in the cytochemical medium as described at Table 1, and equilibrated again in a second linear sucrose gradient (from 1.10 to 1.30 g/ml in density). A major shift of ligand distribution (from ∿ 1.12 to ∿ 1.19 g/ml) is observed only in the presence of H_2O_2, and leads to the concomitant displacement of a minor fraction of liver protein. Adapted from Courtoy et al., 1984.

VALIDITY

The DAB-induced density shift procedure can be of general interest if it is specific, reproducible, and suitable for other ligands and cell types. Its application to the characterization of the composition and properties of endosomes further requires the evaluation of possible effects of the DAB-reaction product on the constituents under study.

The specificity of the density shift was demonstrated by in vitro mixing experiments. Endosomal preparations were prepared from two different rats, injected either with ligand-peroxidase, or with the native ligand, and mixed. When the mixture was submitted to the DAB-shift procedure, the density shift of the endosomes containing the ligand-peroxidase conjugate did not cause a significant displacement of the endosomes containing the unconjugated ligand (Fig. 3a). The DAB shift procedure can thus be used to demonstrate the dissociation of constituents which co-distribute in classical fractionation systems.

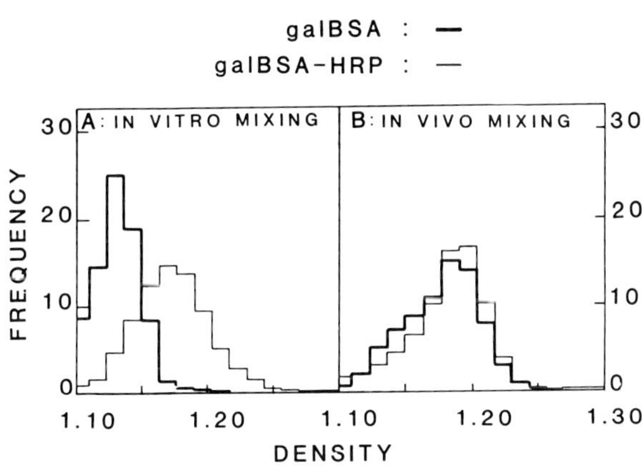

Figure 3. Specificity of the density shift. On the left, galBSA and galBSA-HRP were injected to two different rats. Corresponding LP_1 pools were prepared, mixed in vitro, and submitted to the DAB-shift procedure as at Fig. 2. On the right, both ligands were injected to the same animal and the LP_1 pool was submitted to the DAB-shift procedure. Adapted from Courtoy et al., 1984.

When both conjugated and unconjugated ligands were injected simultaneously to the same rat, so as to be internalized into the same endosomes, the DAB-shift procedure resulted in a concomitant displacement of both distributions. This indicates that the procedure can also be used to assess the true association of constituents (Fig. 3b). This possibility will be exploited in the third application.

However, the specificity of the procedure shift was abolished when free HRP activity was present in the incubation medium at concentrations as low as 30 ng/ml, which roughly corresponded to the content of HRP enclosed in endosomes. This problem is attributed to a better access of DAB to free HRP, compared to membrane-enclosed HRP, and to random cross-linking of organelles by the polycationic reaction product. It is thus essential to clear cell preparations of extracellular HRP, and to minimize the rupture of HRP-containing organelles during pellet resuspension. Further removal of free HRP can be obtained by the first density equilibration. We recently observed that the addition of 7.5 mM ethyl(2-mercaptobenzoato-S) mercury (sodium salt, or thimerosal), inhibited by ∿ 90 % free HRP activity, but did not abrogate the density shift. This suggests that thimerosal does not cross membranes in our experimental conditions. As a consequence, thimerosal largely restored the specificity of the procedure, even in the presence of 30 ng/ml free HRP. In any case, in vitro mixing experiments have been regularly performed to detect organelle aggregation.

In our hands, the procedure proved fairly reproducible, provided great care was taken to avoid particle aggregation, especially by minimizing the occurrence of free HRP activity. The extent of displacement was saturable with HRP concentration, but increased with DAB concentration up to 6 mM. The H_2O_2 concentration and the incubation time were less critical. The procedure has been applied successfully to free HRP (Draye et al., 1983) or to HRP conjugated to pIgA (Quintart et al., 1985), asialoorosomucoid (Stoorvogel and Strous, 1986), or transferrin (Ajioke and Kaplan, 1986). It worked as well on rat liver (Quintart et al., 1983, 1984 and 1985; Kay et al., 1984), as on cultured cells such as fibroblasts (Draye et al., 1983), Hep G_2 cells (Stoorvogel et al., 1986) and Hela cells (Ajioka and Kaplan, 1986).

The effect of the DAB-reaction product on endosomal activities is controversial. In our hands, polymerized DAB interfered with colorimetric assays, but, at the concentration used, did not appreciably affect enzymatic activities associated with several HRP-containing endosomal or lysosomal preparations (Fig. 4). In addition, when oxidized DAB was generated on the pericellular membrane of rat fibroblasts after covalent binding of HRP, neither 5'-nucleotidase, nor alkaline phosphodiesterase I activities were significantly altered (data not shown). In contrast, Ajioka and Kaplan (1986) have reported that incubation with high concentrations of DAB abolished some enzymatic and receptor activities in endosomes.

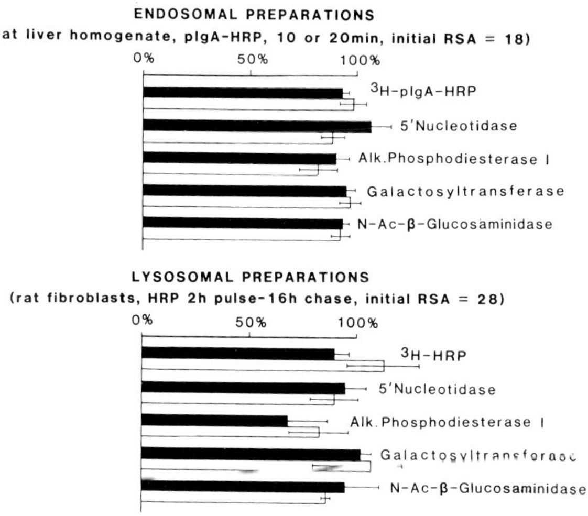

Figure 4. Effect of DAB cytochemistry on enzymatic activities. HRP was introduced into rat hepatocytes endosomes (using pIgA-HRP), or into rat fibroblast lysosomes. Initial RSA are the relative specific activities of the preparations before cytochemistry and reequilibration. Filled bars show activities measured after incubation in DAB and H_2O_2, in % of controls. Open bars show recoveries in the gradients after subsequent equilibration, expressed as % of recoveries in the corresponding control gradients. Shifted endosomes contained up to 20 % of 5'-nucleotidase and 28 % of alkaline phosphodiesterase I of the preparation submitted to the procedure. Shifted lysosomes contained up to 80 % of N-acetyl-β-glucosaminidase of the preparation (mean ± SD; n=4).

APPLICATIONS

We will review three applications of the procedure : the purification of endosomes, the analysis of ligand sorting, and the composition of endosomal compartments.

Purification of Endosomes

The purification of endosomes (Quintart et al., 1984) was followed by the ratio of the yield of radiolabeled ligand-peroxidase conjugate to that of liver protein (Table 2). Because the DAB reaction product interferes with the colorimetric assays of protein, we radiolabeled liver protein by a pulse of [^{14}C]leucine or [^{35}S]methionine given intraperitoneally 36 hours before sacrifice. At 10 minutes post injection, endosomes containing galBSA-HRP occupied less than 1 % of the hepatocyte volume and were mostly recovered in the L fraction (organelles recovered in the tight pellet after sedimentation between 33,000 and 450,000 g x min). After a first sucrose density gradient, the low density pool (L_1) already showed considerable purification, but the contamination was still substantial (Fig. 5a). After cytochemistry and equilibration in a second sucrose gradient, preparations recovered between 1.20 to 1.22 g/ml in density were purified more than 250 fold, with a yield of 10 % of the homogenate. Those preparations were almost pure (Fig. 5b).

TABLE 2. Purification of rat liver endosomes. Yields are expressed as % of the homogenate. For details, see text. Reproduced from Quintart et al., 1984.

	Protein (%)		[^{3}H]galBSA-HRP (%)	Relative specific activities
	Lowry et al.	[^{35}S]Met.		
L fraction	4.02	4.02	23.0	5.7
L_1 pool	0.162	0.168	13.4	82.1
Shifted L_1	-	0.039	10.4	267

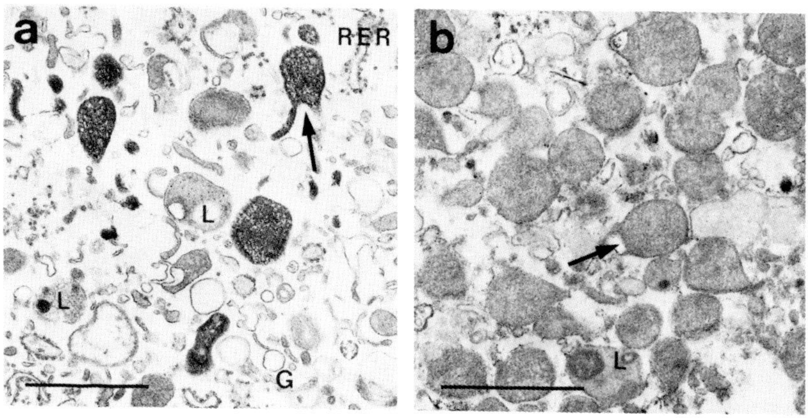

Figure 5. Electron microscopy of endosomal preparations from rat liver. On the left, L_1 pools were fixed, collected on Millipore filters, processed for peroxidase cytochemistry, and examined. Endosomes are indicated by arrows. Major contaminants are the rough endoplasmic reticulum (RER), Golgi stacks (G), and lysosomes (L). On the right, shifted L_1 material was fixed, collected on Millipore, and examined without further staining. Bar is 1 μm. Reproduced from Quintart et al., 1983 and 1984.

Demonstration and Kinetics of Ligand Sorting

Because the DAB shift procedure can test the association and the dissociation of constituents, it can be applied to follow the kinetics of ligand segregation into distinct organelles. We have injected simultaneously galBSA-HRP and polymeric IgA, two ligands recognized by different receptors and transferred to different destinations, i.e. lysosomes and bile canaliculi, respectively. GalBSA-HRP was used as a shift-inducer and the possible association with pIgA was followed by a concomitant displacement. Early after injection, the majority of pIgA followed galBSA-HRP, but this proportion decreased considerably with time (Fig. 6). We concluded that both ligands are initially associated in the same endosomes and subsequently sorted by endosomes into separate structures, also recovered at low densities (Courtoy et al., 1985). The composition of those late endosomes will be analyzed in the last section. This principle has also been applied

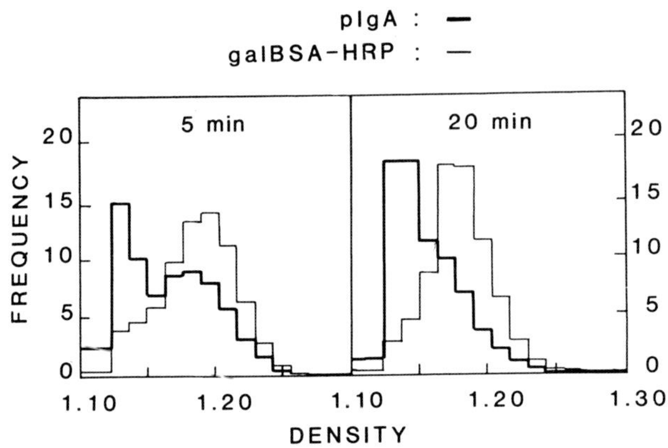

Figure 6. Ligand sorting in rat hepatocytes. Rats were injected with galBSA-HRP together with pIgA at 5 or 20 min before sacrifice. LP_1 pools were prepared and submitted to the DAB-shift procedure. A concomitant displacement of pIgA is used as a criterion of ligand association. Adapted from Courtoy et al., 1985.

to the study of sorting between asialoorosomucoid and transferrin in the Hep G_2 cells. Sorting occurred in a few minutes, but was blocked at 20°C, or by the presence of primaquine (Stoorvogel et al. 1986).

Composition of Endosomal Compartments

The activity of several marker enzymes for plasma membrane (5'-nucleotidase and alkaline phosphodiesterase I), Golgi complex (galactosyltransferase) and lysosomes (N-acetyl-β-glucosaminidase) could be assayed after the DAB procedure. We thus followed their possible displacement, concomitant with the ligand-peroxidase conjugates, as an evidence for association with the corresponding endosomes. We have examined the occurrence of marker enzymes in various endosomal populations by using two different ligands (galBSA-HRP and pIgA-HRP), injected separately at 10 or 20 min before sacrifice. Data for galBSA-HRP at 10 min have been previously reported (Quintart et al., 1984), and were undistinguishable at 20 min. Data for pIgA-HRP are shown at Fig. 7. To evaluate the association of marker enzymes with endosomes, we have measured the activity specifically displaced, in the most

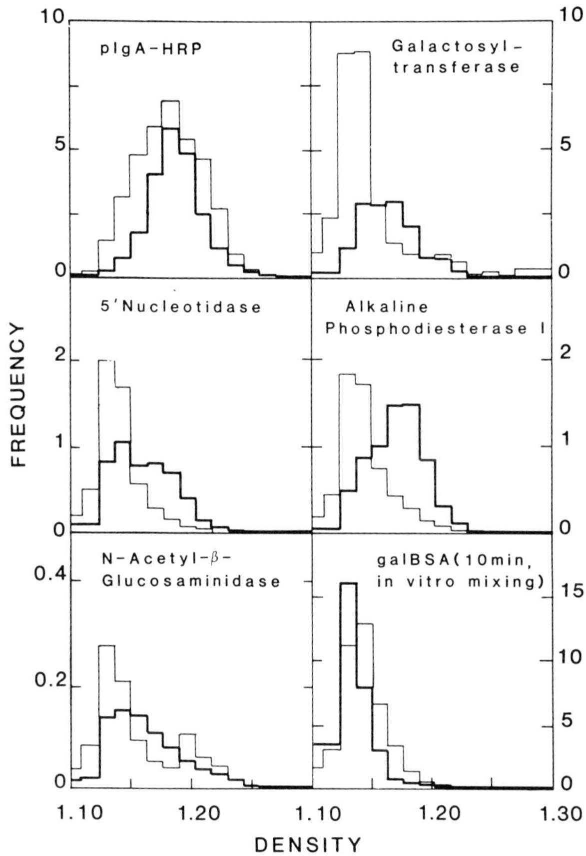

Figure 7. Association of marker enzymes with endosomes. Rats were injected with pIgA-HRP at 10 (thin line) or 20 min (thick line) before sacrifice. LP_1 pools were prepared and submitted to the DAB-shift procedure. Mean distributions in the gradients are presented after normalization of the areas of histograms with respect to the homogenate, and hence reflect the yield of each constituent in the LP_1 pool (n=3 to 5). A shift of marker enzyme concomitant to the ligand is used to calculate association with endosomes, as shown in Table 4. In the lower right panel these LP_1 pools were also mixed, immediately before cytochemistry, with similar preparations obtained from rats injected with galBSA 10 min before sacrifice. The absence of significant displacement of galBSA indicates the specificity of the density shift.

purified preparations, by subtracting the activity recovered at the same density when H_2O_2 was omitted. The yield of marker enzymes was expressed as percentage of the homogenate and divided by the corresponding yield of endosomes (i.e. their ligand content). Values were not considerably changed if all activity lost after cytochemistry was also attributed to the endosomes, to correct for a possible enzyme inactivation (data not shown).

As shown at Table 3, a very low fraction of all activities tested could be associated with the endosomes which contain galBSA-HRP at 10 min postinjection. This Table also shows that similar values are obtained from different starting preparations (L_1 versus LP_1 pools), irrespective of the final purification (from 20 to 270 fold). Nor could any difference be detected between 10 and 20 min postinjection.

Compared to galBSA-HRP, the occurrence of plasma membrane markers in endosomes which carry pIgA-HRP was very similar at 10 min, but considerably increased at 20 min (Table 4). This is consistent with ligands still being sorted at 10 min, and with the occurrence at 20 min of pIgA-HRP in a distinct endosomal compartment, whose membrane composition resembles that of the bile canalicular membrane. This "maturation," and the true association of significant amounts of plasma membrane markers with the endosomes containing pIgA-HRP at 20 min are suggested by correlations between ligand and marker contents (Fig. 8).

Figure 8. Association of alkaline phosphodiesterase I to endosomes. The content of alkaline phosphodiesterase I in several shifted fractions (in % of the homogenate) are compared to those of pIgA-HRP at 10 min (o) and 20 min postinjection (•). By linear linear regression, endosomes carrying pIgA at 10 and 20 min apparently account for 2.3 and 10.0 % of alkaline phosphodiesterase I of the homogenate, respectively. Dashed lines indicate lower limits of reproducibility of assays.

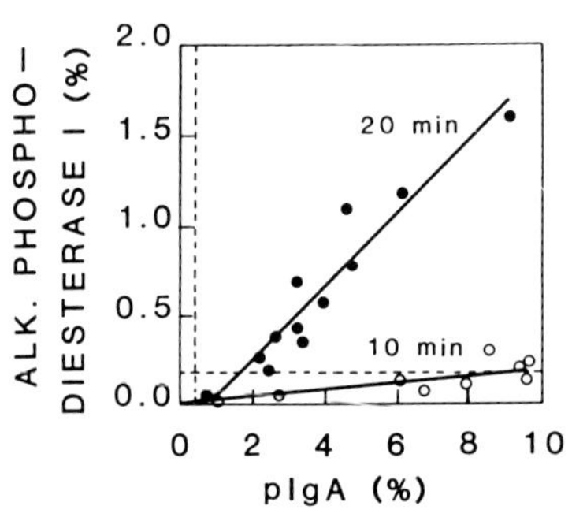

TABLE 3. Characterization of liver endosomes containing galBSA-HRP. Rats were injected with the ligand-HRP conjugate at 10 or 20 min before sacrifice. L_1 or LP_1 pools were submitted to the DAB-shift procedure and the association of constituents with endosomes was calculated on the most purified shifted preparations. Values are expressed as % of the homogenate, presented as means ± standard deviations from 3 to 4 experiments.

	10 min		20 min
	(L_1,n=3)	(LP_1,n=3)	(LP_1,n=4)
Yield			
^{3}H-Ligand	10.4 ±2.6	24.7±3.6	12.0±4.8
^{35}S-Protein	0.04±0.005	1.2±1.1	0.7±0.5
Association of markers with endosomes			
5'-Nucleotidase	0.8±0.1	0.7±0.5	1.2±0.4
Alk. Phosphodiest. I	1.2±0.5	2.0±0.7	2.7±1.3
Galactosyltransferase	2.7±0.1	4.1±2.9	0.4+0.7
N-acetyl-β-glucosamin.	0.9±0.6	0.1±0.1	0.1∓0.1

TABLE 4. Characterization of liver endosomes containing pIgA-HRP (same presentation as in Table 3).

	10 min	20 min
	(LP_1, n=3)	(LP_1, n=3)
Yield		
^{3}H-Ligand	20.6±9.3	21.5±5.8
Association of markers with endosomes		
5'-Nucleotidase	1.0±0.6	5.8±2.6
Alk. Phosphodiesterase I	2.1±0.5	10.2±1.6
Galactosyltransferase	12.4±1.8	7.0±6.0
N-acetyl-β-glucosaminidase	0.5±0.3	0.5±0.3

Finally, the DAB shift procedure also proved useful to study the polypeptide composition of endosomes purified from rat liver (see Bergeron and coworkers, this volume). Ajioka and Kaplan (1986) also took advantage of the selective inactivation of some constituents at high DAB concentration to follow the content of transferrin receptor in the endosomes of Hela cells.

ACKNOWLEDGEMENTS

This work was supported by grants 2.4540.80 and 3.4547.79 from the Belgian Research Council, of which P.C. is Research Associate.

REFERENCES

Ajioka RS, Kaplan J (1986). Intracellular pools of transferrin receptors result from constitutive internalization of unoccupied receptors. Proc Natl Acad Sci USA 83:6445-6449.

Courtoy, PJ, Quintart J, Baudhuin P (1984). Shift of equilibrium density induced by 3,3'-diaminobenzidine cytochemistry : a new procedure for the analysis and purification of peroxidase-containing organelles. J Cell Biol 98:870-876.

Courtoy PJ, Quintart J, Limet JN, Deroe C, Baudhuin P (1985). Polymeric IgA and galactose-specific pathways in rat hepatocytes : evidence for intracellular ligand sorting. In Pastan I, Willingham MC (eds): "Endocytosis," New York: Plenum, pp 163-184.

Draye JP, Courtoy PJ, Quintart J, Baudhuin P (1983). Association of plasma-membrane marker enzymes with the lysosomal membrane in cultured rat fibroblasts. Arch Int Physiol Biochim 91:B97-98.

Graham RC, Karnovsky MJ (1966). The early stages of absorption of injected horseradish peroxidase in the proximal tubules of mouse kidney : ultrastructural cytochemistry by a new technique. J Histochem Cytochem 14:281-302.

Helenius A, Mellman I, Wall D, Hubbard A (1983). Endosomes. Trends Biochem Sci 8:245-250.

Kay DG, Khan MN, Posner BI, Bergeron JJM (1984). *In vivo* uptake of insulin into hepatic Golgi fractions : application of the diaminobenzidine-shift protocol. Biochem Biophys Res Comm 123:1144-1148.

Quintart J, Courtoy PJ, Limet JN, Baudhuin P (1983). Galactose-specific endocytosis in rat liver : biochemical and morphological characterization of a low-density compartment isolated from hepatocytes. Eur J Biochem 131:105-112.

Quintart J, Courtoy PJ, Baudhuin P (1984). Receptor-mediated endocytosis in rat liver : purification and enzymic characterization of low density organelles involved in uptake of galactose-exposing proteins. J Cell Biol 98:877-884.

Quintart J, Courtoy PJ, Baudhuin P (1985). Enzymic characterization of vesicles involved in transcellular transport in rat hepatocytes. Arch Int Physiol Biochim 93:B108-109.

Stoorvogel W, Geuze, HJ, Strous GJ (1986). Sorting of endocytozed transferrin and asialoglycoprotein occurs immediately after internalization in Hep G_2 cells. J Cell Biol "in press".

Cell-Free Analysis of Membrane Traffic, pages 185–194

THE ISOLATION OF ENDOSOMES USING COLLOIDAL GOLD

Julia Beardmore and Colin Hopkins

Department of Biochemistry, Imperial College of Science and Technology, London SW7 2AZ.

INTRODUCTION

Colloidal gold probes have been used to follow the distribution and intracellular sorting of specific receptors in human epithelial cell lines. In these studies epithelial cells in monolayer culture were allowed to endocytose colloidal gold particles complexed to receptor specific monoclonal antibodies for different times at 4°C, 20°C or 37°C. The cells were then fixed and the location of the complexes along the endocytic pathway identified by electron microscopy.

In most of our studies we have used gold complexes carrying the well characterised monoclonal antibody B3/25 which is specific for the human transferrin receptor (Trowbridge and Omary, 1981) to study the endocytic pathway within A431 cells (Hopkins and Trowbridge, 1983; Hopkins, 1983). The endosome system outlined using the gold-B3/25 probe has recently been shown by video enhanced microscopy of living cells to be a widely distributed, highly pleiomorphic organelle which participates in recycling endocytosed transferrin-receptor complexes back to the plasma membrane (De Brabander et al, In Press).

Other recent studies (Miller et al, 1985) have shown that epidermal growth factor (EGF) receptors and EGF are also processed within this system and it is clear therefore, that it is also involved in delivering endocytosed ligand receptor complexes to lysosomes. In view of the central role which the endosome system plays in sorting internalised

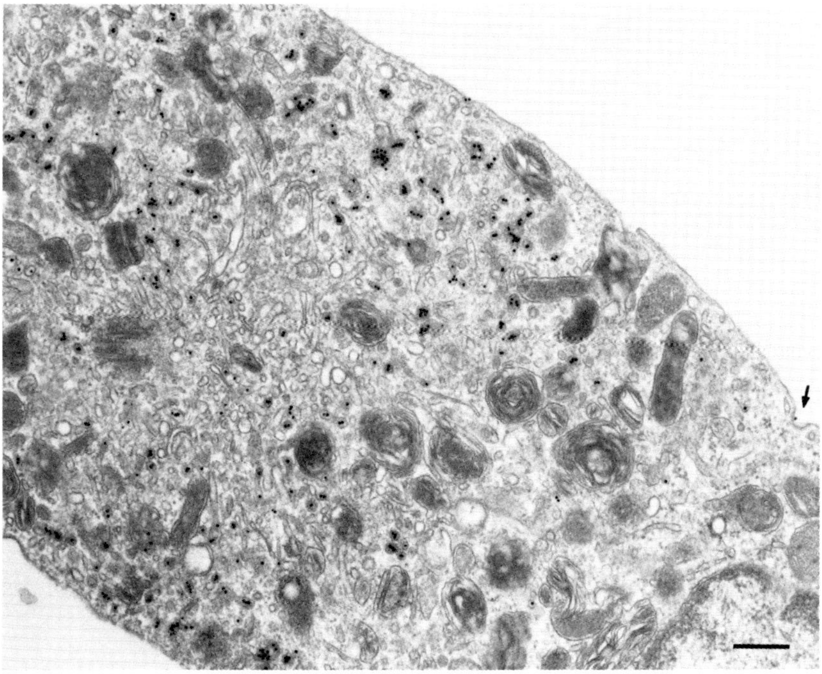

Figure 1. Gold-B3/25 loaded endosome compartments in the intact A431 cell after a 45 minute incubation at 20°C x42000 Cells were fixed in dilute Karnovsky fluid (Karnovsky, 1965), postfixed in 2% osmium tetroxide dehydrated and embedded in Epon. 40nm section. Bar = 0.2µM. Arrow coated pit.

ligands and their receptors, it has become important to devise methods for isolating compartments within the endosome pathway so that they can be characterised biochemically. Previous cell fractionation studies have shown that endosome elements have a similar equilibrium density to other 'light' membrane organelles (plasma membrane, Golgi fractions, endoplasmic reticulum) and morphological studies show that these elements are highly heterogeneous in size and shape. Conventional centrifugation methods of subcellular fractionation are thus unable to yield homogeneous endosome fractions of high purity. We report here an alternative density modification approach (see Beardmore et al., 1987 for a more detailed report).

METHOD

The procedure used for labelling and producing an endosome fraction is as follows:-

1. 9cm confluent dishes of A431 cells are incubated in serum free DMEM for 30 minutes at 37°C and then incubated with ^{125}I-gold-B3/25 for 1 hour at 20°C in binding medium (DMEM, 10mM PIPES, 10μg.ml^{-1} BSA, 0.01% carbowax 20 M, pH7.4).

In some experiments, gold-B3/25 loaded cells are cooled and further incubated on ice with 5ng ^{125}I-EGF for 20 minutes, thus providing a non-enzymatic plasma membrane marker.

To follow the distribution of cellular protein or phospholipid, cells are preincubated for 18 hours with 50μCi L-[^{35}S] methionine or with 3μCi [methyl-^{3}H] choline chloride, respectively.

2. Cells are washed and scraped off the dishes in ice-cold isotonic sucrose containing 10mM acetic acid, 10mM triethanolamine and 1mM $MgCl_2$, pH 7.4 (STEA buffer).

3. A431cells (approximately 20 x 10^6 cells/ml) containing ^{125}I-gold B3/25 in STEA buffer, 4°C. Lysis is achieved by pipetting with a 0-250μl Finn Pipette or Gilson - this is usually 90% efficient as judged by phase light microscopy.

4. The homogenate is incubated with 0.5μg.ml^{-1} trypsin (EC3.4.21.4, Sigma Chemical Co. Ltd., London) for 3 minutes at 37°C followed by the addition of the equivalent activity of SBT1 and the homogenate cooled on ice.

Unbroken cells and nuclei are removed by centrifugation at 800g for 5 minutes at 4°C.

The resulting post nuclear supernatant (PNS) is again centrifuged as above and then a repeat incubation with trypsin performed.

Centrifugation of PNS at 800g, 5 minutes, 4°C.

5. The final PNS, usually containing 70% of the cell associated ^{125}I-gold B3/25 is loaded onto 12ml 26-52% continuous sucrose gradients formed on top of a 0.7ml agar pellet (4% agar in 52% sucrose) which supports a 0.1μ, 13mm

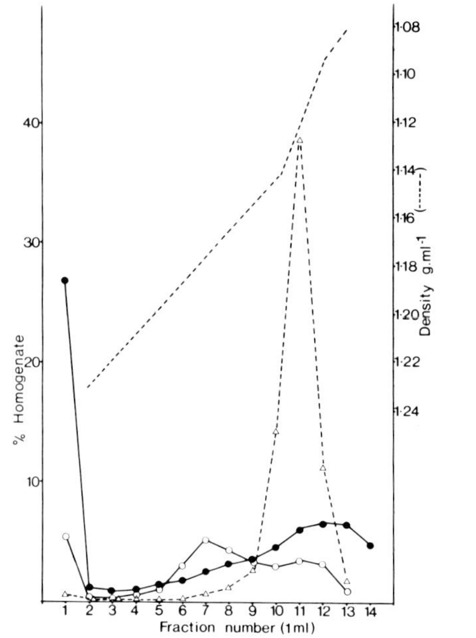

Figure 2. The recovery of ^{125}-labelled B3/25 and ^{125}I-labelled gold-B3/25 on sucrose gradients. Gradient load 1ml. ^{125}I-labelled gold-B3/25, ●—● trypsin treated (n=7), ○—○ no trypsin treatment (n=4), △- -△ ^{125}I-labelled B3/25, trypsin treated (n=3).

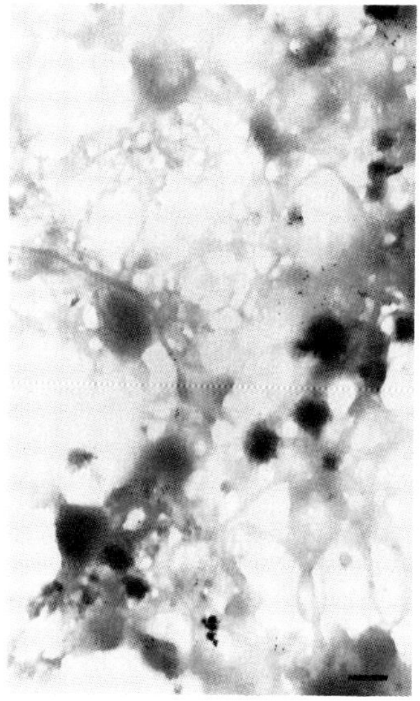

Figure 3. Whole-mount preparation of endosome enriched fraction from non-trypsinised cell homogenate. The filter fraction was resuspended, fixed and applied to poly-L-lysine coated coverslips which were rotary shadowed with platinum/carbon. x 15000 Arrows: gold containing elements. Bar 0.3µm.

Millipore filter.

6. Centrifugation in swinging bucket rotor at 200,000g for 18 hours at 4°C.

7. Gradients are fractionated from just above the filter. The filters supporting the density shifted endosome fraction are either processed for electron microscopy or the material on them recovered for biochemical analysis.

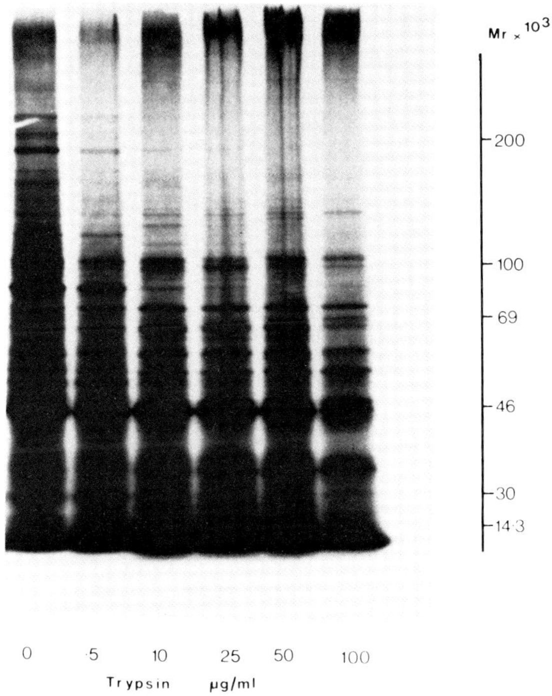

Figure 4. Fluorograph of PNS from L-[^{35}S] methionine labelled cells after increasing treatments with trypsin. 5-12% gradient gel under reducing conditions. Each lane L-[^{35}S] methionine labelled PNS from approximately 1 x 10^6 cells.

RESULTS

We have studied the size of the density shift produced by loading endosomes with gold-B3/25, by comparing the position on gradients of endosomes loaded with ^{125}I-gold-B3/25 (see Figure 2). We concluded that a density shift of at least 0.102 g.ml^{-1} was produced (the filter fraction being at a density > 1.23g.ml^{-1}).

During the development of an adequate lysis procedure for A431 cells, we investigated the effects of trypsin treatment, now used as part of the standard protocol. When fractionation was performed without the addition of trypsin to the homogenate, the yield of ^{125}I-gold-B3/25 on the filter at the bottom of the gradient was substantially less (Figure 2). Examining carbon shadowed samples of the filter fraction by electron microscopy showed that gold containing globular elements were entangled in a meshwork of filamentous material (Figure 3). We presumed this was derived from the cyto-skeleton. The trypsin treatment thus achieved an increase in both the yield and the purity of the endosome fraction and appears to act by degrading cytoplasmic filament proteins, permitting the more efficient dispersal and fractionation of membrane bound elements during centrifugation.

To evaluate the effect of trypsin further we prepared PNS from cells preincubated with L-[^{35}S] methionine and treated them with increasing amounts of trypsin. Figure 4 is a fluorograph of the SDS-PAGE gel of the supernatants. The loading of the lanes varied somewhat but it is clear that at concentrations of trypsin of 10μg ml^{-1} and higher, there is a loss of several protein bands (most noticeably around 220, 190, 120, 85 x 10^3 Mr). At the much lower concentrations of trypsin routinely used in the preparation of our A431 cell homogenate (0.5μg.ml^{-1}) the proteolysis of most cellular proteins is undetectable. The effects of trypsin treatment on cellular proteins may vary from cell type to cell type but our experience suggests that it is necessary for the effective isolation of endosome elements from several epithelial cell lines (Hep 2, KB, WISH).

Figures 5a and 5b show the appearance of a fixed filter supporting the isolated endosome fraction from trypsin treated A431 cells. Clearly the fraction consists of a variety of vesicles and tubules approximately 80% (n=600) of which can be seen to contain gold in a single section. Gold within membrane elements is often seen in close opposition to membrane expecially in larger elements, suggesting that it is still bound to the transferrin receptor via the B3/25 antibody. Apparently free gold comprises less than 25% of the gold on the filter. Serial sections show this to be an overestimate (unpublished observations).

The lack of contaminating cellular organelles in the endosome filter fraction was confirmed by assaying the gradient fractions for marker enzymes for these organelles (Figure 6). When the relative specific activity (RSA) of each marker to

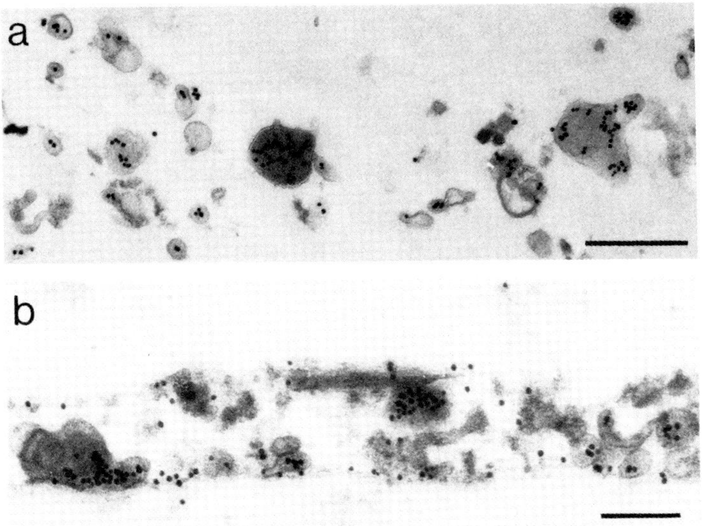

Figure 5. Enriched endosome fraction from gradient filter a) 60nm thick section x75,000, b) 800nm thick section x63000 Bar 0.2μm. Typical cross section views of the filter surface displaying endosomal elements comprising small vesicular structures, tubular elements and larger vesicles.

protein is calculated for the filter fraction, we routinely obtain a value >200 fold for ^{125}I-gold-B3/25, which is an indication of the purification achieved. The only other markers that were slightly enriched in the filter fraction were those for lysosomes with a mean RSA of 2.1 suggesting a 2 fold enrichment in this fraction compared with the PNS.

The endosome fraction from L-[^{35}S]methionine labelled cells was compared with PNS from the same cells and with ^{125}I surface iodinated A431 cells (Hubbard and Cohn, 1975), on 5-12% gradient SDS-acrylamide gels (Figure 7). As expected the endosome fraction is less complex than plasma membrane or PNS, containing at least 10 major polypeptide bands and several minor bands. Three polypeptides (approximately 200, 63, 15 x10^3 Mr) appear to be enriched in this fraction as compared with PNS. Interestingly, the 200 x10^3 Mr protein seems to also be strongly represented in the ^{125}I-plasma membrane preparation.

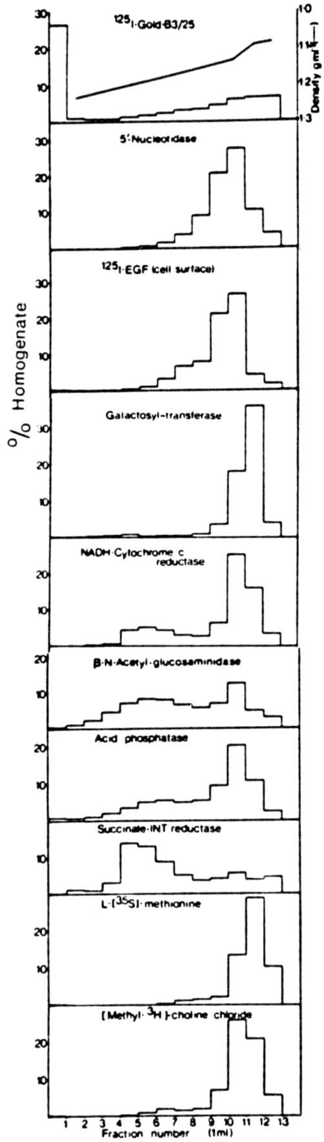

Figure 6. The distribution of marker enzymes and radioactive markers from lysis supernatants on 26-52% sucrose gradients. Gradient load 1ml. Recoveries were 70-100%.

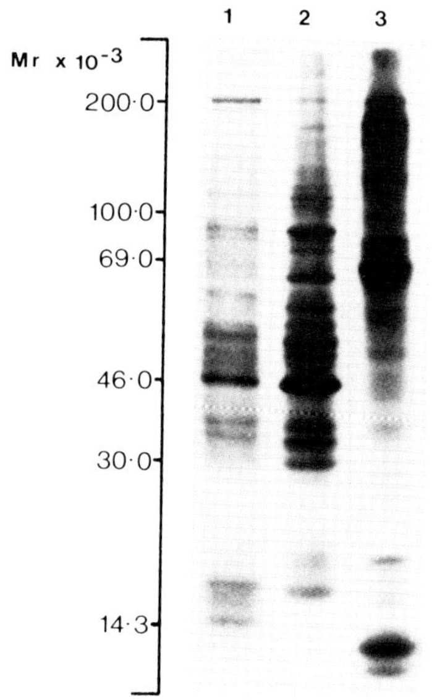

Figure 7. Fluorograph of L-[^{35}S] methionine labelled endosome enriched fraction, PNS and ^{125}I-labelled plasma membrane from A431 cells. A 5% to 12% gradient gel was used under reducing conditions. Lane 1 is L-[^{35}S]methionine labelled endosome fraction (sample load = total filter fraction from 60 x 10^6 cells); Lane 2 is L-[^{35}S]methionine labelled PNS (1 x 10^6 cells); lane 3 is ^{125}I-labelled plasma membrane (3 x 10^6 cells).

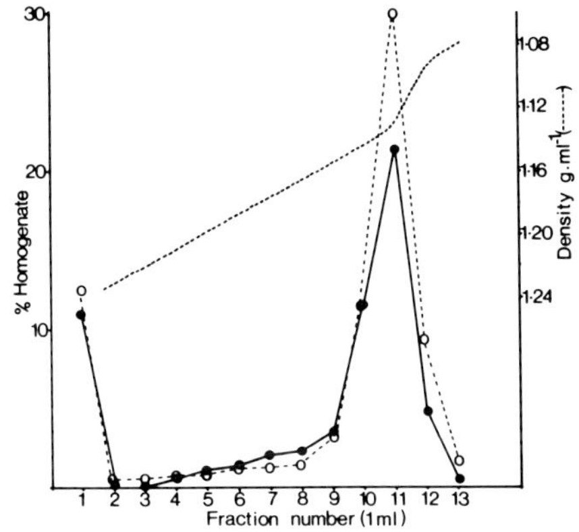

Figure 8. The distribution of ^{125}I-transferrin and ^{125}I-B3/25 on sucrose gradients. Cells were incubated for 1 hour at 20°C in the presence of 100ng ml^{-1} ^{125}I-transferrin or ^{125}I-B3/25 and gold B3/25 and chased for 30 min at 20°C with 10μg ml^{-1} transferrin. Gradient load 1ml. ●——● ^{125}I-labelled B3/25; ○- - -○ ^{125}I-transferrin.

By incubating cells simultaneously with ^{125}I-transferrin or ^{125}I-B3/25 and gold-B3/25 we could show that 12% of ^{125}I-transferrin and 11% of ^{125}I-B3/25 (presumably still bound to the transferrin receptor) is also routed into the highly enriched endosome fraction collected on the filter (Figure 8). It is therefore possible to quantitate the routing of different ligands into the endosome compartment isolated. There is a difference between the percentage of ^{125}I-gold-B3/25 and ^{125}I-transferrin recovered in the filter fraction. We think that a high proportion of the peak of ^{125}I-transferrin in the light gradient fractions is in endosome elements that do not contain sufficient gold to increase their density significantly. Our observation of pink and therefore, gold containing protein bands in these fractions of heavily loaded gradients, adds support to this suggestion.

If this interpretation is correct then there will be opportunities of subfractioning the endosome pathway by selectively loading its subcompartments. It may be possible for example, to selectively isolate cisternal elements (rich in transferrin receptors), from multivesicular bodies (rich in EGF receptors) or even separate microdomains such as clathrin coated pits from their membrane boundaries in which they reside.

CONCLUSION

We have developed a quick and simple method for preparing a highly enriched endosome fraction from an epithelial cell line. The procedure can be readily applied to different cell types as long as suitable gold complexes are available. By varying the incubation conditions with the gold complex, it may be possible to isolate different endosome subfractions allowing further detailed characterizations of the endosome pathway.

REFERENCES

Beardmore J, Howell KE, Miller K, Hopkins CR (1987). Isolation of an endocytic compartment from A431 cells using a density modification procedure employing a receptor-specific monoclonal antibody complexed with colloidal gold. J Cell Sci 87:(in press).

De Brabander M, Nuydens R, Geerts H, Hopkins CR. Dynamic behaviour of the transferrin receptor followed in living A431 cells with Nanovid microscopy. Cell Motility and Cytoskeleton (in press).

Hopkins CR, Trowbridge IS (1983). Internalisation and processing of transferrin and the transferrin receptor in human carcinoma A431 cells. J Cell Biol 97:508-521.

Hopkins CR (1983). Intracellular routing of transferrin and transferrin receptors in epidermoid carcinoma A431 cells. Cell 35:321-330.

Hubbard A, Cohn Z (1975). Externally disposed plasma membrane proteins. I. Enzymatic iodination of mouse L cells. J Cell Biol 64:438-460.

Karnovsky MJ (1965). A formaldehyde-glutaraldehyde fixative of high osmolarity for use in electron microscopy. J Cell Biol 27:(pt. 2), 137a (Abstr.).

Miller K, Beardmore J, Kanety H, Schlessinger J, Hopkins CR (1986). Localisation of the epidermal growth factor (EGF) receptor in EGF-stimulated epidermoid carcinoma (A431) cells. J Cell Biol 102:500-509.

Trowbridge IS, Omary MB (1981). Human cell surface glycoprotein related to cell proliferation is the receptor for transferrin. Proc Natl Acad Sci USA 78:3039-3043.

Cell-Free Analysis of Membrane Traffic, pages 195–209

CHARACTERIZATION OF INTESTINAL MEMBRANE VESICLES WITH FLOW CYTOMETRY

Jean-Pierre Gorvel[1], Zohair Mishal[2] and Suzanne Maroux[1]
1. CBM-CNRS, BP 71, 31, Chemin J. Aiguier, 13402 Marseille Cedex 9, France
2. CTRSC, Service de Fluorometrie Analytique et Separative, 94800 Villejuif, France

INTRODUCTION

The flow fluorometry technique has been used widely to characterize, identify and sort closely related cells and chromosomes (Herzenberg et al. 1976, Loken and Stall 1982, Muirhead et al. 1985). We recently showed that this technique could be useful and efficient as a means of characterizing membrane vesicles as it is for cells (Gorvel et al. 1984a, Moktari et al. 1986). During subcellular fractionation, membranes from different organelles can be characterized by specific markers. However, a small amount of a marker thought to be specific for one type of membrane can very often be detected in another membrane fraction. It is then difficult to determine whether this reflects a cross-contamination or the actual presence of the same marker on two membrane types, in very different amounts. To settle this problem, we applied the technique of flow fluorometry for the first time to membrane vesicles (Moktari et al. 1986).

The plasma membrane of highly differentiated absorptive cells is composed of at least two structurally and functionally distinct domains. The apical domain is formed by the microvilli of luminal brush border and contains a large number of digestive hydrolases (Kenny and Maroux 1982). The basolateral domain is characterized by other specific markers such as Na^+, K^+-ATPase (Gorvel et al. 1982, Reggio et al. 1982) and histocompatibility antigens (Kirby and Parr 1979, Curman et al. 1979, Scott et al. 1980, Gorvel et al. 1984b).

After immunolabeling of jejunal sections, no hydrolases such as aminopeptidase N (APN) could be observed in the basolateral domain (see Figure 2a), whereas small amounts of hydrolase enzymic activity have been observed in all the basolateral membrane fractions obtained after subcellular fractionation (Weiser et al. 1978, Quaroni et al. 1979a,b, Colas and Maroux 1980, Feracci et al. 1985, Moktari et al. 1986). In view of the high sensitivity of the flow fluorometry method, the actual presence of APN in the basolateral membrane vesicles was investigated in the present study with this technique, using monoclonal antibodies directed against various epitopes of the enzyme.

RESULTS

Principles of the Flow Fluorometric Technique

A flow cytometer includes a light source, a sample chamber and an optical and processor assembly (Muirhead et al. 1985). Cells, microorganisms, chromosomes or cell organelles are introduced into the sample chamber and constrained into the center of a fast fluid stream. When a particle crosses the laser beam, the illuminating light is elastically scattered by the particle and the intensity of the scattered light gives information about the size, shape, density and surface morphology (Salzman 1982, Benson et al. 1984, Bohmer et al. 1984). The amount of light scattered at low angles in the forward direction along the axis of the laser beam is related to the particle size (Salzman 1982). Measurements are performed at rates of 10^4 to 10^6 particles per minute.

In most applications of flow cytometry, the cells are characterized and sorted after being specifically labeled by fluorescent probes, usually rhodamin and fluorescein, coupled to antibodies. After excitation by the laser beam, the emissions from several fluorochromes can be analyzed and quantified separately on the same particle by using a series of appropriate dichroic mirrors and filters. A particularly attractive feature of flow cytometry is that flow sorters are able to physically isolate cells or chromosomes from heterogeneous samples. When a droplet containing the particle of interest breaks off from the main stream, the droplet is given a negative or positive charge and is subsequently deflected electrostatically into a container (Muirhead et al. 1985).

Subcellular Fractionation

We have proposed a fast and easy procedure for concomitantly preparing from the same mucosa of A^+ rabbits four fractions enriched in membrane from brush border (BB), basolateral domain of plasma membrane (BL), rough endoplasmic reticulum (RER) and smooth endoplasmic reticulum plus Golgi apparatus (G + SER) (Moktari et al. 1986). A^+ rabbits are genetically defined by the presence of human blood group A determinants in the secreted and membrane bound glycoproteins of several organs (Gorvel et al. 1985). The glycoproteins integrated in the plasma membrane of enterocytes bear this type of determinant (Gorvel et al. 1982). Briefly, rabbits were killed and small intestine without the duodenum excised and washed with cold phosphate saline buffer. After two centrifugations at low speed, the high density membranes containing BB and RER were separated from the low density membranes containing SER + G + BL by centrifugation through a sucrose step gradient. BB and RER were then completely separated by calcium precipitation of RER. The BL was satisfactorily separated from the G + SER membranes by specific immunoprecipitation by anti-blood group A antibodies. The amount of APN in BL amounted to about 6% of the total activity in the homogenate.

Immunodetection of Membrane Markers

Immunolabeling by specific antibodies on ultrathin frozen sections of jejunum and subcellular fractions was first carried out by light and electron microscopy as shown in Figures 1-3. Immunogold labeling of brush border membrane vesicles agreed very well with the immunofluorescence labeling of the brush border _in situ_ (compare Figures a and b). The human blood group A antibody reacted very satisfactorily with the basolateral domain _in situ_ (Figure 1a) and with some large vesicles of the light membrane fraction tentatively identified with the basolateral membrane vesicles (Figure 1c). By contrast, although the anti-rabbit histocompatibility antigen class I strongly labeled

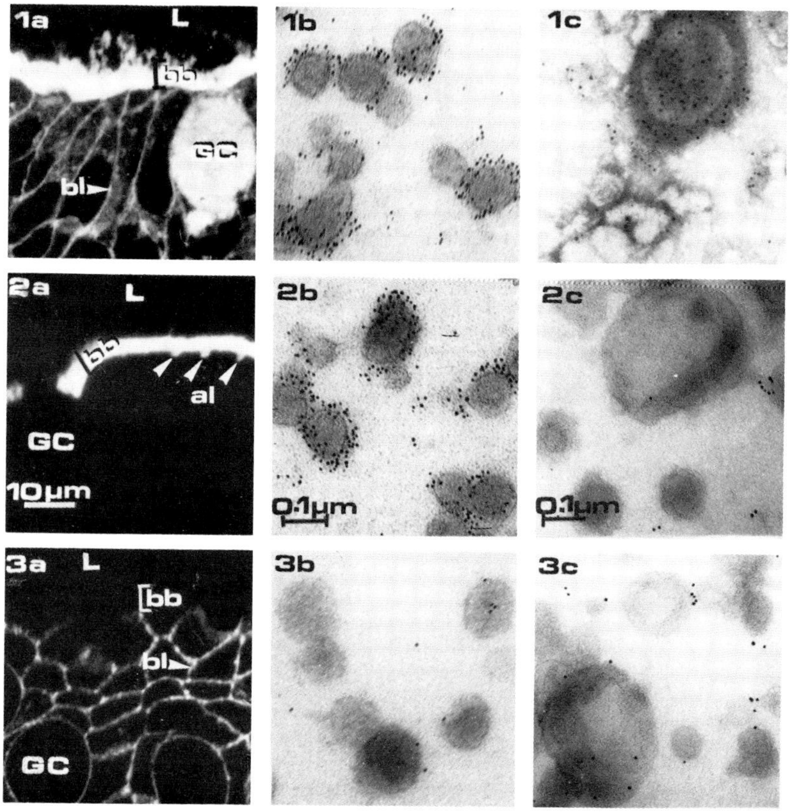

Figures 1-3. Immunofluorescence labeling of ultrathin frozen sections (1800 Å) of jejunum from A^+ rabbit in a) and electron microscopy of negative stained brush border membrane vesicles in b) and light membrane (BL + G + SER) fraction in c) after immunogold labeling with: mouse monoclonal antibody anti-human blood group A determinants (Cl 3.3) in 1; mouse monoclonal antibody anti-aminopeptidase N (1D8.1) in 2; rat monoclonal antibody anti-rabbit histocompatibility antigen class I (8A3.1) in 3. In a, the first antibody (30 µg/ml) was revealed by a second antibody coupled to fluorescein (Feracci et al. 1982). L: lumen; GC: goblet cell; bb: brush border; bl: basolateral plasma membrane; al: apical region of the lateral membrane. In b and c the first antibody (30 µg/ml) was revealed by a double sandwich technique using a rabbit anti-mouse IgGs polyclonal antibody cross-reacting with rat IgGs, and then by protein-A-gold (Gorvel et al. 1986b).

the basolateral domain *in situ* (Figure 3a), no reaction with corresponding membrane vesicles contained in the low density membrane fraction could be observed (Figure 3c). Consequently, the absence of any labeling of this membrane vesicle fraction by anti-APN (Figure 2c) cannot be taken to be proof of the absence of the enzyme from the basolateral domain.

In conclusion, electron microscopy took a long time and did not seem sufficiently sensitive to detect low amounts of antigen on membrane vesicles. Moreover, there is a possibility that a specific class of vesicles may have been lost during washing.

We sought to overcome some of the limitations associated with electron microscopy by using flow fluorometry for the analysis and quantification of membrane vesicle antigens labeled by means of antibodies (Moktari et al. 1986) As flow fluorometry can be applied only to free particle suspensions, this analysis could not be performed on the final BL fraction since it was obtained in the form of an immunoprecipitate. Specific BL labeling was deduced by comparing the labelings of the light membrane fraction before and after BL immunoprecipitation. Figure 4 shows the results obtained. The BB vesicles were labeled by all the antibodies directed against some well characterized brush border markers: human blood group A determinants, aminopeptidase N (see Figures 1a and 2a) and the early expressed 140 K antigen (Gorvel et al. 1986a, b), but not the anti-RLA class I, a specific marker of the basolateral domain (see Figure 3a). As expected, the latter antibody specifically labeled the BL vesicles contained in the low density membrane fraction (BL + G + SER). Anti-human blood group A and anti-APN also reacted with BL vesicles.

Although immunoblots and immunofluorescence showed that the polyclonal anti-APN antibodies were specific (Moktari et al. 1986), the presence of APN on BL vesicles was more closely investigated by monoclonal antibodies directed against various epitopes of the enzyme. Quantitative data are given in Table 1.

The antibodies able to react with the BB fraction labeled around 70% of the vesicles. This was the first strict evaluation of the homogeneity of this membrane preparation. Moreover, about 30% of BL + G + SER were labeled

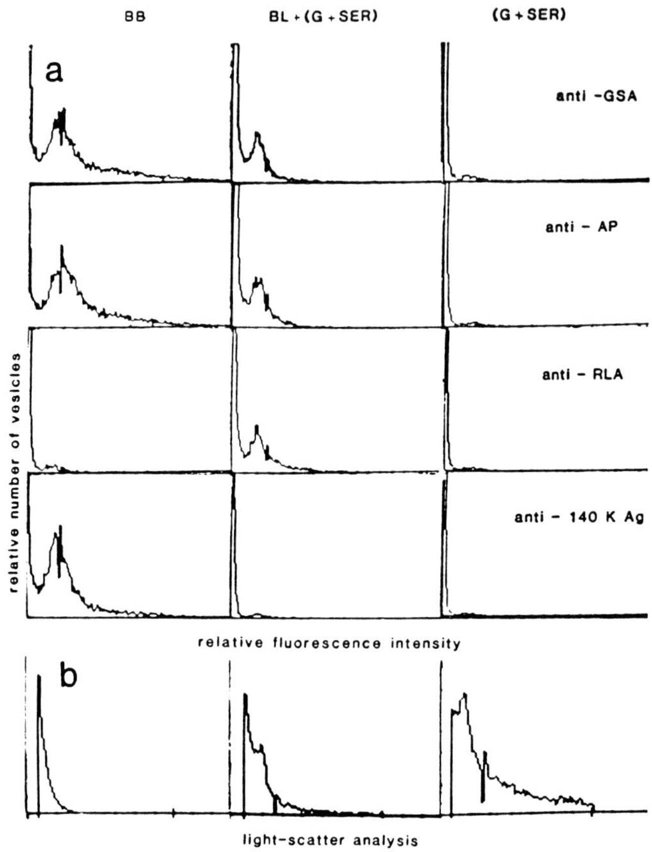

Figure 4. In a), surface labeling of various membrane fractions resulting from subcellular fractionation of A+ rabbit mucosal scraping (Moktari et al. 1986) labeled by several specific antibodies coupled with fluorescein, and assayed by flow fluorometry. Vesicles were stained by direct immunofluorescence with FITC-conjugated with: rabbit polyclonal anti-human blood group A determinants (60 μg/ml);goat polyclonal anti-APN (40 μg/ml); rat monoclonal anti-rabbit lymphocyte antigen of the major histocompatibility complex class I (10 μg/ml); mouse monoclonal anti-140 K antigen (14 μg/ml). Vesicle number was plotted vs. fluorescence intensity (channels 0 to 256). The cell sorter was set at a gain of 50. In b) light-scatter analysis of the three membrane fractions. The number of vesicles was 10,000 in each experiment.

by either anti-RLA, anti-human blood group A determinants or anti-APN, so that the proportions of BL in this fraction could be definitely estimated. Among the interesting information provided by this flow fluorometric analysis, was the estimation of the density of labeling. It should be mentioned that the amount of AP per μm^2 of BL is 20 times less than in BB vesicles. This is in good agreement with the relative specific activities of AP in BL (60) and in BB (1500).

Table 1. Quantitation of surface labeling of membrane fractions by several specific antibodies.

Antibodies	% of labeled vesicles			$\frac{\text{Density of AP in BL}}{\text{Density of AP in BB}}$ X100
	BB	BL+G+SER	G+SER	
Anti-AP				
Polyclonal	74	30	3	4.5
4 H 7.1	65	27	7	2.7
1 D 8.1 7 H 1.3	68	24	3	6
4 C 11.1 3 H I.4	63	19	4	6
4 H 11.1 2 D 12.1 3 G 4.1	70	17	3	3.5
Anti-GSA	78	27	4	-
Anti-RLA	5	38	2	-
Anti-140KAg	74	3	2	-

For quantitative immunofluorescence measurements, the cell sorter was calibrated with a view to establishing the relationship between the number of fluorochromes and the fluorescence emitted, as described by Le Bouteiller et al. (1983). Polyclonal anti-aminopeptidase (anti-AP) and 8 monoclonal anti-AP classified into four groups recognizing different epitopes were tested: 40 μg/ml with polyclonal antibody and 30 μg/ml with monoclonal anti-AP antibodies. Anti-GSA: anti-human blood group A determinants (60

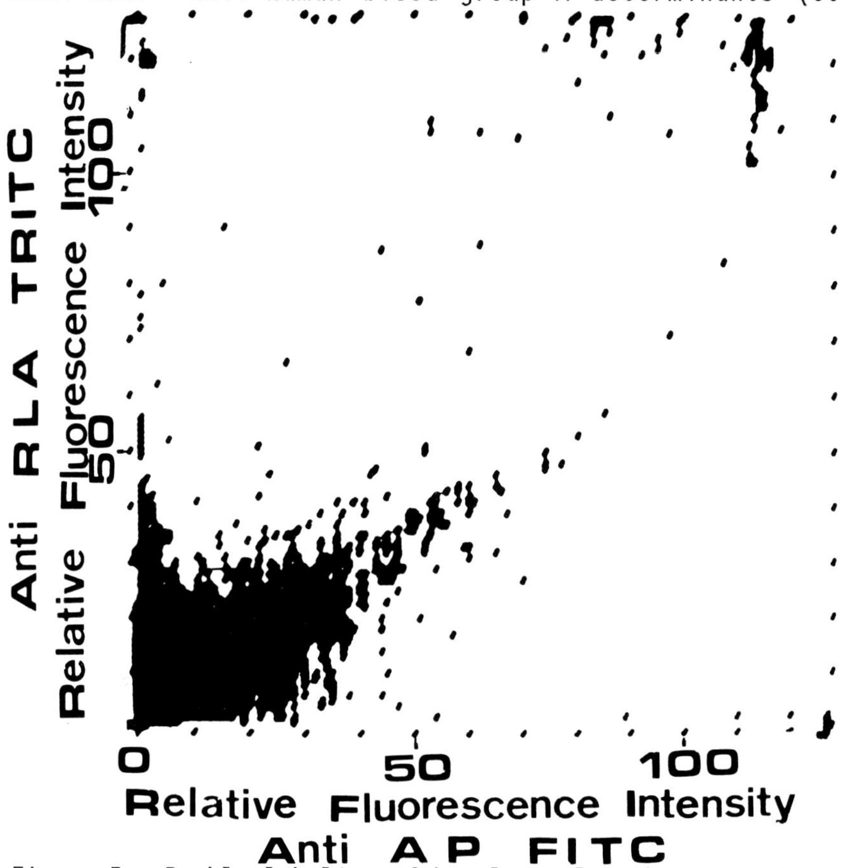

Figure 5. Double labeling of basolateral membrane vesicles with fluorescein-anti-aminopeptidase N (anti-AP-FITC) and rhodamin anti-histocompatibility antigen class I (anti-RLA-TRITC) (Moktari et al. 1986). The relative fluorescence intensity increased linearly to the right (index channel numbers 0-128 with anti-AP-FITC) and to the top (channel numbers 0-128 with anti-RLA-TRITC).

μg/ml). Anti-RLA: anti-rabbit histocompatibility class I antigen (10 μg/ml). Anti- 140 K Ag: anti-rabbit brush border antigen (14 μg/ml) (Gorvel et al. 1986a).

To confirm these results, the presence of both APN and RLA class I on the same vesicles was tested more closely by a dual labeling experiment (Figure 5). This was achieved using anti-RLA coupled to rhodamin and anti-APN coupled to fluorescein.

Quantitative analysis of the diagram showed that all vesicles labeled by anti-APN were labeled by anti-RLA class I. This means that APN was present on BL membrane. The other noteworthy result was the finding that 40% of the labeled vesicles bound RLA exclusively. These vesicles might be basal membrane vesicles.

Physicochemical Studies of Membrane Vesicles by Flow Cytometry

Flow cytometry also can be used to investigate the size and fluidity of membrane preparations. As shown in Figure 6, flow cytometry is a rapid means of analyzing the size of vesicles at a single particle level by forward angle light-scatter analysis. The size homogeneity of a membrane preparation can be checked by looking at the sharpness of the histograms. When BB and BL vesicles prepared as described by Colas and Maroux (1980) were analyzed, BL vesicles were found to be less homogeneous in size than BB. After calibration of the flow cytometer with latex beads or chromosomes, the average diameter of BB and BL vesicles was found to be 0.1 μm and 0.2 μm, respectively. These values were in good agreement with those measured by electron microscopy (Figure 6).

The membrane fluidity of BB and BL membranes prepared as described by Colas and Maroux (1980) were compared by studying the fluorescence polarization of diphenylhexatriene (DPH), a convenient probe for the lipid hydrocarbon chains (Andrich and Vanderkooi 1976). To ensure that histograms corresponded to statistical analysis of membrane vesicles, we combined light-scatter and fluorescence polarization analysis (Table 2).

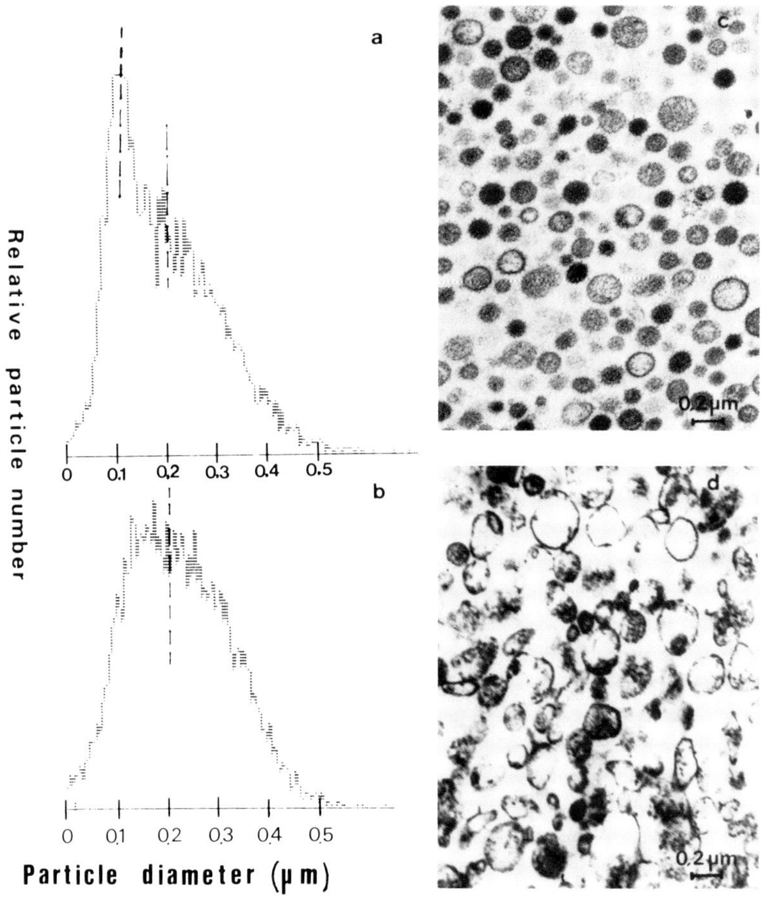

Figure 6. Light-scatter patterns of BB fraction(a) and BL fraction prepared as described by Colas and Maroux (1980) in b were analyzed by flow cytometry. Sections through the pellet of BB and BL are shown in c and d, respectively. The EPICS V cell sorter (Coulter Electronics, Hialeah, Fl., USA) was used with an argon ion laser mode 171-17 (12 W) to analyze vesicles in phosphate saline buffer as they crossed the laser beam. 5000 vesicles were analyzed. The laser excitation power was 400 mW at 363 nm. The flow rate was adjusted to 1000 vesicles per second. The light-scatter from particles was measured within the narrow angle of 0-9° in relation to the incident light.

TABLE 2. Fluorescence polarization of brush border and basolateral membranes assayed by flow cytometry.

Membranes	Fluorescence polarization		
	4°C	18°C	37°C
BB	0.325 (0.005)	0.273 (0.006)	0.252 (0.004)
BL	0.223 (0.006)	0.210 (0.007)	0.200 (0.007)

BB and BL membranes prepared as described by Colas and Maroux (1980) were incubated for 30 min at 4, 18 and 37°C with 2 μM DPH. The power of the laser excitation was 400 mW at 363 nm. The flow rate was adjusted to 1000 vesicles/s and the cell sorter analyzed 5000 vesicles. The fluorescence polarization values corresponded to the median value of histograms (Gorvel et al. 1984a) obtained by plotting the vesicle number as a function of the fluorescence polarization of DPH. The on-line computer calculated the fluorescence polarization of each vesicle according to the equation:

$$P = \frac{I_{pa} - I_{pe}}{I_{pa} + I_{pe}}$$

where I_{pa} and I_{pe} are the fluorescent intensities recorded when fluorescent light crossed two kinds of polarizers, oriented in parallel and perpendicularly, respectively.

First, the values of the fluorescence polarization of DPH were in good agreement with those obtained by spectroscopy (Hauser et al. 1982). The membrane fluidity deduced from the fluorescence polarization was higher in BL than in BB and the membrane fluidity increased with temperature in both fractions. Analysis of membrane fluidity by flow cytometry has distinct advantages: with flow cytometry, measurements can be obtained at the level of a single particle, so that the permanent problem of depolarization due to light-scattering is avoided; disrupted particles, membrane debris and aggregates can be efficiently eliminated by light-scatter gratings (Gorvel et al. 1984a), so that the analysis is conducted on intact vesicles.

CONCLUSION

Aided by parallel advances in instrumentation and analytical methodologies, recent progress in cell and molecular biology has generated important new insights in several domains of cell biology. Flow cytometry is a good example of hybrid technology, since it has evolved as the result of confluent advances in laser technology, fluorochrome chemistry, computer processing and monoclonal antibody production. Flow cytometry holds great promise for subcellular fractionation. This technique provides a rapid means of controlling the homogeneity of membrane fractions and quantifying antigens at the surface vesicles by using fluorescent antibodies. Analysis at a single particle level makes it possible in one step to sort vesicles in order to test their protein and lipid content. We are investigating the sorting of membrane vesicles, which we hope to achieve shortly.

The sensitivity of the flow cytometry technique makes it possible to detect an antigen present in small amounts on the surface of membrane vesicles. When used to characterize the two plasma membrane domains of rabbit enterocytes, this technique definitely proved the presence of APN, a brush border marker, in at least 60% of the basolateral membrane vesicles. We recently obtained strong evidence that the presence of APN in the basolateral domain results from the fact that the newly synthesized enzyme transits through this domain reaching the brush border (Massey et al. 1986).

Dual labeling of basolateral vesicles has shown that although 60% of vesicles bore APN and RLA class I antigen, 40% expressed only the RLA class I antigen. The latter vesicles might correspond to vesicles obtained from the basal membrane domain. With flow cytometry, it might thus be possible to identify, characterize and isolate specialized domains of the basolateral membrane, which has not hitherto been possible with other methods.

This technique does, however, have limitations which are mainly due to the starting material. First, immunoprecipitated vesicles cannot be analyzed with this method, and the maximum concentration of antibodies just before aggregate formation has to be determined by light-scatter analysis. Consequently, the concentration of an abundant antigen is liable to be underestimated. Secondly, the risk of obtaining from the same membrane two types of closed vesicles, some right side out and the others inside out should also be kept in mind. Both types of vesicles cannot be characterized by the same antibodies.

REFERENCES

Andrich MP, Vanderkooi JM (1976). Temperature dependence of 1,6-diphenyl-1,3,5-hexatriene fluorescence in phospholipid artificial membranes. Biochemistry 15:1257-1261.

Benson MC, Mc Dougal DC, Coffey DS (1984). The application of perpendicular and forward angle scatter to assess nuclear and cellular morphology. Cytometry 5:515-522.

Bohmer RM, King NJC (1984). Flow cytometric analysis of immunogold cell surface label. Cytometry 5:543-546.

Colas B, Maroux S (1980). Simultaneous isolation of brush border and basolateral membrane from rabbit enterocytes. Biochim Biophys Acta 600:406-420.

Curman B, Kampe O, Rask L, Peterson PA (1979). Presence of alloreactive Ia antigens on murine intestinal epithelial cells. Scand J Immunol 10:11-15.

Feracci H, Bernadac A, Gorvel JP, Maroux S (1982). Localization by immunofluorescence and histochemical labeling of aminopeptidase. Gastroenterology 82:317-324.

Feracci H, Rigal A, Maroux S (1985). Biosynthesis and intracellular pool of aminopeptidase N in rabbit enterocytes. J Membrane Biol 83:139 146.

Gorvel JP, Wisner-Provost A, Maroux S (1982). Identification of glycoproteins bearing human blood group A determinants in rabbit enterocyte plasma membrane. FEBS Lett 143:17-20.

Gorvel JP, Liabeuf A, Massey D, Liot D, Goridis C, Maroux S (1983). Recognition of sodium- and potassium-dependent adenosine triphosphatase in organs of the mouse by means of a monoclonal antibody. Cell Tissue Res 234:619-632.

Gorvel JP, Mawas C, Maroux S, Mishal Z (1984a). Flow cytometry is a new method for the characterization of intestinal plasma membrane. Biochem J 221:453-457.

Gorvel JP, Sarles J, Maroux S, Olive D, Mawas C (1984b). Cellular localization of class I (HLA-A,B,C) and class II

(HLA-DR and DQ) MHC antigens on the epithelial cells of normal human jejunum. Biol Cell 52:249-252.
Gorvel JP, Rigal A, Sarles J, Maroux S (1985). Aminopeptidase N- and human blood group A-antigenicity along the digestive tract and associated glands in the rabbit. Cell Tissue Res 239:241-248.
Gorvel JP, Rigal A, Olive D, Mawas C, Maroux S (1986a). Identification of an early expressed marker of the luminal membrane of rabbit small intestinal columnar cells. Presence of a homologous antigen in kidney proximal tubules and glomureli. Biol Cell 56:121-126.
Gorvel JP, Rigal A, Bernadac A, Maroux S (1986b). Further characterization of an early expressed glycoprotein of the rabbit small intestinal brush border. Its interaction with some hydrolases. Eur J Biochem 159:499-505.
Hauser H, Gains N, Semenza G, Spiess M (1982). Orientation and motion of spin-labels in rabbit small intestinal brush border vesicle membranes. Biochemistry 21:5621-5628.
Herzenberg LA, Sweet RG, Herzenberg LA (1976). Fluorescence activating cell sorting. Sci Am 234:108-117.
Kenny AJ, Maroux S (1982). Topology of microvillar membrane hydrolases of kidney and intestine. Physiol Rev 62:91-128.
Kirby WN, Parr EL (1979). The occurrence and distribution of H-2 antigens on mouse intestinal epithelial cells. J Histochem Cytochem 27:746-750.
Le Bouteiller P, Mishal Z, Lemmonier FM, Kourilsky FM (1983). Quantification by flow cytometry of HLA class I molecules at the surface of murine cells transformed by cloned HLA genes. J Immunol Methods 61:301-315.
Loken MR, Stall AM (1982). Flow cytometry as an analytical and preparative tool in immunology. J Immunol Methods 50:R85-R112.
Massey D, Feracci H, Gorvel JP, Rigal A, Maroux S (1986). Evidence for the transit of aminopeptidase N through the basolateral membrane before it reaches the brush border of enterocytes. J Membrane Biol (in press).
Moktari S, Feracci H, Gorvel JP, Mishal Z, Rigal A, Maroux S (1986). Subcellular fractionation and subcellular localization of aminopeptidase N in the rabbit enterocytes. J Membrane Biol 89:53-63.
Muirhead KA, Horan PK, Poste G (1985). Flow cytometry: present and future. Biotechnology 3:337-356.

Quaroni A, Kirsch K, Weiser MM (1979a). Synthesis of membrane glycoproteins in rat small intestinal villus cells. Redistribution of L-(1,5,6,-^{3}H)-fucose labelled membrane glycoproteins among Golgi, lateral basal and microvillus membrane in vivo. Biochem J 182:203-212.

Quaroni A, Kirsch K, Weiser MM (1979b). Synthesis of membrane glycoproteins in rat small intestinal villus cells. Effect of cochicine on the redistribution of L-(1,5,6-^{3}H)-fucose labelled membrane glycoproteins among Golgi, lateral and microvillus membrane in vivo. Biochem J 182:213-221.

Reggio H, Coudrier E, Louvard D (1982). Surface and cytoplasmic domains in polarized epithelial cells. In Liss AR (ed): "Membranes in growth and development," NY: pp 89-105.

Salzman GC (1982). Light scattering analysis of single cells. In Catsimpoclas H (ed): "Cell Analysis" Vol. 1, NY: Plenum Publishing Corp., pp 111-143.

Scott H, Silheim BG, Brantzaeg P, Thorsby E (1980). HLA-DR- like antigens in the epithelium of the human small intestine. Scand J Immunol 12:77-82.

Weiser MM, Neumeier MM, Quaroni A, Kirsch K (1978). Synthesis of plasmalaminal glycoproteins in intestinal epithelial cells. Separation of Golgi membranes from villus and cryptal surface membranes, glycosyl transferase activity of surface membrane. J Cell Biol 77:722-734.

Cell-Free Analysis of Membrane Traffic, pages 211–225

PURIFICATION OF MEMBRANES ENRICHED IN ROUGH ENDOPLASMIC RETICULUM DOMAIN BY USING A DENSITY PERTURBANT

Alain Amar-Costesec, Danièle Godelaine and Michael Hortsch[+]
International Institute of Cellular and Molecular Pathology, UCL, Brussels and [+]European Molecular Biology Laboratory, Heidelberg

INTRODUCTION

By electron microscopic examination, hepatocytes are found to contain a large membranous network called the endoplasmic reticulum (ER), which exists in two morphologically distinct forms : the rough ER, composed of parallel layers of flat cisternae bearing ribosomes on the surface facing the cytoplasm, and the smooth ER which has the appearance of convoluted tubules protruding from the extremities of the cisternae or accumulating in some regions of the cytoplasm to form a closely meshed network.

ER plays important roles in cellular physiology : the rough ER is the site for the synthesis of secretory proteins as well as some organelle proteins, whereas the transformation of lipophilic xenobiotics into more polar molecules takes place in the smooth ER. To carry out these and numerous other functions, the ER possesses a wide variety of enzymes and other protein constituents. Its biochemical complexity is reflected by the composition of the microsomal fraction (Amar-Costesec et al., 1974a), which contains the majority of the smooth and rough ER-derived vesicles produced during homogenization of the liver.

Subfractionation of microsomes by isopycnic centrifugation led to the unexpected finding that rough and smooth vesicles do not separate from each other, but instead they overlap extensively. In fact, fragmentation of the ER gives rise to rough vesicles with a varying number of bound ribosomes. These vesicles equilibrate along the gradient according to the number of ribosomes bound per vesicle (Wibo et al., 1971; Beaufay et al., 1974b).

This experimental approach also led to a classification of ER constituents based upon their density distributions. NADH- and NADPH-cytochrome c reductases, cytochromes b5 and P-450 (Beaufay et al., 1974b), epoxide hydrolase, cholesterol 7α-hydroxylase and hydroxymethylglutaryl-coenzyme A reductase (Mitropoulos et al., 1978) peak at a density of 1.14 which is the modal equilibrium density of smooth vesicles. Alternatively, ribophorins (Marcantonio et al., 1984) and high affinity ribosome binding sites (Amar-Costesec et al., 1984) peak at a density of 1.22. Finally, glucose 6-phosphatase, glucuronosyltransferases, esterases and nucleoside diphosphatase displayed a more regular, almost bimodal distribution pattern with two shoulders at densities of 1.14 and 1.22. These three groups of constituents were termed b, d and c, respectively. Because the distinction between rough and smooth ER, which is essentially based on morphological observations, is insufficient to account for this heterogeneity, we presented our interpretation of the molecular organisation of the ER (Amar-Costesec, 1978a,b ; Amar-Costesec and Beaufay, 1981, reviewed in Beaufay et al., 1981). According to our hypothesis, the ER constituents would be distributed between three biochemically distinct domains. Domains b and d would share the cytoplasmic aspect of the membrane, each domain representing about one-half of the total surface area (Amar-Costesec and Beaufay, 1981) and containing the corresponding groups of constituents. At the lumenal side, group c enzymes almost totally occupy this surface. Most of our present knowledge of the biochemical composition and molecular organisation of the ER concerns domains b and c while very little is known about domain d (the bona-fide rough domain). How many integral membrane proteins of the ER are segregated within this domain ? Is it exclusively devoted to protein synthesis by membrane-bound ribosomes, or is it also involved in other functions as has been suggested by some authors (see for instance Venkatesan et al., 1980) ? In order to answer the first question, we have compared the protein patterns of various membrane preparations by two-dimensional gel electrophoresis (in collaboration with O. Arroyo, K. Willard-Gallo, H. Beaufay) in the hope that we can recognize polypeptides that are specific for the rough ER. The membrane preparations differ from each other by the degree of their enrichment in d domain. One of the difficulties we have encountered during this work has been the contamination of the membranes by peroxisomal cores whose equilibrium density is close to that of the roughest ER vesicles. We describe here how the use of a density perturbant allows the removal of this contamination.

MATERIALS AND METHODS

Materials

The following products were purchased from the indicated sources : sucrose grade I, chymostatin, pepstatin, leupeptin, antipain, leucyl leucyl leucine,aprotinin, phenyl methyl sulfonyl fluoride (PMSF) : Sigma Chemical Co., St. Louis, MO ; the protease inhibitor solutions were freshly prepared by dissolution in dimethyl sulfoxide (chymostatin, pepstatin, leucyl leucyl leucine), in water (leupeptin, antipain, aprotinin), or in ethanol (PMSF) ; imidazol puriss. p.a. : Fluka A.G., Buchs, Switzerland ; tissue grinder type C : Arthur H. Thomas Co., Philadelphia, PA.

Isolation of Postmitochondrial Supernatants and Microsomes

All sucrose solutions were buffered at pH 7.4 with 3 mM imidazole-HCl. In some experiments, the homogenization medium contained : chymostatin, pepstatin, leupeptin, antipain, leucyl leucyl leucine, each at a final concentration of 5 µg/ml ; 1 µg/ml aprotinin, 40 µg/ml PMSF and 3 mM dithiotreitol. All operations were conducted at 0-4°C.

Female rats of the Wistar strain weighing 180-200 g were fasted for 18 hrs, with water ad libitum, before being killed by decapitation. After sacrifice of the animals, the livers were quickly removed, immersed in cold 250 mM sucrose and weighed. The livers were chopped into pieces, diluted with 3 ml 250 mM sucrose per g, and one liver at a time was homogenized with one stroke (in about 15 sec) of the teflon pestle driven at 1,300 rpm. When used, PMSF was added directly to the homogenizer containing the sucrose and the tissue pieces. The homogenate was centrifuged for 10 min at 1,700 rpm (DRP 6000 centrifuge, rotor n° 252, International Equipment Co., Needham Heights, MA). The nuclear pellet was washed twice by resuspension in 3 ml sucrose per g liver and centrifugation for 10 min at 1,400 rpm in the same rotor as above. The pooled supernatants, forming the postnuclear supernatant, were centrifuged for 6 min 42 sec at 25,000 rpm (Beckman rotor n° 40, Beckman Instruments Inc., Spinco Division, Palo Alto, CA). The large granule fraction was washed twice by resuspension in 3 ml sucrose per g liver and centrifugation (conditions as above). The pooled supernatants constitute the postmitochondrial supernatant.

The microsomal fraction was obtained by centrifugation of the postmitochondrial supernatant for 30 min at 39,000 rpm (Beckman rotor n° 40) or 54 min at 29,000 rpm (Beckman rotor n° 30),when large volumes of postmitochondrial supernatant have to be centrifuged.

Isolation of Rough Microsomes

Except when otherwise stated, all sucrose solutions are given as % w/w. Three rotors of different capacities were used for preparing rough microsomes, depending on the amount of material to be treated.

E-40 rotor. The method used was essentially that described by Beaufay et al. (1974b). To the periphery of the E-40 rotor (Beaufay, 1966),spinning between 5,000 and 7,000 rpm the following samples were injected in succession : a 10 ml-sample, containing the microsomes from 9-10 g liver, a 32 ml-linear sucrose gradient between densities 1.10 (23.2 %) and 1.25 (52.2 %) and a 6 ml-sucrose cushion of density 1.34 (67.3 %) (for details see Beaufay et al., 1974b and Beaufay and Amar-Costesec, 1976). After centrifugation for 140 min at 40,000 rpm, the rotor was decelerated to 10,000 rpm and 12 subfractions of 3-4 ml were collected. Subfractions with a density of 1.19 or more were pooled, except for the most dense subfractions which basically contain the cushion. This preparation (average density = 1.25) is defined as the rough microsomes.

Beckman SW 25-2 rotor. A 20 ml-linear sucrose gradient between densities 1.07 (16.6 %) and 1.20 (43.1 %), and a 10 ml-sucrose layer of density 1.20 were pumped successively into one to three centrifuge tubes ; a 30 ml-sample containing the microsomes or the postmitochondrial supernatant from approximately 10 g liver was layered on top of each sucrose gradient. The tubes were centrifuged overnight at 24,000 rpm. The first 50 ml, which corresponds to the volume of the sample and the gradient, was carefully removed using a peristaltic pump. The pellet was recovered using a glass rod and resuspended in the viscous supernatant to give the rough microsomes.

Beckman Ti-14 rotor. A step gradient was made of the following sucrose solutions pumped in sequence to the edge of the rotor spinning at 4,000 rpm : 50 ml of density 1.10 (23.2 %), 50 ml of 1.12 (27.4 %), 100 ml of 1.20, 80 ml of 1.27 (55.7 %) and 180 ml of 1.30 (60.7 %) as a cushion. The microsomal suspension containing the particles from 30 g or more of liver, suspended in 75 ml of 250 mM sucrose, was loaded in the center of the rotor followed by an overlay of 120 ml of 250 mM sucrose. This gradient was centrifuged for 4 hrs at 45,000 rpm. The rotor was decelerated to 5,000 rpm and unloaded by pumping sucrose with a density of 1.30 into the periphery. At periodic intervals, the density of the liquid displaced from the rotor was measured. When the density reached 1.20, the rough microsomes were collected as a single fraction in a total volume of 90 ml with an average density of 1.25.

Floatation of Stripped Rough Microsomes

E-40 rotor. To 14 ml of rough microsomes, containing the particles isolated from 9 g liver, were added under continuous stirring 5 ml of a sucrose solution of density 1.25 containing 200 µmoles Na pyrophosphate (PPi) and 3 mM imidazole-HCl, pH 8.2. To the periphery of the E-40 rotor, spinning between 5,000 and 7,000 rpm, we injected in succession : a 32 ml-sucrose gradient, linear between densities 1.05 and 1.18, 10 ml of the PPi-treated rough microsomes (density 1.25) and a 6 ml-cushion (density 1.34). After centrifugation for 140 min at 40,000 rpm, 12 subfractions were collected.

Beckman Ti-14 rotor. To 90 ml of rough microsomes (density 1.23), containing the particles isolated from 16 g liver, were added under continuous stirring 10 ml of a sucrose solution of density 1.25 containing 100 mM Na PPi, pH 8.2. To the center of the T-14 rotor - filled up with sucrose of density 1.30 and spinning at 4,000 rpm -, were pumped in succession a 90 ml-aliquot of the PPi-treated rough microsomes and the following sucrose solutions : 100 ml of density 1.18 and 350 ml of density 1.034. After an overnight centrifugation at 45,000 rpm, the rotor was decelerated at 5,000 rpm. A first fraction of 300 ml was collected, followed by 26 fractions of 15 ml.

Determination of Densities and Volumes of Gradient Subfractions

Gradient subfractions were collected in tared tubes and weighed. The densities were measured in a gradient produced from o-dichlorobenzene and petroleum ether (b.p. 80-

100°C) as previously described by Beaufay and Amar-Costesec (1976). The gradient was established in a glass tube and kept at 4°C. A 10 µl-aliquot of each subfraction was introduced in the gradient calibrated with KBr standard solutions of densities ranging from 1.010 to 1.3. The volumes were calculated from the weights and densities of the subfractions.

Biochemical Determinations

The enzymes listed in Table 1 were used to monitor the purification of the rough ER-derived vesicles. They were assayed according to the referenced methods.

Protein was assayed according to Lowry et al. (1951) with bovine serum albumin as a standard. Phospholipid was measured as described by Beaufay et al. (1974a). RNA was estimated by the Schmidt-Tannhauser procedure as modified by Fleck and Munro (1962).

The contribution of each subcellular organelle (mitochondria, peroxisomes, ER, plasma membrane, Golgi complex, outer mitochondrial membrane) to the protein content of the isolated fractions was calculated as described by Leighton et al. (1968) using the percentages given in Table 1.

RESULTS AND DISCUSSION

Although three different rotors were used to obtain rough microsomes from either microsomal fractions or post-mitochondrial supernatants, no significant differences in composition were observed between the resulting preparations. Rough microsomes were isolated as the membrane fraction that passes the 1.20 density barrier after centrifugation in a linear sucrose gradient (Fig. 1).

During this step, the membrane fragments that are derived from the outer mitochondrial envelope, the plasma membrane, and the Golgi complex were mostly eliminated along with the upper part of the gradient. As shown in Table 2, the recovery of the relevant marker enzymes declines markedly in rough microsomes by comparison with microsomes. Galactosyltransferase activity is almost totally absent, whereas alkaline phosphodiesterase I and monoamine oxidase activities are present but at considerably lower levels than in microsomes. Part of the monoamine oxidase activity probably reflects the presence of intact mitochondria in the region of

TABLE 1. Percentage of liver protein accounted for by various subcellular organelles and enzyme assays used to monitor the purification of membranes enriched in rough ER domain

Subcellular organelle	Percent of liver protein*	Marker enzyme	Reference of the assay method
	%		
ER	20	NADPH cytochrome *c* reductase NADH cytochrome *c* reductase Glucose 6-phosphatase Esterase	Beaufay *et al.* (1974a)
Mitochondria	20	Cytochrome oxidase	Beaufay *et al.* (1974a) as modified by Amar-Costesec (1981)
Outer mitochondrial envelope	3	Monoamine oxidase	Wurtman & Axelrod (1963) as modified by Amar-Costesec (1981)
Peroxisomes	2.5	Catalase	Baudhuin *et al.* (1964)
Peroxisomal cores		Urate oxidase	de Duve *et al.* (1955)
Plasma membranes	3	Alkaline phosphodiesterase I	Beaufay *et al.* (1974a)
Golgi complex	1	Galactosyltransferase	Beaufay *et al.* (1974a)

*Data from Leighton *et al.* (1968) and Amar-Costesec *et al.* (1974).

TABLE 2. Comparison of the biochemical composition of total and rough microsomes

Constituent	Percent of liver content in	
	Microsomes	Rough microsomes
Protein	20.6 ± 1.5 (9)	5.85 ± 1.76 (9)
RNA	53.4 ± 8.1 (9)	30.7 ± 2.2 (9)
NADPH cytochrome *c* reductase	77.5	13.8
Glucose 6-phosphatase	64.8 ± 6.9 (9)	20.7 ± 6.6 (9)
Cytochrome oxidase	7.23 ± 2.19 (6)	1.49 ± 1.27 (6)
Catalase	4.10 ± 0.84 (3)	2.70 ± 1.18 (3)
Urate oxidase	20.4	7.85
Monoamine oxidase	23.7 ± 4.8 (5)	4.0 (2)
Galactosyltransferase	72.2 ± 5.6 (9)	0.53 ± 0.33 (9)
Alkaline phosphodiesterase I	61.6 ± 7.7 (5)	2.57 ± 1.66 (5)

Number of experiments given in parentheses.

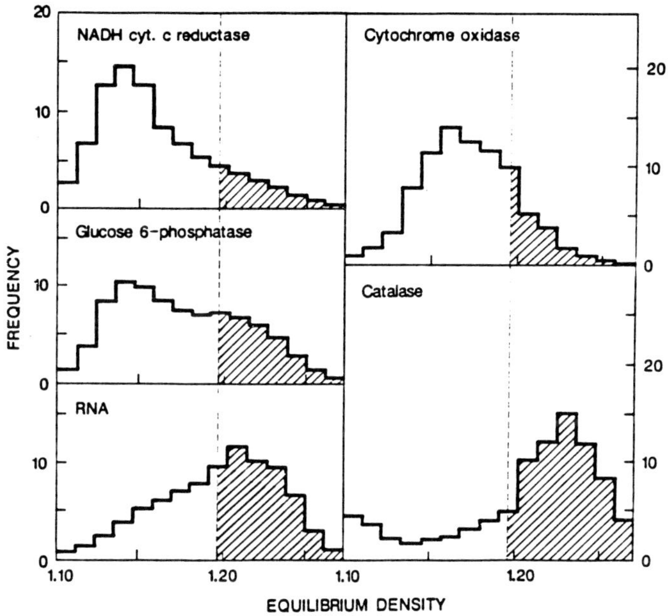

Figure 1. Isopycnic equilibration of total microsomes in the E-40 rotor. The rough microsomes (shaded area) were selected as the particles passing density 1.196 (dashed line). The results of this experiment and of that shown in Fig. 3 are included in the data presented in Table 3.

the gradient. Together, these three membrane components contribute to less than 4 % of the rough microsomal protein content, a value estimated from : (i) the recoveries given in Table 2 for monoamine oxidase (4.0 %), alkaline phosphodiesterase I (2.6 %) and galactosyltransferase (0.5 %) in rough microsomes ; (ii) the percentage of liver protein accounted for by outer mitochondrial envelope (3 %), plasma membranes (3 %) and Golgi complex (1 %) (Table 1) ; and (iii) the protein content per g liver (242.7 $\pm$ 20.4 mg, mean of 9 experiments).

Mitochondria and peroxisomes have modal densities of 1.18 and 1.23, respectively. Thus, these contaminants are still present in the fraction that passes the 1.20 density barrier, as indicated by the percentages of cytochrome oxidase and catalase (2 % and 2.7 %, respectively, see also Fig. 1). The presence of peroxisomal cores is even more important as revealed by the higher content in urate oxidase (7.9 %). These contaminants, easily recognizable at the electron microscope (Fig. 2), were eliminated thanks to the "PPi shift": since the equilibrium density of rough vesicles is primarily determined by the number of ribosomes bound to their external surface, the PPi treatment, which detaches a substantial portion of membrane-bound ribosomes, results in a marked decrease in the equilibrium density of these vesicles (Amar-Costesec et al., 1974b).

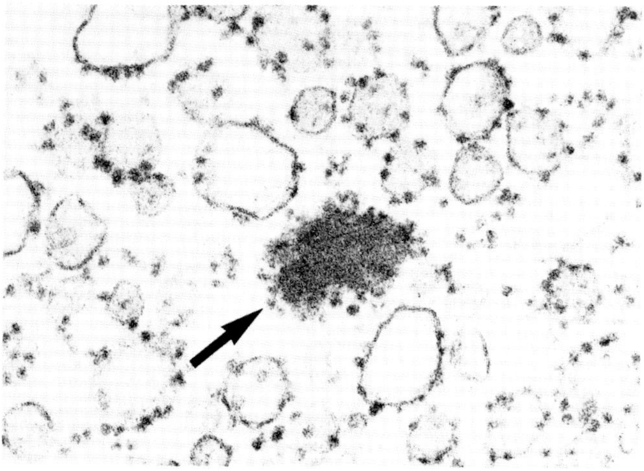

Figure 2. Peroxisomal cores (arrow) contaminate the rough microsomes (x 76,500).

Fig. 3 shows the results of a typical experiment in which rough microsomes were first treated with PPi and then floated in a sucrose density gradient. Mitochondria, which peaks now at a density of 1.20, and peroxisomes, whose equilibrium density is not affected by the treatment, remain at the bottom of the gradient. The subfractions with a density of less than 1.19 contain the bulk of ER vesicles, stripped of bound ribosomes, as shown by the density distribution profiles of NADH cytochrome *c* reductase and esterase. The results given in Table 3 indicate that the percentage of cytochrome oxidase, catalase and urate oxidase in the floated preparation is 5-, 21- and 26-times smaller, respectively, than in rough microsomes. The floated membranes account for 15-18 % of the total ER with a 6-fold enrichment over the homogenate. The yield is 7 mg protein per g liver. These membranes, which have conserved the functional capacities and the polypeptide composition of the original rough ER membrane (results not shown) were used to isolate and characterize the rough ER domain.

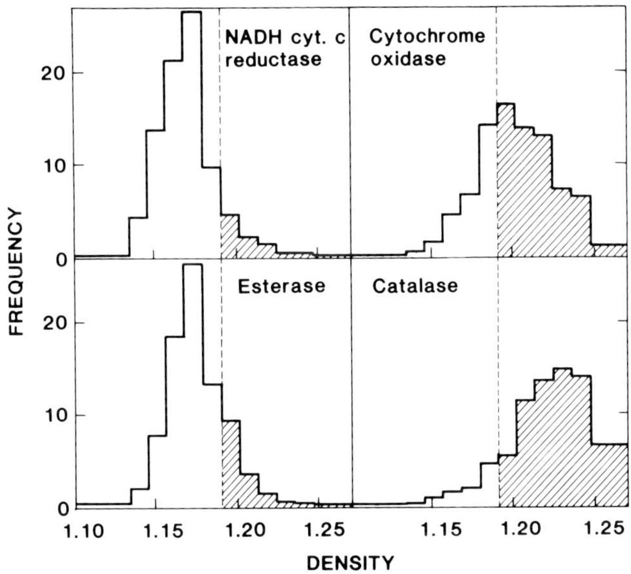

Figure 3. Removal of mitochondrial and peroxisomal contaminants by floatation of PPi-treated rough microsomes in a sucrose gradient. The subfractions with a density greater than 1.19 (shaded area) were discarded.

ACKNOWLEDGEMENTS

We thank Mrs. Ch. Turu for expert technical assistance, Dr. M. Wibo for performing the electron microscopy, Dr. K. Willard for many improvements to the manuscript, Dr. P. Courtoy for his help in preparing Fig.2 and Ms. M. Gabarz-Kaufmann for careful typing of the manuscript. This research was supported by grants from the Fonds de la Recherche Collective and the Services de Programmation de la Politique Scientifique.

TABLE 3. Comparison of the biochemical composition of various subcellular preparations

Constituent	Homogenate	Microsomes	Rough microsomes	Floated subfractions (2)	Nbr of exp.
	mg/g liver	% (1)	%	%	
Protein	243.8 ± 26.1	20.0 ± 1.2	6.14 ± 0.95	2.88 ± 1.39	4
	U/g liver				
NADH cytochrome _c_ reductase	126.9 ± 12.2	60.9 ± 1.7	18.5 ± 4.7	12.5 ± 2.4	4
Glucose 6-phosphatase	28.9	78.2	36.7	14.5	
Esterase	239.5 ± 63.7	72.9 ± 5.1	23.6 ± 4.2	18.3 ± 2.9	4
Cytochrome oxidase	27.8 ± 2.1	8.9 ± 1.3	2.02 ± 1.29	0.46 ± 0.28	4
Catalase	46.2 ± 13.5	3.86 ± 0.91	2.70 ± 1.18	0.13 ± 0.04	3
Urate oxidase	1.59	20.4	7.85	0.49	
Monoamine oxidase	0.57	23.3	9.2	0.86	2
Alkaline phosphodiesterase I	17.0 ± 2.6	60.2 ± 10.0	3.28 ± 1.88	1.74 ± 1.58	3
	mU/g liver				
Galactosyltransferase	15.4	74.3	0.50	0.50	

(1) Percentages refer to the sum of contents in the various fractions and subfractions. They are given as means ± standard deviations.

(2) Corresponding to the shaded area in Fig. 2.

REFERENCES

Amar-Costesec A (1978a). The molecular organization of endoplasmic reticulum in liver : A plausible model to account for the enzymic heterogeneity. Arch Internat Physiol Biochim. 86:193-194.

Amar-Costesec A (1978b). The distribution of enzymes in the plane of endoplasmic reticulum membranes of rat liver. Arch Internat Physiol Biochim 86:843-844.

Amar-Costesec A (1981). Analytical study of microsomes and isolated subcellular membranes from rat liver. VII. Distribution of protein-bound sialic acid. J Cell Biol 89: 62-69.

Amar-Costesec A, Beaufay H (1981).A structural basis of enzymic heterogeneity within liver endoplasmic reticulum. J Theor Biol 89:217-230.

Amar-Costesec A, Beaufay H, Wibo M, Thinès-Sempoux D, Feytmans E, Robbi M, Berthet J (1974a). Analytical study of microsomes and isolated subcellular membranes from rat liver. II. Preparation and composition of the microsomal fraction. J Cell Biol 61:201-212.

Amar-Costesec A, Todd JA, Kreibich G (1984). Segregation of the polypeptide translocation apparatus to regions of the endoplasmic reticulum containing ribophorins and ribosomes. I. Functional tests on rat liver microsomal subfractions. J Cell Biol 99:2247-2253.

Amar-Costesec A, Wibo M, Thinès-Sempoux D, Beaufay H, Berthet J (1974b). Analytical study of microsomes and isolated subcellular membranes from rat liver. IV. Biochemical, physical and morphological modifications of microsomal components induced by digitonin, EDTA and pyrophosphate. J Cell Biol 62:717-745.

Baudhuin P, Beaufay H, Rahman-Li Y, Sellinger OZ, Wattiaux R, Jacques P, de Duve C (1964). Tissue fractionation studies. 17. Intracellular distribution of monoamine oxidase, aspartate aminotransferase, alanine aminotransferase, D-amino acid oxidase and catalase in rat-liver tissue. Biochem J 92:179-184.

Beaufay H (1966). "La centrifugation en gradient de densité. Applications à l'étude des organites subcellulaires". Thèse d'Agrégation de l'Enseignement Supérieur (Ed. : Ceuterick, Louvain, Belgique, S.A.) Université Catholique de Louvain, pp 132.

Beaufay H, Amar-Costesec A (1976). "Cell fractionation techniques". (Ed. : ED Korn) New York-London : Plenum, 6:pp 1-100.
Beaufay H, Amar-Costesec A, de Duve C (1981). "The topology of enzymes in microsomes from liver". (Eds. : CP Lee, G. Schatz, G. Dallner) Reading (Mass) : Addison-Wesley Publishing Company : pp 629-653.
Beaufay H, Amar-Costesec A, Feytmans E, Thinès-Sempoux D, Wibo M, Robbi M, Berthet J (1974a). Analytical study of microsomes and isolated subcellular membranes from rat liver. I. Biochemical methods. J Cell Biol 61:188-200.
Beaufay H, Amar-Costesec A, Thinès-Sempoux D, Wibo M, Robbi M, Berthet J (1974b). Analytical study of microsomes and isolated subcellular membranes from rat liver. III. Subfractionation of the microsomal fraction by isopycnic and differential centrifugation in density gradients. J Cell Biol 61:213-231.
de Duve C, Pressman BC, Gianetto R, Wattiaux R, Appelmans F (1955). Tissue fractionation studies. 6. Intracellular distribution patterns of enzymes in rat-liver tissue. Biochem J 60:604-617.
Fleck A, Munro HN (1962). The precision of ultraviolet absorption measurements in the Schmidt-Thannhauser procedure for nucleic acid estimation. Biochim Biophys Acta 55:571-583.
Godelaine D, Beaufay H, Wibo M, Amar-Costesec A (1979). The dolichol pathway of protein glycosylation in rat liver. Stimulation by GTP of the incorporation of N-acetylglucosamine in endogenous lipids and proteins of rough microsomes treated with pyrophosphate. Eur J Biochem 96:17-26.
Leighton F, Poole B, Beaufay H, Baudhuin P, Coffey JW, Fowler S, de Duve C (1968). The large-scale separation of peroxisomes, mitochondria and lysosomes from the livers of rats injected with Triton WR-1339. Improved isolation procedures, automated analysis, biochemical and morphological properties of fractions. J Cell Biol 37:482-513.
Lowry OH, Rosebrough NJ, Farr AL, Randall RJ (1951). Protein measurement with the Folin phenol reagent. J Biol Chem 193:265-275.
Marcantonio E, Amar-Costesec A, Kreibich G (1984b). Segregation of the polypeptide translocation apparatus to regions of the endoplasmic reticulum containing ribophorins and ribosomes. II. Rat liver microsomal subfractions contain equimolar amounts of ribophorins and membrane-bound ribosomes. J Cell Biol 99:2254-2259.

Mitropoulos KA, Venkatesan S, Balasubramanian S, Peters TJ (1978). The submicrosomal localisation of 3-hydroxy-3-methylglutaryl-coenzyme-A reductase, cholesterol 7α-hydroxylase and cholesterol in rat liver. Eur J Biochem 82: 419-429.

Venkatesan S, Mitropoulos KA, Balasubramanian S, Peters TJ (1980). Biochemical evidence for the heterogeneity of membranes from rat liver endoplasmic reticulum. Studies on the localization of acyl-CoA:cholesterol acyltransferase. Eur J Cell Biol 21:167-174.

Wurtman RJ, Axelrod J (1963). A sensitive and specific assay for the estimation of monoamine oxidase. Biochem Pharmacol 12:1439-1441.

Cell-Free Analysis of Membrane Traffic, pages 227–235

SEPARATION OF CELL ORGANELLES AND MEMBRANE VESICLES BY PHASE PARTITION

Per-Åke Albertsson

Department of Biochemistry, University of Lund, Box 124, S-221 00 Lund, Sweden

Cell organelles and membrane vesicles can be separated by partition between two immiscible, aqueous polymer phases (Albertsson et al.,1982; Albertsson 1986). By this method, cell particles are separated according to their surface properties, and partition is therefore a useful complement to centrifugation methods. The separation can be based on several different surface properties of the cell particles, such as electrical charge properties, hydrophobicity or biospecific recognition between receptors on the surface and ligands on the polymers. The method can be made highly selective, it preserves biological activity and it can be applied on a small scale as well on a large scale.

The basic aspects of the method and various applications are treated in two recently published books (Albertsson, 1986; Walter et al., 1985) and in recent review articles (Albertsson et al.,1982, 1987; Walter and Johansson, 1986).

Cell particles, such as cell organelles, membrane vesicles or whole cells, are distributed between one of the liquid phases and the interface. If the distribution is very different for two particles they can be separated by one or a few partition steps. If they are closer to each other in their behaviour, one has to use a multistage procedure such as countercurrent distribution (Albertsson 1986). The phase systems mainly used so far are composed of dextran, polyethylene glycol (PEG), and water. Other polymer-polymer two-phase systems and their general properties have been described by Albertsson (1986).

The most prominent feature of these phase systems is that both phases are rich in water. In a dextran - PEG system the upper PEG phase contains 90 to 95 % water and the lower dextran phase contains 80 to 90 % water. Sucrose, sorbitol, salts, and low molecular weight compounds are distributed almost equally between the phases and can be included in the phase system (without gross change in the phase diagram) to give an optimal environment for the stability of cells and cell organelles. The contribution to the tonicity of the polymers alone is about 30 m Osm (Bamberger et al 1985). The surface tension of the interface is very small, 0.0001 to 0.1 mN/m (Albertsson, 1986).

The polymer two-phase systems have a relatively long settling time (from 3 min to hours), due to the high viscosity of the phases and the low density differences. The settling time may be reduced to less than a minute by low speed centrifugation, provided the partitioned particles do not sediment.

Generally the aqueous polymer two-phase systems act mildly toward biological material. This is due to the high water content, low interfacial tension between the phases and the presence of the polymers which protect enzymes and cell membranes against denaturation.

FACTORS DETERMINING PARTITION

Partition of cell particles depends on several factors (Albertsson, 1986): the surface properties of the cell particles including surface charges, hydrophobicity of the surface and specific receptors on the membrane surface; the ionic composition of the phase system, the type and molecular weight of the polymers and their concentration.

The different factors which determine partition can be exploited separately or in combination to achieve an effective separation. We can also enlarge some of the factors so that they will dominate the partition behavior. With respect to the partitioned substance the following types of partition can be distinguished:

1. Size-dependent partition: molecular size or the surface area of the particles is the dominating factor.
2. Electrochemical partition: electrical potential between the phases is used to separate molecules or particles according to their charge.
3. Hydrophobic affinity partition: hydrophobic properties of a phase system are used for separation according to the hydrophobicity of molecules or particles.
4. Biospecific affinity partition: the affinity between sites on the molecules or particles and ligands attached to the phase polymers is used for separation.
5. Conformation dependent partition: conformation of the molecules and particles is the determining factor.
6. Chiral partition: enantiomeric forms are separated.

Formally , we can split the logarithm of the partition coefficient into several terms:

$$\ln K = \ln K^{o} + \ln K_{el} + \ln K_{hfob} + \ln K_{biosp}$$

$$+\ln K_{size} + \ln K_{conf}$$

where el, hfob, biosp. size, and conf stand for electrochemical, hydrophobic, biospecific, size, and conformational contributions to the partition coefficient and $\ln K^{o}$ includes other factors.

The different factors are more-or-less independent of each other though probably none is completely independent of the others. For example, when we introduce a hydrophobic group on one of the phase-forming polymers, we may slightly influence both the distribution of ions and the electrical potential. When the molecular weight of a macromolecule is increased, its net charge may also increase. When the conformation of a macromolecule is changed, new groups of atoms are exposed to the surrounding phase.

Ionic Composition of the Phase System

Different ions have different chemical affinities for the two phases. An electrical potential is therefore created between the two phases when a salt is included in the phase system (1). For cells and cell organelles which usually carry a net negative charge, the positively charged ions cause the cell particles to favour the upper phase in the following order: $Cs^+ \sim K^+ < Na^+ < Li^+$. The negative ions have the same effect in the follwing order: $Cl^- \sim H_2PO_4^-$ < $HPO_4$2- SO_4^{2-} citrate. Therefore, to get a maximal distribution into the upper phase, one should therefore use Li_2-SO_4 and to get a maximal distribution into the lower phase one should use NaCl or KCl.

Polymer Concentration

An increase in polymer concentration of the dextran PEG system will favour the collection of particles at the interface.

APPLICATIONS

Phase partition can be used for several purposes. It can be used for

a) purifying a desired cell organelle or membrane fragment.

b) a study of the heterogeneity of cells or cell organelles

c) gaining information on the surface properties of membranes.

Fragmentation of membranes followed by separation using centrifugation, which separates according to size, and phase partition, which separates according to surface properties, can also be used to get information on the domain organization of membranes. By plotting the concentration of one component versus the concentration of another component for different separated fragments, one obtains characteristic plots which can be compared with similar plots from model membranes (Albertsson, 1987). Such a comparison can be used to build a model of a membrane with respect to its domain organization. This approach has been used successfully on thylakoids (Melis et al.; Albertsson, 1987 a,b).

Plasma Membranes

Plasma membranes from several species, both animal and plant, have a strong tendency to favour the upper phase of a dextran-polyethylene glycol system. By changing the polymer concentration at a given salt composition or by changing the salt composition at a given polymer composition (Albertsson et al., 1982), one can select a phase system where plasma membranes are found in the upper phase while most other intracellular membrane particles such as mitochondria, endoplasmic reticulum derived vesicles, chloroplast and chloroplast fragments, are found at the interface or in the bottom phase. This has been utilized for the preparation of plasma membranes from several plant species (Larsson, 1985). Of great importance is the fact

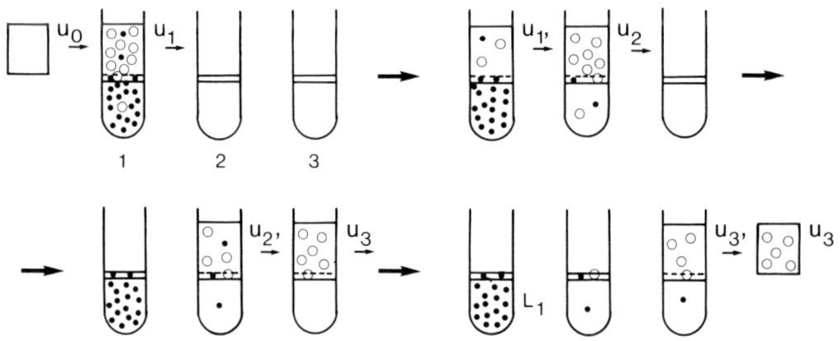

Figure 1. A batch-procedure involving three partition steps for isolation of plasma membranes. A micosomal preparation is used as the starting material. In the first step the plasma membranes (O) are partitioned into the upper phase while other membrane vesicles are mainly found at the interface or in the bottom phase. By the second and third partition steps, the plasma membrane vesicles are further purified. Composition of the phase system: 6,2% (w/w) Dextran 500, 6,2% (w/w) polyethylene glycol 4.000, 0,33 M sucrose, 3 mM KCl, 5 mM potassium phosphate. From Kjellbom and Larsson (1984).

that the procedure works very well for the isolation of plasma membranes from green cells (Kjellbom and Larsson, 1984), where the removal of thylakoid membrane vesicles from the plasma membranes is a great problem. Fig. 1 shows the separation scheme for the isolation of plasma membranes from spinach or barley leaves.

A similar procedure can be used with animal cells. In a systematic study Gierow et al. (1986) studied the partition of rat liver plasma membranes obtained by centrifugation. Different membrane markers were used to assay the partition of different organelles and membrane vesicles. The affinity for the upper phase was highest for the plasma membrane and lowest for the endoplasmic reticulum. The affinity for the upper phase increased in the following order: endoplasmic reticulum, mitochondria, lysosomes, Golgi membranes and plasma membranes. Moreover, by countercurrent distribution these workers could demonstrate a considerable heterogeneity among the plasma membrane vesicles. A several-fold increase in the ratio of leucine aminopeptidase (a bile-canalicular marker) and 5' -nucleotidase to asialo-orosomucoid binding (a blood-sinusoidal marker) was obtained in one fraction, whereas another fraction gave a 2-3 fold increase in the ratio of blood-sinusoidal to bile-canalicular markers.

Mitochondria

Mitochondria from rat liver can be separated into two subpopulations by countercurrent distribution (Albertsson 1986). Mitochondria from developing young rats have been compared with those from adult rats. It was found that the ratio between the two subpopulations changed with age. Mitochondria from 1.5 day-old rats lacked one of the subpopulations which dominated in the full grown rat.

Rat brain mitochondria have been separated from contaminating synaptosomes (Lopez-Perez et al., 1981). In this case the high affinity of the synaptosomes for the upper phase, due to the properties of their plasma membrane, was utilized for an efficient separation from the mitochondria.

Partition by a few batch-steps has been used to purify leaf mitochondria from other organelles including chloroplast fragments (Bergman et al. 1980).

Neural Membranes

Countercurrent distribution has been used to fractionate neural membrane vesicles from calf brain. A partial separation of opiate receptor activity from the acetylcholinesterase activity was obtained (Johansson et al., 1984).

Chloroplasts

A crude chloroplast preparation obtained by differential centrifugation can be resolved into three different populations by countercurrent distribution (Albertsson and Larsson 1976; Larsson et al.; 1971). Inside-out thylakoid vesicles can be separated from right-side-out vesicles and unbroken thylakoids by partition, using only two or three partition steps.

Sonication of inside-out thylakoid vesicles yields smaller vesicles which can be separated by countercurrent distribution. All these studies on the partition of fragmented thylakoids have given us a new insight of the structure of the thylakoid membrane with regard the lateral and transverse organization (Andersson et al.,1985).

CONCLUDING REMARKS

Partitioning in aqueous polymer two-phase systems has now been applied to such a large member of different cell organelles and membrane vesicles that it can be considered a general method for preparative separation and analysis of cell particles.

It is an advantage if phase partition, which is a surface-dependent method, is used as a complement to centrifugation methods which separate according to size and density. The only efficient procedures for isolation of clean intact chloroplasts or mitochondria from green leaves involve a combination, in sequence, of phase partition with differential and/or density gradient centrifugation.

Partition is particularly useful for large-scale separation. It should therefore be useful for the large-scale preparation of plasma membranes as starting material for the isolation of receptors.

TABLE 1. Examples of applications of phase partition on cell organelles and membrane vesicles.

Material	References
chloroplast	Albertsson and Larsson, 1976 Larsson et al., 1971
thylakoid vesicles, inside-out thylakoids	Andersson et al., 1985 Albertsson, 1985 Melis et al., 1986
multiorganelle complexes	Albertsson and Larsson, 1976
mitochondria	
from green leaves	Bergman et al., 1980
from rat liver	Albertsson, 1986
from rat brain	López-Pérez et al., 1981
submitochondrial partides	Kay et al., 1985
synaptosomes	López-Pérez et al., 1981
nerve cell membrane vesicles	Johansson et al., 1984
plasma membranes, plant	Kjellbom and Larsson, 1984 Larsson, 1985
plasma membranes, animal	Gierow et al., 1986

REFERENCES

Albertsson PÅ (1985). Subfractionation of inside-out thykaloid vesicles and the localization of cytochrome f. Physiol Vég 23:731-739.

Albertsson PÅ (1986). Partition of Cell Particles and Macromolecules. John Wiley & Sons Inc., New York, N.Y.

Albertsson PÅ (1987). Analysis of the domain structure of membrane by fragmentation and separation in aqueous polymer two-phase systems. Quart Rev Biophysics, in press.

Albertsson PÅ (1987). The domain organization of the chloroplast photosynthetic membrane. Chemica Scripta, in press.

Albertsson PÅ, Andersson B, Larsson C, Akerlund HE (1982). Phase partition - A method for purification and analysis of cell organelles and membrane vesicles. Meth Biochem Anal 28:115-150.

Albertsson PÅ, Larsson C (1976). Properties of chloroplasts isolated by phase partition. Mol Cell Biochem 11:183-189.

Andersson B, Sundby C, Akerlund HE, Albertsson PÅ (1985). Inside out thylakoid vesicles. An important tool for the characterization of the photosynthetic membrane. Physiol Plant 65:322-330.

Bergman A, Gardeström P, Ericson J (1980). Method to obtain chlorophyll-free preparation of intact mitochondria from spinach leaves. Plant Physiol 66:442-445.

Gierow P, Sommarin M, Larsson C, Jergil B (1986). Fractionation of rat liver plasma membrane regions by two-phase partitioning. Biochem J 235:685-691.

Johansson G, Akerlund HE, Olde B (1984). Liquid-liquid extraction of membranes from calf brain using conventional and centrifugal counter-current distribution techniques. J Chromatogr 311:277-289.

Kay CJ, Ericson J, Gardeström P, Palmer J, Möller IM (1985). Generation and purification of submitochondrial particles of different polarities from plant mitochondria. FEBS Lett 193:169-174.

Kjellbom P, Larsson C (1984). Preparation and polypeptide composition of chlorophyll-free plasma membranes from leaves of light-grown spinach and barley. Physiol Plant 62:501-509.

Larsson C (1985). Plasma Membranes in Modern Methods of Plant Analysis, New Series. In Linskens HF, Jackson JF (eds): Springer Verlag, Berlin 1:85-104.

Larsson C, Collin C, Albertsson PÅ (1971). Characterization of three classes of chloroplasts obtained by counter-current distribution. Biochim Biophys Acta 245:425 438.

López-Pérez MJ, Paris G, Larsson C (1981). Highly purified mitochondria from rat brain prepared by phase partition. Biochim Biophys Acta 635:359-368.

Melis A, Svensson P, Albertsson PÅ (1986). The domain organization of the chloroplasts thylakoid membrane. Localization of photosystem I and of the cytochrome b_6-f complex. Biochim Biophys Acta 850:402-412.

Walter H, Brooks DE, Fisher D (1985). Partitioning in Aqueous Two-phase Systems. Academic Press, Inc. Orlando, Fl, USA.

Walter H, Johansson G (1986). Partitioning in aqueous two-phase systems: An overview. Analyt Biochem 155:215-242.

Cell-Free Analysis of Membrane Traffic, pages 237–241

ISOLATION OF HIGHLY PURIFIED PLASMA MEMBRANE FROM HEPATOMAS AND CULTURED CELLS BY TWO-PHASE PARTITION

Placido Navas, D. David Nowack and D. James Morré

Departamento de Biologia Celular (P.N.), Universidad de Córdoba, 14071 Córdoba, Spain and Department of Medicinal Chemistry (D.D.N., D.J.M), Purdue University, West Lafayette, Indiana, USA

Studies of cell surfaces comparing normal and altered cells have benefited from investigations of isolated plasma membranes (Wallach, 1975). Non-uniform homogenization, contamination by organelles other than plasma membrane and changes in culture conditions as well as the transformation process itself tend to complicate the isolation of plasma membranes from cultured cells and hepatomas compared to tissues of origin.

The objective of this research was to explore aqueous two-phase polymer partition as a simple, rapid procedure to obtain plasma membranes from cells and hepatomas in high yield and purity. The procedure of Albertsson et al. (1982) was followed with minor modifications.

Hepatomas were homogenized in a glass/glass tissue homogenizer for 6 min at 600 rpm in a medium (1:2 w/v) containing 37 mM Tris-maleate, pH 6.4, 0.5 M sucrose, 1% dextran, 5 mM $MgCl_2$ and 5 mM β-mercaptoethanol and centrifuged for 15 min at 5000 x g. The bottom 2/3 of this pellet was resuspended in 5 ml of 1 mM bicarbonate in a conical glass/glass homogenizer for 2 min at 600 rpm with an additional 5 ml of bicarbonate being added after suspension. This mixture was centrifuged for 15 min at 5,000 x g. The top 1/2 of this pellet was used for the two-phase partition.

For cultured cells, six 150 cm^2 flasks were harvested by rinsing with 10 ml of phosphate-buffered saline (PBS),

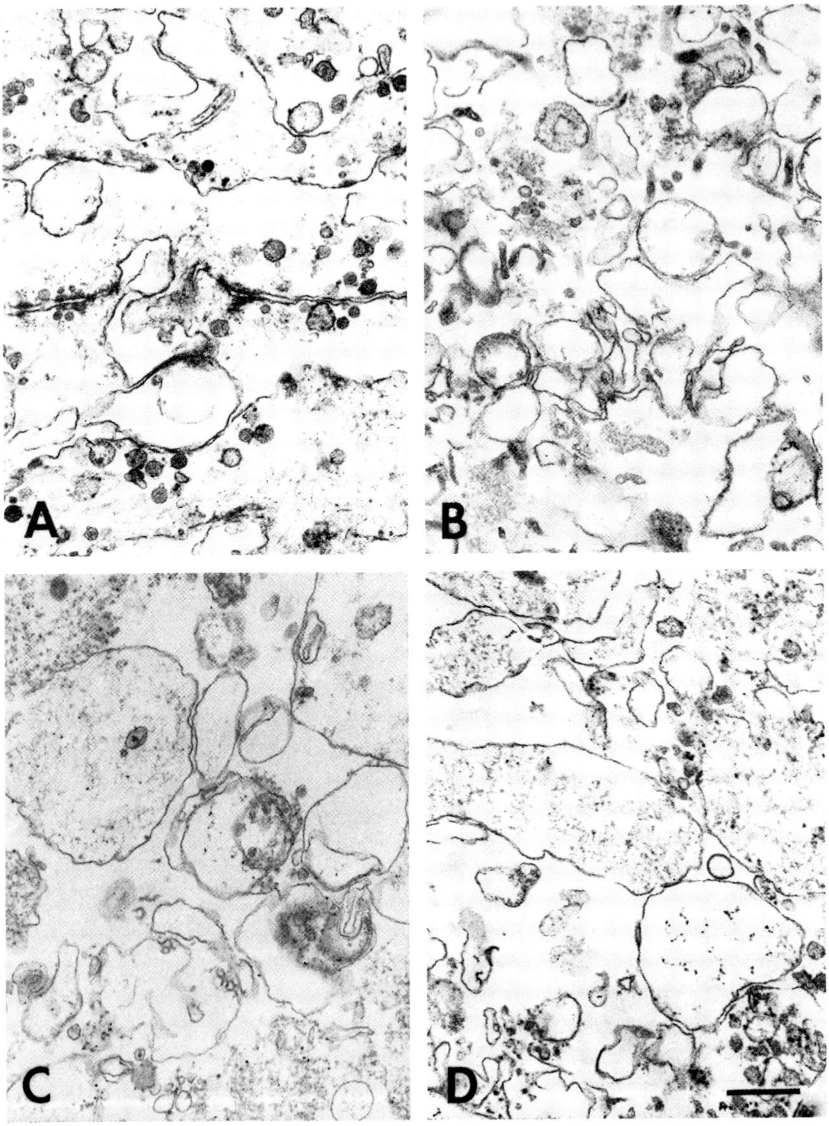

Figure 1. Electron micrographs of plasma membranes prepared by aqueous two phase partitioning from (A) rat liver, (B) WI-38 tissue culture cells, (C) hepatoma designated RLT-N, and (D) RLT-28 tissue culture cells. Bar = 0.5 μm.

pH 7.2, followed by scraping in 10 ml of PBS. The cells were collected by centrifugation for 10 min at 1500 x g resuspended and swollen in 15 ml of 10 mM HEPES, containing 15 mM KCl 1.5 mM Mg acetate, pH 7.2 (medium A) for 10 min. The cells were pelleted again for 10 min at 3,000 rpm and resuspended in 2 ml of medium A. This suspension was homogenized in a Dounce teflon/glass homogenizer for 2 min at 600 rpm with 20 up/down strokes. Next, 2 ml of 9-fold concentrated medium A were added and this mixture was centrifuged for 10 min at 1,000 rpm. The resulting supernatant was centrifuged for 30 min at 16,000 x g and the pellet resuspended in 1 ml of 5 mM potassium phosphate, pH 7.2, as the starting material for the two-phase partition.

The two-phase partition system contained 6.6% (w/w) Dextran T500, 6.6% (w/w) Polyethylene Glycol 3350, 0.25 M sucrose, and 5 mM potassium phosphate, pH 7.2. Two tubes, each containing 7 g of the above mixture, were prepared. To one tube, one g of suspended membranes was added. To the second tube, one g of buffer only was added. The tubes were shaken vigorously 40 times in the cold (4°C) and the phases were separated by centrifugation for 5 min at 750 rpm in a swinging bucket rotor. The two upper phases were exchanged, the tubes shaken, and phases separated as before. The upper phases of both tubes were pooled, not taking the membranes at the interphase, and pelleted following dilution with 5 volumes of 1 mM bicarbonate by centrifugation for 30 min at 10,000 rpm.

The K^+-stimulated, ouabain-inhibited, p-nitrophenyl phosphatase (EC 3.1.3.1) assay of Kashiwampta et al. (1979) was used to assess the yield of plasma membrane. Other assays included succinate-INT reductase (EC 1.3.9.1) (Pennington, 1961) for mitochondria; NADH-cytochrome C reductase (EC 1.6.2.3) (Mahler, 1955) for endoplasmic reticulum; glucose-6-phosphatase (EC 3.1.3.9) (Nordlie and Arion, 1966) for endoplasmic reticulum and galactosyltransferase (EC 2.4.1.13) (Palmiter, 1969) for Golgi apparatus.

Some characteristics of plasma membrane fractions prepared by aqueous two-phase partitioning from liver hepatoma and cultured cells (Table 1) illustrate the generality of the two-phase method for preparation of surface membranes as well as the relative purity of preparations obtained. Even with liver, the results are of

TABLE 1. Marker Enzyme Characteristics of Isolated Plasma Membranes. RLT-N designates a liver tumor originally induced by feeding diethylnitrosamine. The units of specific activity are micromoles/h per mg protein for K^+-stimulated, ouabain inhibited, p-nitrophenyl phosphatase, succinate-INT reductase, glucose-6-phosphatase, and nanomoles/h per mg protein for NADPH cytochrome c reductase and galactosyltransferase.

Marker activity	Fraction	Specific activity: Liver	RLT-N hepatoma	RLT-28 cells
K^+-stimulated, ouabain-inhibited p-nitrophenyl-phosphatase (plasma membrane)	Homogenate	0.06 ± 0.03	0.26 ± 0.05	0.11 ± 0.01
	Plasma Memb	1.8 ± 0.5	1.49 ± 0.16	0.84 ± 0.18
Succinate-INT reductase (mitochondria)	Homogenate	1.9 ± 0.4	0.91 ± 0.5	2.2 ± 0.7
	Plasma Memb	0.4 ± 0.07	0.31 ± 0.17	0.02 ± 0.07
Glucose-6-phosphatase (endoplasmic reticulum)	Homogenate	6.7 ± 0.3	3.0 ± 0.5	6.6
	Plasma Memb	0.05 ± 0.02	0.04	0.8 ± 0.1
NADPH-cytochrome c reductase (endoplasmic reticulum)	Homogenate	21.0 ± 4.0	1.4 ± 0.4	13.0 ± 0.7
	Plasma Memb	6.7 ± 0.9	0.75 ± 0.5	1.0 ± 0.2
Galactosyltransferase (Golgi apparatus)	Homogenate	2.7 ± 0.5	5.02	3.1
	Plasma Memb	1.7	1.96	44.0

exceptional quality and improved over gradient methods at least to the extent that the procedure appears less domain specific (Figure 1A). Contaminating membranes enter the lower phase. The latter provides a source of material for isolation of other membranes free of plasma membrane. After washing to remove the dextran, the membranes may be subjected to free-flow electrophoresis to yield fractions enriched in rough endoplasmic reticulum or in mitochondria and mitochondrial fragments.

The plasma membrane isolates from aqueous two-phase partition frequently are mixtures of right-side out and inside-out vesicles. These vesicles of differing orientation may be resolved by free-flow electrophoresis (Heidrich and Leutner, 1974). The right-side out vesicles migrate further toward the cathode than do the wrong-side out vesicles of lower negative charge density.

Thus, the two-phase partition procedure provides a rapid and facile alternative to gradient procedures for isolation of plasma membranes from cultured cells (Baumbach et al., 1983) and hepatomas (Walter and Morré, 1983). It is easy, reproducible and rapid. With only minor modifications, it may be applied to a wide range of cell and tissue types as well as under different growth or culture conditions. By combination with free-flow electrophoresis, fractions of plasma membrane vesicles of known absolute orientation are obtained.

REFERENCES

Albertsson PÅ, Andersson B, Larsson C, Akerlund HE (1982). Phase partition-a method for purification and analysis of cell organelles and membrane vesicles. Meth Biochem Anal 28:115-150.

Baumbach GA, Horst MN, Olymphio MA, Aldrich HC, Roberts RM (1983). Isolation and characterization of plasma membrane from cultured cancer cells. In Reid E, Cook GMW, Morré DJ (eds): "Cancer Cell Organelles," New York:

Heidrich HG, Leutner G (1974). Two types of vesicles from the erythrocyte ghost membrane differing in surface charge. Separation and characterization by preparative free-flow electrophoresis. Eur J Biochem 41:37-43.John Wiley and Sons, pp 351-370.

Kashiwampta S, Goto S, Samba R, Suzuki F (1979). Inhibition by bilirubin of (Na^{+}-K^{+})-activated adenosine triphosphatase and K^{+}-activated p-nitrophenyl phosphatase activities of nal-treated microsomes from young rat cerebrum. J Biol Chem 254:4577-4584.

Mahler HR (1955). Butyryl coenzyme A dehydrogenase. Meth Enzymol 2:553-559.

Nordlie RL, Arion WJ (1966). Glucose-6-phosphatase. Meth Enzymol 9:619-625.

Palmiter RD (1969). Properties of lactose synthetase from mammary gland: Role of a proposed third component. Biochem Biophys Acta 178:35-46.

Pennington RJ (1961). Biochemistry of dystrophic muscle. Mitochondrial succinate-tetrazolium reductase and adenosine triphosphatase. Biochem J 80:649-654.

Wallach D (1975). Membrane molecular biology of neoplastic cells. Elsevier Scientific Company, New York.

Walter VP, Morré DJ (1983). Isolation of plasma membranes from transplanted rat hepatomas. In Reid E, Cook GMW, Morré DJ (eds): "Cancer Cell Organelles," New York: John Wiley and Sons, pp 396-401.

Cell-Free Analysis of Membrane Traffic, pages 243–245

USE OF BIOTINYLATED LECTINS AND IMMOBILIZED STREPTAVIDIN FOR THE SPECIFIC ISOLATION OF SURFACE GLYCOCONJUGATES

J. William Buckie and Geoffrey M.W. Cook

Department of Pharmacology (J.W.B.), and Department of Anatomy (G.M.W.C.), University of Cambridge, Cambridge, England.

INTRODUCTION

We have recently devised an indirect affinity chromatography procedure utilising biotinylated concanavalin A for the analysis of glycoproteins at the cell surface (Buckie and Cook, 1986a). Intact cells, metabolically labelled with [^{35}S] methionine, are treated on ice with biotin-ε-aminocaproyl-concanavalin A and after solubilization in non-ionic detergent, in the presence of protease inhibitors, the biotinylated-lectin/glycoprotein complexes which form only at the cell periphery, are retrieved on immobilized streptavidin. Radiolabelled material may then be eluted from this latter support with 2% SDS at 100°C and analysed by two-dimensional gel electrophoresis.

In our hands, this procedure with intact lymphoid cells has resulted in the identification of more components than may be retrieved from isolated plasma membrane fractions (Warley and Cook, 1973) with immobilized concanavalin A. It is likely that on isolating plasma membrane fractions, material is lost and that a method which employs intact viable cells may more closely approximate the receptors' availability for ligand binding _in vivo_ (glycoprotein-concanavalin A interactions in this instance). Certainly in the analysis of membrane traffic, the use of biotinylated ligands with intact cells may provide a convenient technique for the characterization of materials representative of the intact cell periphery.

THE ADVANTAGES OF STREPTAVIDIN

The avidin-biotin system has been modified as an analytical tool in molecular biology for the characterization and isolation of important biomolecules. However, for membrane analysis, streptavidin with a high affinity for biotin (Ka $\sim 10^{14}$ M^{-1}), has certain advantages over avidin. Unlike avidin, streptavidin is not a glycoprotein (Chaiet and Wolf, 1964) and therefore avoids the likely interaction with the carbohydrate-binding site on the biotinylated lectin. In addition, with a lower pI than avidin (6.5 as opposed to 10.5 for avidin), the possibility that net negatively charged membrane glycoproteins and proteins will bind non specifically by ionic interactions is greatly decreased. In our hands, Sepharose 4B containing approximately 5.0 mg of streptavidin/ml of swollen gel has proved ideal for the retrieval of biotinylated concanavalin A - glycoprotein complexes.

Biotinylation of lectins is readily performed under mild conditions (Bayer et al., 1982) and the appropriate reagents are available commercially (we have routinely used biotin-ϵ-aminocaproic acid, N-hydroxy-succinimide ester from Calbiochem-Behring, Cambridge, U.K.). The degree of biotinylation is however crucial for the successful isolation of glycoconjugates; increasing the level of biotinylation from 5 to approximately 30 biotin residues per tetramer of concanavalin A raised the binding to immobilized streptavidin from 15% to 93%. This latter result compares favourably with the findings of Bayer et al. (1976) who used immobilized avidin with biotinylated concanavalin A.

CONCLUSIONS

Clearly the use of biotinylated concanavalin A can be extended to a variety of cells, and one is not limited to the use of $[^{35}S]$ methionine labelled material. Cells metabolically labelled with radioactive sugars have been used with the above method to study changes in glycan structures at the cell surface (Buckie and Cook, 1986b). In addition, there is a wide range of lectins available in purified form and with biotinylation, would no doubt further extend the range of surface glycoconjugates that can be isolated with this methodology.

REFERENCES

Bayer EA, Skutelsky E, Wilchek M (1976). Affinity cytochemistry: The localization of lectin and antibody receptors on erythrocytes via the avidin-biotin complex. FEBS Lett 68: 240-244.

Bayer EA, Skutelsky E, Wilchek M (1982). The ultrastructural visualisation of cell surface glycoconjugates. Methods Enzmol 83: 195-215.

Buckie JW, Cook GMW (1986a). Specific isolation of surface glycoproteins from intact cells by biotinylated concanavalin A and immobilized streptavidin. Anal Biochem 156: 463-472.

Buckie JW, Cook GMW (1986b). An indirect affinity chromatography procedure for the specific isolation of surface glycoconjugates. Biochem Soc Trans 14: 706.

Chaiet L, Wolf FJ (1964). The properties of streptavidin, a biotin-binding protein produced by streptomycetes. Arch Biochem Biophys 106: 1-5.

Warley A, Cook GMW (1973). The isolation and characterization of plasma membranes from normal and leukaemic cells of mice. Biochim Biophys Acta 323: 55-68.

Cell-Free Analysis of Membrane Traffic, pages 247–250

ANALYTICAL DIFFERENTIAL CENTRIFUGATION OF CELL ORGANELLES

Torgeir Flatmark

Department of Biochemistry, University of Bergen, N-5000 Bergen, Norway

INTRODUCTION

Analytical differential centrifugation in homogenous media is a powerful approach in the assignment of the subcellular localization of enzymes/components and in tissue fractionation studies. Thus, in order to design specific fractionation schemes for the optimal recovery of a given organelle, its average sedimentation coefficient ($\bar{s}$-value) and size distribution, should be known. A simple and convenient method has been developed for the determination of $\bar{s}$-values of subcellular particles (Slinde and Flatmark, 1973), based on a logarithmic expression of the rate equation of parallel sedimentation in horizontal cylindrical tubes.

THEORETICAL CONSIDERATIONS

Considering an ideal and parallel sedimentation of subcellular particles, suspended in a homogenous medium, the fraction (Y) of a chosen $\bar{s}$-particle, which has sedimented to the bottom at a certain time integral of rpm^2, will follow a sediterm, given by Eq. (1) (Slinde and Flatmark, 1973),

$$Y = [(R_{max} - R_y)/(R_{max} - R_{min})]\,100 \qquad (1)$$

where R_{max} and R_{min} are the distances from the axis of rotation to the bottom and the surface of the fluid column, respectively (in cm); the value of R_{min} (= R_y), giving the quantity of the actual $\bar{s}$-particle that has sedimented to the bottom of the tube (R_{max}), is calculated from Eq. (2)(De Duve and Berthet, 1953),

$$R_{min} = R_{max}\ 10^{(-\bar{S} \int_o^t rpm^2 dt/3.5x10^{13})} \qquad (2)$$

The average sedimentation coefficient ($\bar{s}$, in Svedberg units) is given by Eq. (3) (Slinde and Flatmark, 1973),

$$\log[1 - (1-R_{min}/R_{max})Y/100] = -(\bar{S} \int_o^t rpm^2 dt/3.5x10^{13}) \qquad (3)$$

If the term on the left hand side is plotted against the time integral of rpm^2, a straight line is obtained (Fig. 1), and the $\bar{s}$-value estimated from the slope.

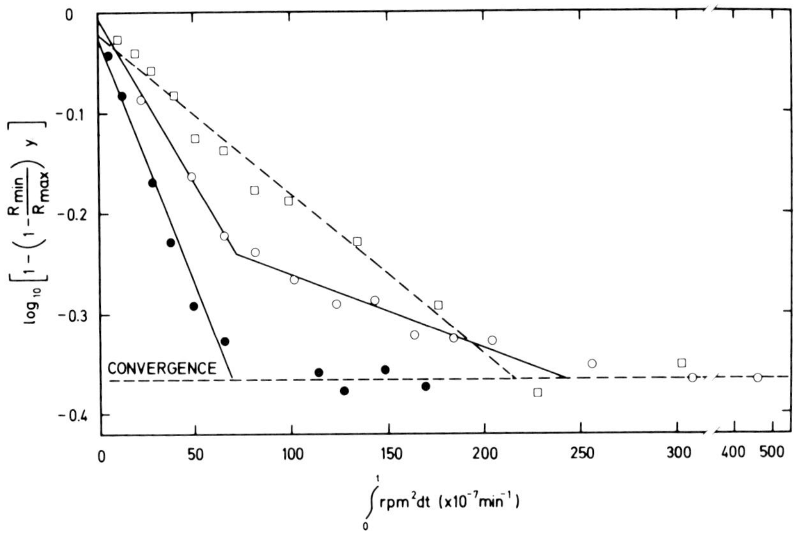

Figure 1. Sedimentation profile of two populations of mitochondria using malate dehydrogenase as the marker enzyme. (●), sediterm of rat liver mitochondria with $\bar{s}$ = 17200 ± 1400 S; 0.16 mg protein/ml and convergence level 89 %. (□), sediterm of guinea pig ileal mitochondria with $\bar{s}$ = 5640 ± 340 S; 0.4 mg protein/ml and convergence level 90 %. (o), a mixture of isolated liver mitochondria (64 % of total activity) and ileal mitochondria (36 % of total activity); the $\bar{s}$-values were estimated to be 17800 ± 1300 S and 5300 ± 300 S for liver and ileal mitochondria, respectively. From Slinde et al., 1978.

DETERMINATION OF $\bar{s}$-VALUES OF CELL ORGANELLES

The theory has been confirmed experimentally for different intact organelles in a number of tissues such as mitochondria (Slinde and Flatmark, 1973; Slinde and Flatmark, 1975; Slinde et al., 1978; Flatmark et al., 1980; Kryvi et al., 1986), lysosomes (Berge et al., 1984), normal peroxisomes (Flatmark et al., 1981; Christiansen et al., 1981), microperoxisomes (Normann and Flatmark, 1982; Norseth et al., 1982), microsomes (Norseth et al., 1982) and storage organelles (Bødtker-Næss et al., 1978). Thus, the enzymatic activities of e.g mitochondria and peroxisomes follow predicted sediterms expected for particles approaching a monodisperse population of particles. Generally, relatively small standard errors in the $\bar{s}$-values (< 10 %) are observed. Marked variations in the $\bar{s}$-values of mitochondria and peroxisomes have, however, been observed in rat liver as a result of changes, e.g. in the dietary regime (Kryvi et al., 1986).

THE EFFECT OF POLYDISPERSITY

The sedimentation of e.g. rat liver mitochondria, which contain enzymes with a single location, gives a linear sediterm (Fig. 1) and an $\bar{s}$-value with a small standard error. If the enzyme/component is localized to more than a single monodisperse population of particles, the experimentally obtained sediterm should theoretically consist of a number of linear segments. This has been experimentally shown to be the case for a mixed population of mitochondria with different $\bar{s}$-values (Fig. 1)(Slinde et al., 1978), catecholamine storage granules (Bødtker-Næss et al., 1981), microsomes (Norseth et al., 1982) and lysosomes (Berge et al., 1985).

REFERENCES

Berge RK, Flatmark T, Osmundsen H (1984). Enhancement of long-chain acyl-CoA hydrolase activity in peroxisomes and mitochondria of rat liver by peroxisomal proliferators. Eur J Biochem 141:637-644.

Bødtker-Næss V, Slinde E, Terland O, Flatmark T (1978). On the polydispersity of bovine adrenal chromaffin granules Analytical differential centrifugation in isotonic homogeneous medium. Biochim Biophys Acta 541:124-134.

Christiansen EN, Flatmark T, Kryvi H (1981). Effects of marine oil diet on peroxisomes and mitochondria of rat liver. A combined biochemical and morphometric study. Eur J Cell Biol 26:11-20.

De Duve C, Berthet J (1953). Reproducibility of differential centrifugation experiments in tissue fractionation. Nature (London) 172:1142.

Flatmark T, Kryvi H, Tangerås A (1980). Induction of megamitochondria by cuprizone (biscyclohexanone oxaldihydrazone). Evidence for an inhibition of the mitochondrial division process. Eur J Cell Biol 23:141-148.

Flatmark T, Christiansen EN, Kryvi H (1981). Polydispersity of rat liver peroxisomes induced by the hypolipidemic and carcinogenic agent clofibrate. Eur J Cell Biol 24:62-69.

Kryvi H, Christiansen E, Tangerås A, Flatmark T (1986). Effects of essential fatty acid deficiency on mitochondria and peroxisomes in rat hepatocytes with special reference to a partially hydrogenated fish oil diet. Biochim Biophys Acta 881:87-92.

Normann PT, Flatmark T (1982). Microperoxisomes and mitochondria of brown adipose tissue. Hydrodynamic parameters, isolation and capacity of long-chain fatty acid oxidation. Biochim Biophys Acta 712:621-627.

Norseth J, Normann PT, Flatmark T (1982). Hydrodynamic parameters and isolation of mitochondria, microperoxisomes and microsomes of rat heart. Biochim Biophys Acta 719:569-579.

Slinde E, Flatmark T (1973). Determination of sedimentation coefficients of subcelluar particles of rat liver homogenates. A theoretical and experimental approach. Anal Biochem 56:324-340.

Slinde E, Morild E, Flatmark T (1975). A general and rational approach to the optimal recovery of mitochondria by differential centrifugation in homogenous media. Anal Biochem 66:151-158.

Slinde E, Flatmark T (1975). Sedimentation coefficient and buoyant density of brown adipose tissue mitochondria from guinea pigs. Anal Biochem 65:581-585.

Slinde E, Morild E, Flatmark T (1978). On the polydispersity of mitochondria in tissue homogenates and the determination of the average sedimentation coefficient of mixed populations of mitochondria. Anal Biochem 90:516-524.

Section II
MEMBRANE TRAFFIC IN PLANTS

Cell-Free Analysis of Membrane Traffic, pages 253–254

Membrane Traffic in Plants

Plants present additional challenges to the study of membrane traffic imparted by a rigid exoskeleton, the cell wall, adjacent to the plasma membrane, and a large central storage organelle, the vacuole surrounded by a vacuole membrane or tonoplast. Furthermore, while exocytosis has been studied widely in plants, endocytosis has received attention only with protoplasts from which walls have been removed enzymatically. This treatment is necessary to permit access to the outer plasma membrane surface of externally supplied endocytic tracers. Operation of endocytic events in normal walled plant cells remains to be established.

The vacuole is by far the largest organelle in mature plant cells. While not exhibiting any known biosynthetic capacities, it is important to the bidirectional traffic of ions and molecules between the cytoplasm and the inner vacuolar compartment. Thus, it is appropriate that the first chapter of this section by Boudet and Alibert summarizes the use of isolated vacuoles and tonoplast vesicles in cell-free systems as experimental tools in order to study solute traffic across the tonoplast. Methods for the isolation of vacuoles and the preparation of tonoplast vesicles as well as a case study on the transport of O-coumaric acid glucoside through the tonoplast of sweet clover (*Melilotus alba*) are presented. These authors conclude that such '*in vitro*' studies complement those performed '*in situ*' and that a combination of the two approaches should be used more often for a better understanding of the mechanisms of solute traffic between vacuole and cytoplasm.

Methods for tonoplast and plasma membrane isolation are reviewed by Scherer and vom Dorp who emphasize the need for better identification methods for application to the diversity of membrane types encountered in plant cells. Scherer and vom Dorp used filipin labeling and freeze fracture electron microscopy to confirm the identification of tonoplast membrane based on measurements of membrane thickness. A number of the studies reported in this section make use of the free-flow electrophoresis procedure (see also Part IA) (Scherer & vom Dorp, Sandelius & Morré, Kiefer et al. and Auderset et al.). This method has the advantage of allowing for the simultaneous preparation of both the plasma membrane and the tonoplast in high yield and fraction purity.

The controversy as to whether or not endocytosis occurs in plant cells is discussed by Dengler et al. who report on the use of the fluorescent phospholipid NBD-PC [1 acyl-2(N-4-nitrobenzo-2-oxa-1,3-diazole)-aminocaproyl phosphatidyl-choline] to study endocytosis in protoplasts isolated from carrot cells. NBD-PC has previously been used for _in vivo_ studies with animal cells. The use of this reagent is complicated by the presence of phospholipase A_2 activity, though the authors describe how they have begun to discriminate between appearance of NBD-fluorescence due to phospholipase activity versus that due to uptake.

Melkonian et al. remind us that secretion of scales in the green flagellated algae represents an excellent model for the study of outbound membrane traffic. This system is amenable to ready experimental manipulation but remains refractory to decisive cell fractionation.

The section of membrane traffic in plants concludes with a note by Moreau and his colleagues on an experimental approach to the intermembrane transfer of lipids and fatty acids to the plasma membrane _in vivo_ using etiolated seedlings. Compared to proteins, little is known about the intracellular movement of lipids during cell membrane biogenesis and the reader interested in this problem may find their method of interest.

Cell-Free Analysis of Membrane Traffic, pages 255–267

ISOLATED VACUOLES AND TONOPLAST VESICLES AS EXPERIMENTAL TOOLS TO STUDY SOLUTE TRAFFIC TO AND FROM VACUOLES

A.M. Boudet* and G. Alibert**

*Centre de Physiologie Végétale, UA CNRS 241, Université Paul Sabatier, 118, route de Narbonne 31062 Toulouse Cédex - France **Laboratoire de Biotechnologie et Amélioration des Plantes E.N.S.A.T., 145, avenue de Muret, 31076 Toulouse Cédex, France

INTRODUCTION

The vacuole is the largest organelle in mature plant cells. This conspicuous compartment limited by a single membrane can represent up to 90% of the volume of the protoplast. So, the vacuole can be considered as an internal medium of the plant cell interacting with the cytoplasm and playing an important role in different cellular functions (Fig. 1). Some of them seem to be associated to specific strategies of plant cells, i.e. maintenance of internal osmotic pressure, irreversible storage of secondary metabolites or transient storage of nutritive solutes. Other roles including a lysosomal function, the involvement in cytoplasmic homeostasis and possibly in transduction of extracellular stimuli, have more or less their counterpart in other compartments of eucaryotic cells, e.g. lysosomes, endoplasmic reticulum.

The plant vacuole does not exhibit any biosynthetic capacities, but most of its functions are related to the bidirectional traffic of ions and molecules through the vacuolar membrane, i.e. the tonoplast. Therefore it is of the utmost importance to know the transport mechanisms occurring at this membrane and to understand their regulation in order to better integrate the vacuole in the overall functioning of plant cells. In this chapter we would like to summarize to what extent isolated vacuoles and tonoplast vesicles can be used as experimental tools in order to study solute traffic through the tonoplast.

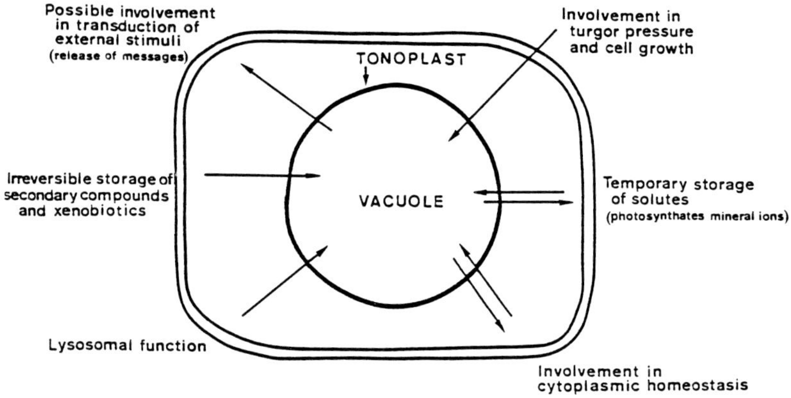

Figure 1. Different functions of plant vacuoles

ISOLATION OF VACUOLES

Due to their large size, vacuoles are very fragile and cannot be obtained through the conventional procedures generally used to isolate other organelles. The most widely used techniques involve the gentle disruption of plant protoplasts. Thus plasmalemma can be ruptured by osmotic mechanical or chemical treatments and the released vacuoles can be recovered and purified by different centrifugation steps. The method we have designed in our lab (Boudet et al. 1981, Alibert et al. 1982a) is briefly described here (Fig. 2). For critical evaluation of the method, see Boudet and Alibert, 1987.

Protoplasts are layered onto a density Ficoll gradient containing successively a polycationic compound DEAE Dextran and a polyanionic compound Dextran sulfate. The polycation binds negative charges of the plasmalemma and facilitates its disorganization. Through the combined action of DEAE Dextran and centrifuge forces, the vacuoles

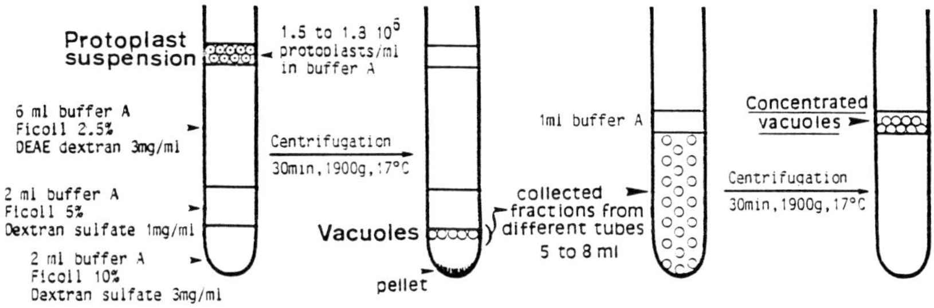

Figure 2. Procedure for obtention and concentration of vacuoles from Acer pseudoplatanus protoplasts. Buffer A: Tris Mes 25 mM, pH 6.5, 0.7 M mannitol.

are released from the protoplasts; then they sediment through a Dextran sulfate-containing layer which binds the excess of DEAE Dextran. They are collected, counted and can be further concentrated by flotation. Starting from Acer pseudoplatanus protoplasts, the final yield of these concentrated fractions amounts to about 15% on an αmannosidase basis, and several ml of vacuole suspensions containing up to 5.10^6 vacuoles/ml can be obtained at once (Fig. 3a).

PREPARATION OF TONOPLAST VESICLES

Tonoplast vesicles are obtained according to two different basic procedures. Starting from crude microsomal preparations in which membranes essentially appear as small vesicles, a tonoplast enriched fraction may be obtained by centrifugation on a discontinuous gradient (Bennet et al. 1984, Poole et al. 1984, Schumaker and Sze 1985). The NO_3^- sensitive ATPase activity and the density of the vesicles are used as markers in order to identify tonoplast enriched fractions. The free-flow electrophoresis technique can also be used to obtain such a purification (Auderset et al. 1986).

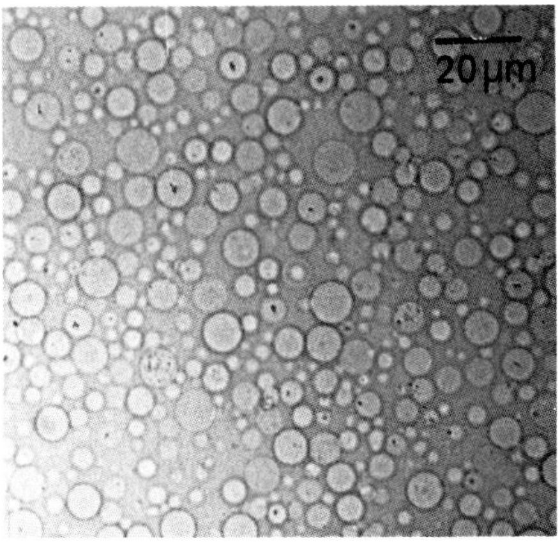

Figure 3a. Isolated vacuoles from <u>Acer pseudoplatanus</u>

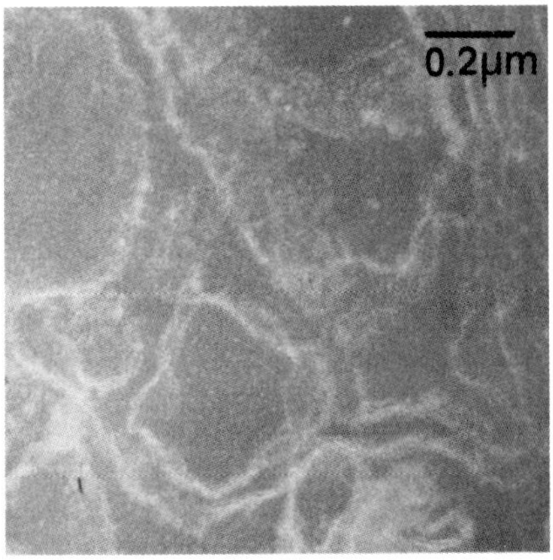

Figure 3b. Electron microscope picture of tonoplast vesicles from <u>Melilotus alba</u> (the average size of the vesicles is about 100 times less than that of the intact vacuoles)

Tonoplast vesicles may also be prepared from isolated vacuoles; in this case more purified preparations are obtained but at a lower scale (Briskin and Leonard 1980, Bush and Sze 1986). This procedure has been used by us to isolate tonoplast vesicles from <u>Melilotus alba</u> vacuoles (Rataboul et al. 1985) (Fig. 3b) and <u>Acer pseudoplatanus</u> (Alibert et al. 1986). Vacuoles were lysed by osmotic shock. The suspension was then centrifuged for 1 min at 2,900 x g and the supernatant centrifuged again 1 h at 100,000 x g. The 100,000 pellet (corresponding to the tonoplast fraction) was resuspended in cold deionized water, and lyophylised.

Lyophylised membranes were resuspended in 25 mM Tris-Mes buffer (pH 6.5) 0.3 M Mannitol 5 mM β-mercaptoethanol. The suspension was transferred to a Potter homogeniser fitted with a Teflon piston and the membranes were homogenised by 20 strokes of the piston to obtain the tonoplast vesicles. When radioactive solutes were added to the homogenisation medium before Potterisation, some of the labeled molecules were trapped inside the vesicles' lumen.

GENERAL PROCEDURES FOR VACUOLAR TRANSPORT STUDIES USING VACUOLES AND TONOPLAST VESICLES

Isolated vacuoles and tonoplast vesicles have opened new possibilities for the study of solute movements through the tonoplast; however, investigations into the corresponding mechanisms and their regulation are still very limited. The main reasons are the difficulty of obtaining relatively large amounts of pure vacuoles, the fragility of these organelles, and the loss of their natural cellular environment after their isolation.

With the exception of studies on proton translocation and the characterization of ionic channels by patch clamp, most of the available results dealing with the transport of solutes arise from the same experimental approach. Basically, the vacuoles (or vesicles) are incubated with a labelled solute [for different lengths of time] and then separated from the incubation medium by centrifugation or filtration.

Vacuoles are usually centrifuged through a Ficoll gradient (Alibert et al. 1982b, Martinoia et al. 1986) or a silicon oil layer (Doll et al. 1979). They have been in some experiments allowed to deposit across a discontinuous Ficoll gradient at 1 g (Buser-Suter et al. 1982). Such a procedure was necessary in this case because the vacuoles were broken upon centrifugation.

Tonoplast vesicles can be collected by filtration (Sze 1985), but such a method is not applicable to vacuoles which burst during filtration. Unlabelled solutes are usually included in the washing solution in order to wash out the radioactivity absorbed at the vesicle surface or filter. In both cases the radioactivity associated to the vesicles on the filter or to the centrifugation pellet is determined by scintillation counting. If necessary, the total intravacuolar (vesicular) volume can be measured rapidly and accurately by centrifuging vacuoles (vesicles) incubated in the presence of ^3H_2O and a ^{14}C labelled non-permeating polymer (Dextran ^{14}C). The ^3H_2O and ^{14}C contents of the resulting pellet measure respectively the total pellet volume and the volume external to the vacuoles. Thus, the intravacuolar volume can be deduced by difference (Komor et al. 1982).

One of the conclusions of these different uptake studies is that a H^+ pumping process fueled by ATP provides the driving force for the transport at the tonoplast of various solutes through antiport or symport mechanisms as predicted by Mitchell's chemiosmotic hypothesis (Marin et al. 1981a and 1981b, Smith et al. 1984). However, in many cases the experimental data do not fit with this general scheme. Uptake of solutes in the absence of any energy source might be related to a preexisting pH gradient between the vacuolar space and the medium, but other undefined mechanisms could also occur.

TRANSPORT THROUGH THE TONOPLAST OF THE O-COUMARIC ACID GLUCOSIDE (o-CG) IN MELILOTUS ALBA: A CASE STUDY

As many other natural substances, this specific plant secondary compound which can reach up to 6% of the dry weight in sweet clover leaves, accumulates in vacuoles (Oba et al. 1981) as a glucoside. Its complete retention within isolated vacuoles indicates that the tonoplast does not become leaky during isolation, and we have studied its uptake by isolated vacuoles (Rataboul et al. 1985). After incubation in presence of the labelled glucoside, the vacuoles were separated on a Ficoll gradient and their radioactivity was determined. The results reported on Figure 4 show that these isolated vacuoles are able to take up the glucoside in a time and temperature dependent manner against a concentration gradient. In our conditions the addition of Mg ATP has no effect. Yet, an interesting observation was that vacuoles isolated from *Acer pseudoplatanus* cells which do not contain o-CG were unable, in the same experimental conditions, to take up the glucoside. Consequently, it seems that the transport system characterized in *Melilotus alba* vacuoles might be restricted to species which accumulate o-CG.

Due to the difficult handling of these vacuoles, we have used tonoplast vesicles to solve another problem: the irreversible sequestration of o-CG in the vacuole. Among different mechanisms, it has been suggested that conformational changes of solutes occurring inside the vacuoles could give rise to non-diffusible forms (Mattern et al., 1983).

In *Melilotus alba* leaves o-CG is synthetized from

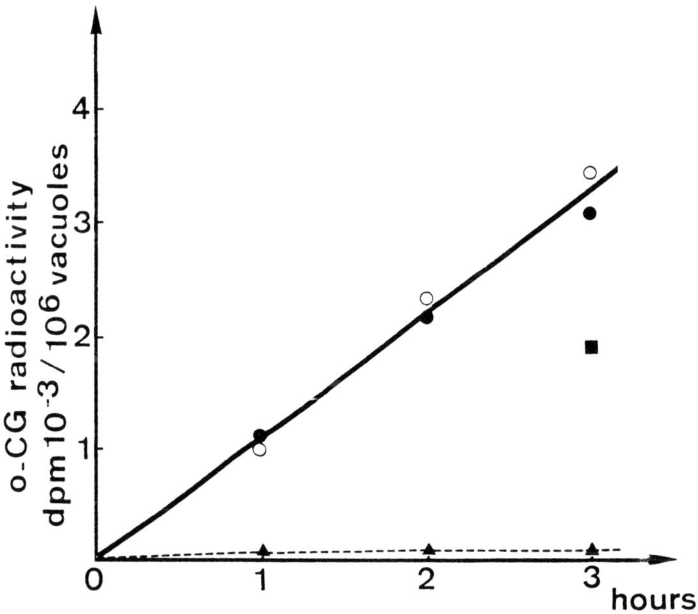

Figure 4. Uptake of ^{14}C o-CG by isolated vacuoles. Experiment performed at pH8 and 20°C (o-CG external concentration 7.6 mM)

Sweet clover vacuoles: ● - ● with ATP Mg; ○ - ○ without ATP Mg; ■ without ATP Mg at 10°C

Acer pseudoplatanus vacuoles ▲ -- ▲

transcinnamic acid as trans-isomer and is then accumulated in vacuoles as cis isomer. In separate experiments, we have prepared tonoplast vesicles in media respectively containing the two ^{14}C o-CG isomers. These different vesicles were also loaded with ^{3}H labelled inulin, a non-diffusible compound used as reference for the quantitation of the vesicles. We have then studied a possible efflux of the isomers during the passage of the vesicles through a gel filtration column (sepharose 4B). Table 1 shows that the radioactivity of the cis isomer is not changed after chromatography on sepharose but is considerably decreased for the trans isomer which seems to be the only form to cross the tonoplast quite easily. This selective impermeability of the membrane to the cis isomer may provide the vacuole with an efficient trapping mechanism responsible for the irreversible sequestration of the cis o-CG.

TABLE 1. Permeability of tonoplast vesicles to trans and cis isomers of O-Coumaric acid glucoside.

^{14}C o-CG isomer	^{14}C o-CG remaining in the vesicles after sieving on sepharose (in %)
Exp. 1	
trans	58
cis	100
Exp. 2	
trans	37
cis	108

It can be difficult to use the results of studies with isolated vacuoles or tonoplast vesicles for the understanding of "in vivo" regulation of transport processes. So vacuolar uptake studies "in situ" have also been performed (Thom et al. 1982). The basic procedure involves the feeding of protoplasts with the compound under study or one of its precursors in a labelled form, the rapid isolation of vacuoles thereafter and the determination of the radioactivity of the vacuolar compound.

Figure 5 shows the kinetics of o-CG synthesis (protoplast) and o-CG vacuolar accumulation (vacuole) after feeding Phenylalanine (a precursor of o-CG) to *Melilotus alba* protoplasts. Such an approach allowed us to show that the "in vivo" vacuolar uptake of o-CG depends on a light induced acidification of the vacuole (Rataboul et al. 1985).

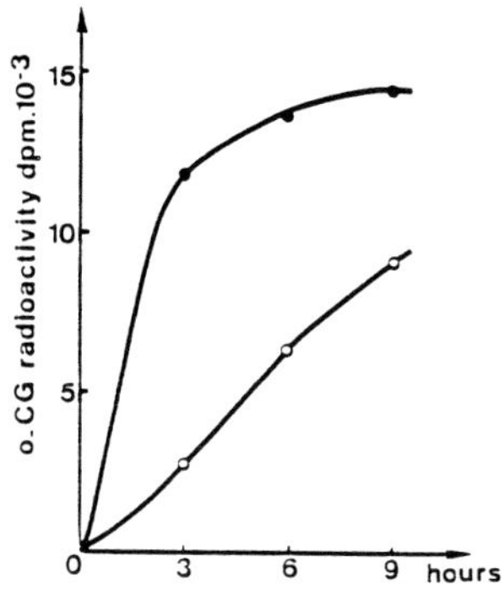

Figure 5. Time course change of o-CG radioactivity in protoplasts (● - ●) and vacuoles (○ - ○).

ADVANTAGES AND DRAWBACKS OF THE DIFFERENT APPROACHES

Isolated vacuoles can be used to study vacuolar transport directly. Furthermore, tonoplast vesicles avoid the possible complications caused by the vacuolar contents. Both systems present advantages and drawbacks in common. They both provide simplified "in vitro" systems allowing a careful control of different effectors in the incubation medium. The common drawbacks are the possible damage of transport systems during isolation and the loss of cytosolic environment. Other specific characteristics of these systems are summarized in Table 2. The major limitation in the use of vacuoles lies in their fragility. These large organelles are very difficult to handle and may be easily disrupted. For this reason different groups have rather used tonoplast vesicles which are also particularly suited for efflux studies after their loading with labelled solutes.

All the exchanges through the tonoplast seem to be regulated by interactions between the cytosol and the vacuole, thus if the "in vitro" studies may be useful in order to identify the basic characteristics of transport mechanisms, they provide little information on their regulation. "In situ" studies of vacuolar uptake on protoplasts represent a convenient system for such a purpose. They can also reveal, by comparison, damages caused to specific transport systems in isolated vacuoles. This system is promising, but complex, and some

TABLE 2

Isolated vacuoles	Advantages	. natural model of membrane vesicles . purity . right side out
	Drawbacks	. fragility . possible exchanges between internal and external solutes
Tonoplast vesicles	Advantages	. stability . possibility of loading (efflux studies...) . control of the internal environment...
	Drawbacks	. right side out or inside out ? . contamination by other membranes . structural heterogeneity
"In situ" vacuolar uptake studies (protoplasts)	Advantages	. conservation of the natural cytosol/vacuole interface . possibility of manipulating environmental conditions
	Drawbacks	. complex interpretation . possible interferences with metabolism and passages through different membranes

difficulties may result from the fact that the labelled solutes fed to the protoplast can be metabolized along different pathways or directed to different internal stores.

Finally, it should be kept in mind that the protoplast is an artificially altered plant cell. The isoosmotic condition required to prepare and to store protoplasts is a serious interference with normal metabolism and could also perturb intracellular transport in certain tissues.

CONCLUSION

Despite the important functions of vacuoles in plant cells, investigation into the mechanisms of solute transport across the tonoplast is still in its infancy. For example, apart from the H^+ pumping ATPase nothing is known about the molecular characteristics of other transport systems and carriers on the tonoplast. "In vitro" techniques may still provide important information on the basic characteristics of transport systems and molecular properties of the corresponding carriers. In parallel, studies using protoplasts and in specific cases (Pi) nuclear magnetic resonance will help to understand the dynamics and the regulation of these transports, i.e. the activities "in situ" of the vacuoles embedded in the cytoplasm.

The two approaches complement each other and should be used more often in combination for a better understanding of the mechanisms of solute traffic through the tonoplast.

REFERENCES

Alibert G, Carrasco A, Boudet AM (1982a). Changes in biochemical composition of vacuoles isolated from _Acer pseudoplatanus_ L. during cell culture. Biochim Biophys Acta 721:22-29.

Alibert G, Boudet AM, Rataboul P (1982b). Transport of O-Coumaric acid glucoside in isolated vacuoles of sweet clover. In Marmé D, Marre E, Hertel R (eds): "Plasmalemma and tonoplast: their functions in the plant cell," pp 193-200.

Alibert G, Carrasco A, Citharel B (1986). Biochemical characteristics of tonoplast preparation isolated from Acer pseudoplatanus cell suspension cultures. Physiol Vég 24:85-96.

Auderset G, Sandelius AS, Penel C, Brightman A, Greppin H, Morré DJ (1986). Isolation of plasma membrane and tonoplast fractions from spinach leaves by preparative free-flow electrophoresis and effect of photoinduction. Physiol Plant 68:1-12.

Bennet AB, O'Neill SD, Spanswick RM (1984). H^+-ATPase from storage tissue of Beta vulgaris l. Identification and characterization of an anion-sensitive H^+-ATPase. Plant Physiol 74:538-544.

Boudet AM, Alibert G (1987). The isolation of vacuoles and tonoplast from protoplasts. Meth Enzymol (in press).

Boudet AM, Canut H, Alibert G (1981). Isolation and characterization of vacuoles from Melilotus alba mesophyl. Plant Physiol 68:1354-1358.

Briskin D, Leonard RT (1980). Isolation of tonoplast vesicles from tobacco protoplasts. Plant Physiol 66:684-687.

Buser-Suter Ch, Wiemken A, Matile Ph (1982). A malic acid permease in isolated vacuoles of a Crassulacean acid metabolism plant. Plant Physiol 69:456-459.

Bush DR, Sze H (1986). Calcium transport in tonoplast and endoplasmic reticulum vesicles isolated from cultured carrot cells. Plant Physiol 80:549-555.

Doll S, Rodier F, Willembrink J (1979). Accumulation of sucrose in vacuoles isolated from red beet tissue. Planta 144:407-411.

Komor E, Thom M, Maretzki A (1982). Vacuoles from sugar cane suspension cultures. III. Proton motive potential difference. Plant Physiol 69:1326-1330.

Marin B, Smith JAC, Luttge U (1981a). The electrochemical proton gradient acid; its influence on citrate uptake in tonoplast vesicles of Hevea brasiliensis. Planta 153:486-493.

Marin B, Marin-Lanza M, Komor E (1981b). The proton motive potential difference across the vacuo-lysosomal membrane of Hevea brasiliensis (rubber tree) and its modification by membrane bound adenosine triphosphatase. Biochem J 198:365-372.

Martinoia E, Schramm MJ, Kaiser G, Kaiser WM, Heber U (1986). Transport of anions in isolated vacuoles. Plant Physiol 80:895-905.

Mattern U, Heller W, Himmelspach K (1983). Conformational changes of apigenin 7-O (6-O malonylglucoside), a vacuolar pigment from parsley, with solvent composition and proton concentration. Eur J Biochem 133:439-448.

Oba K, Conn EE, Canut H, Boudet AM (1981). Subcellular localization of 2-(β-D-glucosyloxy)-cinnamic acids and the related β-glucosidase in leaves of Melilotus alba Desr. Plant Physiol 68:1359-1363.

Poole RJ, Briskin DP, Kratky Z, Johnstone RM (1984). Density gradient localization of plasma membrane and tonoplast from storage tissue of growing and dormant red beet. Characterization of proton transport and ATPase in tonoplast vesicles. Plant Physiol 74:549-556.

Rataboul P, Alibert G, Boller T, Boudet AM (1985). Intracellular transport and vacuolar accumulation of O-Coumaric acid glucoside in Melilotus alba mesophyll cell protoplasts. Biochim Biophys Acta 816:25-36.

Schumaker KS, Sze H (1985). A Ca^{2+}/H^+ antiport system driven by the proton electrochemical gradient of a tonoplast H^+-ATPase from oat roots. Plant Physiol 79:1111-1117.

Smith JA, Uribe EG, Ball E, Heller S, Luttge U (1984). Characterization of the vacuolar ATPase activity of the Crassulacean-acid metabolism plant kalanchoe daignemontiana. Eur J Biochem 141:415-420.

Sze H (1985). H^+-translocating ATPases: advances using membrane vesicles. Ann Rev Plant Physiol 36:175-208.

Thom M, Komor E, Maretzki A (1982). Vacuoles from sugar cane suspension cultures. II. Characterization of sugar uptake. Plant Physiol 69:1320-1325.

Cell-Free Analysis of Membrane Traffic, pages 269–279

ISOLATION BY FREE-FLOW ELECTROPHORESIS AND IDENTIFICATION OF TONOPLAST AND PLASMA MEMBRANE FOR STUDIES OF PLANT MEMBRANE TRAFFIC

Günther F. E. Scherer and Barbara vom Dorp

Botanisches Institut, Universität Bonn, Venusbergweg 22, D-5300 Bonn 1, FRG

INTRODUCTION

Successful purification of subcellular components is clearly one precondition for progress in studies of membrane traffic at the molecular level. Studies with isolated components eventually lead to increased knowledge about markers for these compartments. The latter then provide a basis for cytochemical labels. Molecular studies on transport at a level of sophistication comparable with animal systems, are completely lacking for plants. Part of this deficiency comes from lack of highly purified cell components from plants. Studies of both exocytotic and endocytotic membrane traffic in plants would be facilitated greatly by the availability of purified fractions of cell surface and vacuolar (tonoplast) membranes.

Fractionation of plant membranes has been hampered by the many different compartments that need to be separated: nuclear membranes, endoplasmic reticulum, Golgi apparatus (dictyosomes), tonoplast, plasma membranes, plastids, microbodies, coated and secretory vesicles, and mitochondria. In addition, homogenization often leads to the destruction of recognizable components, so that many smooth vesicles originate even from compartments with distinct morphology such as mitochondria, plastids, dictyosomes, and, through the loss of ribosomes, rough ER (Quail, 1979). Hence, besides isolation techniques, identification methods are needed to verify the establishment of markers and the purity of fractions. Two new methods for the identification of tonoplast have been developed recently (Sandelius et al., 1986; vom Dorp et al., 1986). In our experiments we have combined these two approaches to facilitate studies of membrane trafficking in plants.

MATERIALS AND METHODS

Plant Material

Segments about 1 cm long were cut just below the cotyledons of dark grown (4 - 5 days) zucchini (*Cucurbita pepo* L., var. Cocozelle) or soybeans (*Glycine max* L.). Cress (*Lepidium sativum* L.) roots were grown for 33 - 37 h on a wire mesh to about 2 cm length.

Membranes

Membranes were prepared as described (Scherer and Fischer, 1985). Fractions in ethanolamine-containing buffers were pelleted for 30 min at 40,000 g_{max} and resuspended in chamber buffer for electrophoresis.

Free-flow Electrophoresis

The electrophoresis medium (chamber buffer) contained 0.25 M sucrose, 2 mM KCl, 10 mM triethanolamine, and 10 mM acetic acid, titrated to pH 7.5 (NaOH). For details see Sandelius et al. (1986, this volume).

Analyses

Marker enzymes were measured as described (Scherer, 1984; vom Dorp et al., 1986). Vanadate-inhibited ATPase was measured as the difference of activities plus or minus 0.2 mM orthovanadate pH 6.0 in 50 mM MES/BTP, 0.05% Triton X100 (by weight) in the presence of 50 mM KNO_3. Nitrate-inhibited ATPase was measured as the difference of activities at pH 7.0 in 50 mM MES/BTP plus or minus 50 mM KNO_3/BTP. Assays contained 3 mM $MgSO_4$ and 3 mM ATP and were incubated at 30°C for 70 min. Ice-cold phosphate reagent (Scherer, 1984) was added and color developed for 7 min at 37°C and read at 750 nm. Electron microscopy was as described (vom Dorp et al., 1986).

Results

Different membranes are resolved by free-flow electrophoresis with either microsomes or membrane-enriched fractions as starting material (Fig. 1,2). The protein profile of microsomes after free-flow electrophoresis showed a high central peak. In contrast, the protein profile of the tonoplast-enriched fraction was shifted toward the anode, whereas that of the plasma membrane-enriched fraction was shifted toward the cathode. Since the composition of these enriched fractions had been determined (Scherer and Fischer, 1985; vom Dorp et al., 1986), this result verified other studies where purified tonoplast and plasma membrane fractions were obtained at the outer flanks of the electrophoretic separations (Sandelius et al., 1986).

Free-flow electrophoresis fractions were analyzed by freeze fracture electron microscopy after filipin treatment (vom Dorp et al., 1986) to determine the content of tonoplast, plasma membrane, and other membranes. Tonoplast vesicles clearly showed areas free of filipin-induced lesions covered only (Fig. 3) with intramembranous particles. This feature distinguished them from plasma membrane vesicles which were completely covered by filipin-induced lesions. This difference in morphology, and the fact that all other membranes were unlabeled by filipin treatment, correlates with marker enzymes to help identify tonoplast and plasma membranes in membrane fractions from cress roots (vom Dorp et al. 1986).

When enzyme activities were measured (Fig. 4), markers for plasma membrane, UDPG-sterolglucosyltransferase and vanadate-inhibited ATPase, were increased in specific activities towards the cathode and, more significantly, almost absent in the subfractions of the free-flow electrophoresis of the tonoplast-enriched material. This means that the outermost subfractions for tonoplast and plasma-membrane were the least contaminated. Nitrate-inhibited ATPase, a tonoplast marker, showed the opposite trend with greater specific activity at the anode side of the profile.

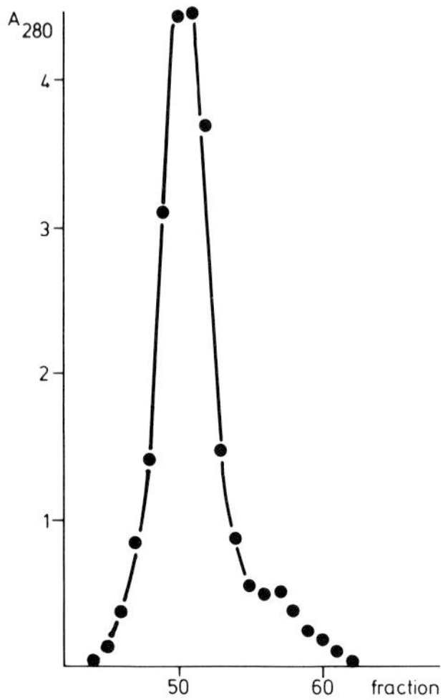

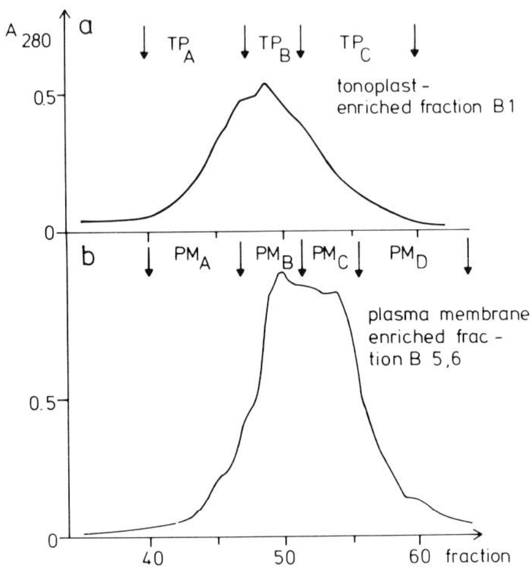

Figure 1 (left). Free-flow electrophoretic separations of soybean microsomes.

Figure 2 (right). Free-flow electrophoresis of zucchini membranes. (a) A tonoplast-enriched fraction (=B1 of Scherer and Fischer, 1985) had been prepared prior to electrophoresis by combined sucrose and glycerol gradient centrifugation (b) A plasma membrane-enriched fraction was subjected to electrophoresis (=B6,7 of Scherer and Fischer, 1985). Membranes were analyzed for marker activities (Fig. 4). Pooled subfractions are indicated by arrows. Membranes for experiments of Figure 3 and Table 1 were prepared and fractions pooled in a similar manner.

Identification of tonoplast and plasma membrane by the filipin-label procedure compared favorably with the identification of plasma membrane by PTA-staining and identification of tonoplast by measurement of membrane thickness (Table 1, and compare Auderset et al., 1986 and Sandelius et al., 1986). With all three methods of identification, a clear separation of tonoplast toward the anode and of plasma membrane toward the cathode was shown. When glycerol gradient centrifugation and free-flow electro-

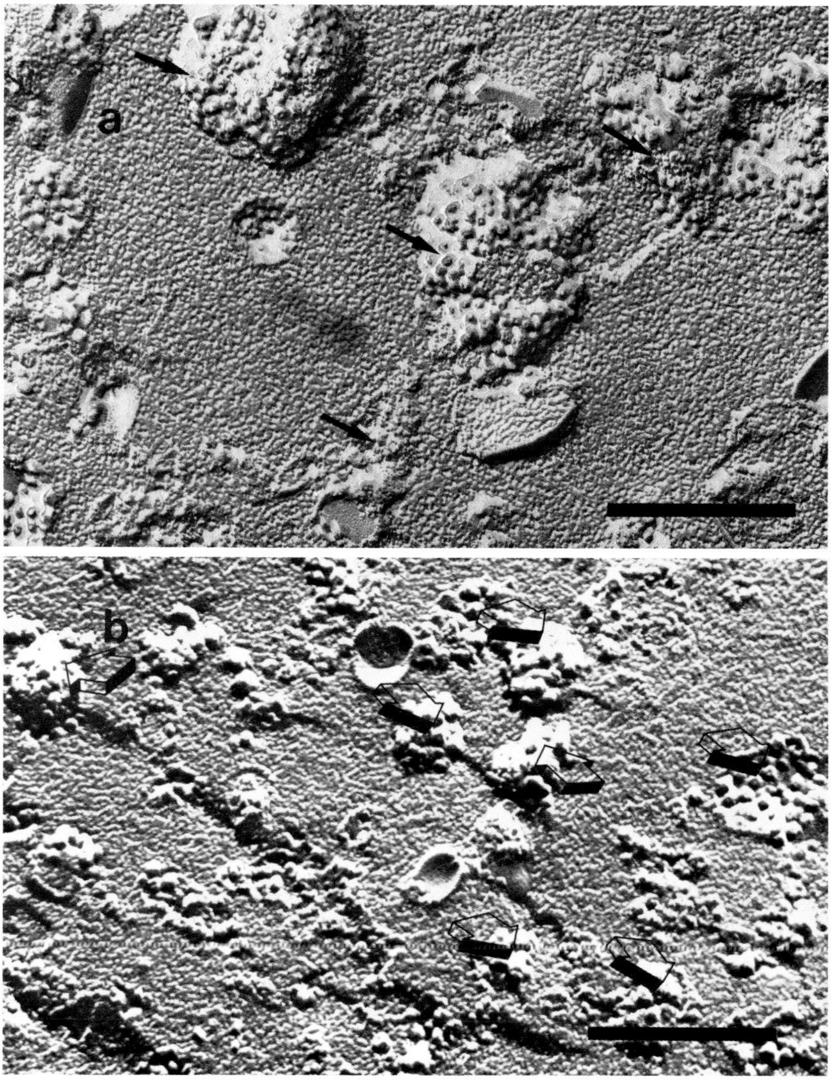

Fig. 3. Survey electron micrograph of highly purified tonoplast (a) and plasma membrane (b) fractions from soybean. The fractions correspond to the experiment at the bottom of Table 1. Tonoplast vesicles (filled arrows) exhibited large indented filipin-induced lesions but also areas with intramembranous particles whereas plasma membrane vesicles were completely covered with filipin-induced lesions (open arrows). Bar = 0.5 μm.

TABLE 1. Identification of tonoplast and plasma membrane vesicles by the filipin-label procedure of vom Dorp et al. (1986). Glycerol gradient centrifugation, free-flow electrophoresis (FFE), or both procedures in sequence were compared.

	Plant Material[a]	Tonoplast	Plasma Membrane	Others
Homogenate	Zucchini	7 %	4 %	89 %
	Cress	8 %	5 %	87 %
B1 (tonoplast-enriched)	Cress	41 %	12 %	47 %
	Cress	35 %	6 %	59 %
	Zucchini	38 %	7 %	55 %
B6, 7 (plasma membrane-enriched)	Cress	25 %	31 %	54 %
	Cress	22 %	40 %	38 %
	Zucchini	13 %	37 %	50 %
FFE of microsomes				
Anode:tonoplast	Zucchini	52 %	9 %	39 %
Middle fraction	Zucchini	5 %	8 %	87 %
Cathode: plasma membrane	Zucchini	9 %	57 %	34 %
Tonoplast subfractions after FFE of B1	Cress	73 %	8 %	19 %
	Soybean	84 %	4 %	12 %
Plasma membrane subfractions after FFE of B6, 7	Cress	3 %	84 %	13 %
	Soybean	8 %	89 %	3 %

[a] Zucchini and soybean tissues were hypocotyls, while cress tissue was roots. All were from dark-grown seedlings.

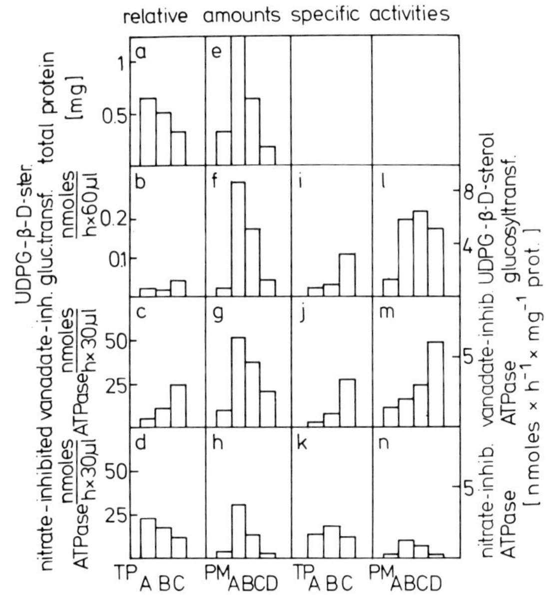

Fig. 4. Marker enzyme activities in pooled zucchini subfractions from free-flow electrophoresis separations of tonoplast and plasma membrane-enriched fractions prepared by glycerol gradient centrifugation. The protein profiles of the free-flow electrophoresis separations and the pooling into the tonoplast, TP_A, TP_B, TP_C, and into the plasma membrane subfractions PM_A, PM_B, PM_C, and PM_D are shown in Fig. 2. UDPG sterolglucosyltransferase is a marker for plasma membrane in zucchini (Scherer, 1984), vanadate-inhibited ATPase is for plasma membrane, and nitrate-inhibited ATPase is for tonoplast (Sze, 1985).

phoresis were applied, sequentially highly purified fractions with about 85% content of either membrane with little cross-contamination were obtained (Table 1). PTA-staining and filipin-label for plasma membranes had been compared previously (vom Dorp et al., 1986). The present results provide strong evidence that this method of identification is valid also for tonoplast.

DISCUSSION

Fractionation of plant cells solely by procedures based on sucrose density gradient centrifugation has been proven difficult. New methods, not based on membrane density as the principal property used for the separation, have proven useful. Among these new approaches is the purification of plasma membrane by partitioning into polyethyleneglycol/dextran phases (Kjellbom and Larsson, 1984). The disadvantage may be that only right-side-out plasma membrane is purified since it is not active in proton pumping. Another technique applied to membrane separation is isoelectric focusing (Griffing and Quatrano, 1984; Steele and Stout, 1986). One drawback here might be that membranes are collected at a pH of 4.5-5.5 which activates phospholipases associated with plant membranes so that membrane lipids would be destroyed rapidly (Scherer and Morré, 1978a/b). More recently, free-flow electrophoresis, originally developed by Hannig and Heidrich (1977), has been used successfully for the purification of tonoplast and plasma membranes (Auderset et al., 1986; Sandelius et al., 1986) from plants.

Membrane identification remains as a serious problem in plant membrane biochemistry. Smooth vesicles may originate from a number of compartments and do not give any "first sight" clues to their morphological identity. Only the PTA staining method for plasma membrane (Roland et al., 1972) has found wide acceptance. For the identification of tonoplast, one approach was its isolation from intact vacuoles, the purity of which can be inspected by light microscopy. In such a preparation the identification of nitrate-inhibited ATPase as a possible marker was achieved (Walker and Leigh, 1981). This marker was shown subsequently to be identical with the nitrate-inhibited H^+-ATPase of tonoplast by similar procedures (Bennett et al., 1984; Mandala and Taiz, 1985) and is consistent with biochemical arguments for a tonoplast location (Churchill and Sze, 1983; O'Neill et al., 1983). More recently, a direct identification of tonoplast in the electron microscope was based on the presence of thick membranes not stained by the PTA method (Auderset et al., 1986; Sandelius et al., 1986). Here we add the identification of tonoplast and plasma membrane by a filipin-label method where the two membranes are differentially labeled to verify and extend previous findings (Auderset et al.; 1986; Sandelius et al., 1986; vom Dorp et al., 1986).

The identification of tonoplast vesicles based on filipin-labeling and freeze fracture electron microscopy confirms the identification of tonoplast based on measurements of membrane thickness. Tonoplast was found at the anode flank of the free-flow separation profiles (Fig. 1, 2) and constitutes a tonoplast fraction of high purity (Table 1). Marker enzyme measurements (Fig. 4) substantiate the assumptions that had been made about nitrate-inhibited ATPase in the tonoplast (Walker and Leigh, 1981; Marin et al., 1982; Churchill and Sze, 1983; O'Neill et al., 1983; Mandala and Taiz, 1985; Scherer and Fischer, 1985; Sandelius et al., 1986), although nitrate inhibition of ATPase was found also in plant mitochondria (Wang and Sze, 1985) and plasma membranes (Anthon and Spanswick, 1986). The presence of UDPG-sterolglucosyltransferase in plasma membrane (Hartmann-Bouillon et al., 1979; Scherer, 1984) has been confirmed although in soybean it is located also in tonoplast (Sandelius et al., 1986). Vanadate-inhibited ATPase as a marker for plasma membranes is to some degree questionable because vanadate also inhibits acid phosphatase (Gallagher and Leonard, 1982; Scherer, 1984). Acid phosphatase is a ubiquitous contaminant in plant membrane preparations and cleaves ATP, thereby constituting another species of vanadate-inhibited ATP hydrolysis. Free-flow electrophoresis, on the other hand, may help to deplete membranes of soluble proteins. With zucchini membranes, vanadate-inhibited ATPase was found in all fractions but was highest in the plasma membrane (Fig. 4).

Since the application of the new techniques in plant membrane fractionation described above is still rather recent, refinement of these techniques can be expected. Perhaps it will be possible to combine centrifugation and free-flow electrophoresis (Table 1) to provide other membranes in addition to tonoplast and plasma membrane, as will be necessary for biochemically oriented studies of membrane traffic in plants.

ACKNOWLEDGMENTS

We thank the Deutsche Forschungsgemeinschaft for a travel grant to G.S. and Keri Safranski, Purdue University, for assistance.

REFERENCES

Anthon GE, Spanswick RM (1986). Purification and properties of the H^+-translocating ATPase from the plasma membrane of tomato roots. Plant Physiol 81:1080-1085.

Auderset G, Sandelius AS, Penel C, Brightman A, Greppin H, Morré DJ (1986). Isolation of plasma membrane and tonoplast fractions from spinach leaves by preparative free-flow electrophoresis and effect of photoinduction. Physiol Plant 68:1-12.

Bennett AB, O'Neill SD, Spanswick RM (1984). H^+-ATPase activity from storage tissue of <u>Beta vulgaris</u>. Plant Physiol 74:538-544.

Churchill KA, Sze H (1983). Anion-sensitive H^+-pumping in membrane vesicles from oat roots. Plant Physiol 71:610-617.

Gallagher SR, Leonard RT (1982). Effect of vanadate, molybdate, and azide on membrane-associated ATPase and soluble phosphatase activities of corn roots. Plant Physiol 70:1335-1340.

Griffing LR, Quatrano RS (1984). Isoelectric focussing of plant cell membranes. Proc Natl Acad Sci USA 81:4804-4808.

Hannig K, Heidrich H-G (1977). Continuous free-flow electrophoresis and its application in biology. In Bloemendahl H (ed): "Cell Separation Methods," New York: Elsevier/North Holland Biomedical Press, pp 95-116.

Hartmann-Bouillon MA, Benveniste P, Roland JC (1979). Sterol metabolism in purified membrane fractions from maize coleoptiles. Biol Cellul 35:183-194.

Kjellbom P, Larsson C (1984). Preparation and polypeptide composition of chlorophyll-free plasma membranes from leaves of light-grown spinach and barley. Physiol Plant 62:501-509.

Mandala S, Taiz L (1985). Proton transport in isolated vacuoles from corn coleoptiles. Plant Physiol 78:104-109.

Marin KB, Cretin H, D'Auzac D (1982). Energization of solute transport and accumulation at the tonoplast in <u>Hevea</u> latex. Physiol Veg 20:584-588.

Quail PH (1979). Plant cell fractionation. Ann Rev Plant Physiol 30:425-484.
O'Neill SD, Bennett AB, Spanswick RM (1983). Characterization of an NO_3^--sensitive ATPase from corn roots. Plant Physiol 72:837-846.
Roland JC, Lembi CA, Morré DJ (1972). Phosphotungstic acid-chromic acid as a selective electron-dense stain for plasma membrane of plant cells. Stain Technol 47:195-200.
Sandelius AS, Penel C, Auderset G, Brightman A, Millard M, Morré DJ (1986). Isolation of highly purified fractions of plasma membrane and tonoplast from the same homogenate of soybean hypocotyls by free-flow electrophoresis. Plant Physiol 81:177-185.
Scherer GFE (1984). Subcellular localization of H^+-ATPase from pumpkin hypocotyls (Cucurbita maxima L.). Planta 160:348-356.
Scherer GFE, Fischer G (1985). Separation of tonoplast and plasma membrane H^+-ATPase from zucchini hypocotyls by consecutive sucrose and glycerol gradient centrifugation. Protoplasma 129:109-119.
Scherer GFE, Morré DJ (1978a). In vitro stimulation by 2,4-dichlorophenoxyacetic acid of an ATPase and inhibition of phosphatidate phosphatase of plant membranes. Biochem Biophys Res Comm 84:238-247.
Scherer GFE, Morré DJ (1978b). Action and inhibition of plant phospholipases. Plant Physiol 62:933-937.
Steele ME, Stout RG (1986). Isoelectric focussing of oat root membranes. Plant Physiol 82:327-329.
Sze H (1985). H^+-translocating ATPases: advances using membrane vesicles. Ann Rev Plant Physiol 36:175-208.
vom Dorp B, Volkmann D, Scherer GFE (1986). Identification of tonoplast and plasma membrane in membrane fractions from garden cress (Lepidium sativum L.) with and without filipin treatment. Planta 168:151-160.
Walker RR, Leigh RA (1981). Characterization of salt stimulated ATPase activity associated with vacuoles isolated from storage roots of red beet (Beta vulgaris L.). Planta 153:140-149.
Wang Y, Sze H (1985). Similarities and differences between the tonoplast-type and the mitochondrial H^+-ATPase of oat roots. J Biol Chem 260:10434-10443.

Cell-Free Analysis of Membrane Traffic, pages 281–284

FREE-FLOW ELECTROPHORESIS FOR PREPARATION OF PLANT MEMBRANES

Anna Stina Sandelius and D. J. Morré

Department of Plant Physiology, Botanical Institute, University of Göteborg, S-41319 Göteborg, Sweden and Department of Medicinal Chemistry, Purdue University, West Lafayette, IN 47907 USA

Fractionation of mature plant cells relative to those of animals is complicated by the presence of essentially two surface membranes - the plasma membrane at the cell surface and a tonoplast membrane surrounding the vacuole. These two membranes have been difficult to isolate in quantity, each free of contamination by the other, from the same homogenates. We have used preparative free-flow electrophoresis (Hannig and Heidrich, 1977) to prepare and resolve these two membrane types in a manner that provides functional integrity (sealed, enzymatically active vesicles), high yield and >90% fraction purity.

The method of homogenization may influence the relative abundance of the different membranes in each of the different fractions (H. Canut, personal communication). The medium is not critical but cell walls and debris are removed both by filtration through a porous cellulose fabric, Miracloth (Chicopee Mills, NY) and by a preliminary centrifugation at 6,000 g for 10 min. The starting material applied to the electrophoresis is a so-called "microsomal" fraction obtained by centrifugation of the 6,000 g supernatant for 30 min at 40,000 x g. The conditions of the electrophoretic separation are described by Sandelius et al., (1986).

Using stem segments of dark-grown seedlings of soybeans, the typical pattern of separation shown in Figure 1 was obtained. The shoulder toward the anode designated A has the characteristics of tonoplast. Membranes are 9-11

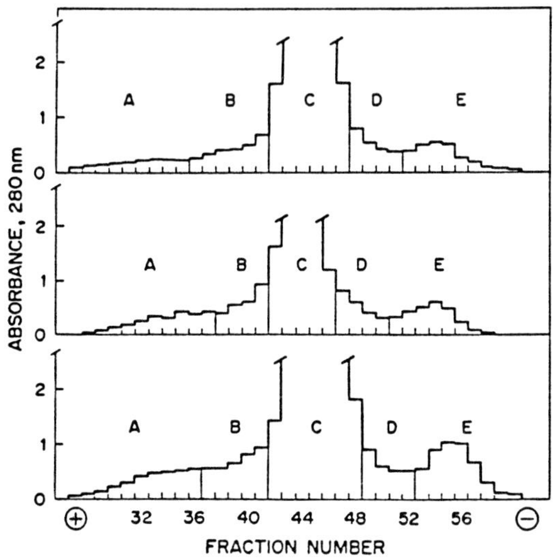

Figure 1. Free-flow electrophoretic separation of a crude microsomal fraction from hypocotyls of dark-grown seedlings of soybean monitored by the absorbance at 280 nm. The letter designations refer to pooled fractions. From Sandelius et al. 1986.

nm in thickness but do not stain with phosphotungstic acid at low pH (Fig. 2). They are characterized by an ATPase inhibited 80% by 50 mM nitrate (data not shown) as is thought to be characteristic for the tonoplast membrane (Walker and Leigh, 1981). The second shoulder, toward the cathode, has membranes 11-12 nm in thickness that react with phosphotungstic acid at low pH as is characteristic of plant plasma membranes (Roland et al. 1972) (Fig. 2). The membranes also bind naphthylphthalamic acid, an inhibitor of the polar transport of growth hormones of the auxin type in plants (Fig. 2) and contain an ATPase inhibited about 50% by Na_2VO_4 but not by nitrate as is characteristic of plasma membranes (Sandelius et al. 1986).

Table I shows the distribution among free-flow electrophoresis fractions A-E of the plasma membrane markers mentioned and of two enzyme activities considered to be localized predominantly in plasma membranes. Of the plasma membrane recovered from the free-flow electrophoresis separation, 30-50% is recovered in Fraction E with ≥90% fraction purity. The remainder of the plasma membrane derived vesicles co-migrate with internal membranes in Fraction C. The basis for the separation is that the two vesicle populations are of different sidedness (H. Canut, private communication).

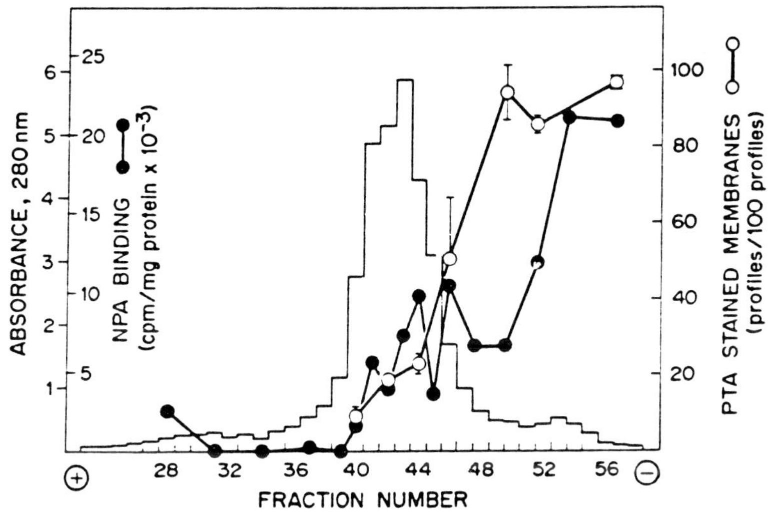

Figure 2. Free-flow electrophoresis separation showing the location of two plasma membrane markers in plants, the binding of naphthylphthalamic acid (NPA), an inhibitor of polar transport of growth hormone (•), and reaction with phosphotungstic acid at low pH (o). Both markers co-localize with the least electronegative fractions toward the cathode and are low or absent from the more electronegative functions to the left containing the tonoplast. From Sandelius et al. (1986).

Endoplasmic reticulum fragments and mitochondria, as well as etioplast fragments, are found in the central main peak of the separation (Fraction C of Figure 1). The Golgi apparatus has not been localized with certainty in these separations, but both fractions B and D contain some enzyme activities (fucosyl transferase, latent nucleoside diphosphatase) characteristic of this cell component in plants.

While there is no obvious advantage of free-flow electrophoresis over aqueous two-phase partitioning (Albertsson, this volume, Kjellbom and Larsson, 1984) for the isolation of plasma membranes, the advantage is the simultaneous preparation of both plasma membrane and tonoplast. Additionally, the method is rapid, easy and amenable to continuous preparative operation.

TABLE 1: Distribution in percent of recovered protein and of plasma membrane markers of free-flow electrophoresis separations of microsomal fractions from soybean hypocotyls. From Sandelius et al. 1986.

Fraction:	A	B	C	D	E
			(%)		
Sterol glyceride synthesis	9	14	21	7	50
Glucan synthetase II	>1	3	39	25	32
NPA binding	2	1	46	9	40
Class I membranes (11-12 nm)	<1	2	57	7	33
Reactivity with phosphotungstic acid at low pH	<1	3	57	8	32
Protein	7	8	71	6	8

REFERENCES

Hannig K, Heidrich H-G (1977). Continuous free-flow electrophoresis and its application in biology. In Bloemendal, H (ed): "Cell Separation Methods, Part IV, Electrophoretic Methods," New York: Elsevier/North Holland Biomedical Press, pp 95-116.

Kjellbom P, Larsson C (1984). Preparation and polypeptide composition of chlorophyll-free plasma membranes from leaves of light-grown spinach and barley. Physiol Plant 62:501-509.

Roland JC, Lembi CA, Morré DJ (1972). Phosphotungstic acid-chromic acid as a selective electron-dense stain for plasma membrane of plant cells. Stain Tech 47:195-200.

Sandelius AS, Penel C, Auderset G, Brightman A, Millard M, Morré DJ (1986). Isolation of highly purified fractions of plasma membrane and tonoplast from the same homogenate of soybean hypocotyls by free-flow electrophoresis. Plant Physiol 81:177-185.

Walker RR, Leigh RA (1981). Characterization of a salt-stimulated ATPase activity associated with vacuoles isolated from storage roots of red beet (_Beta vulgaris_ L.). Planta 153:140-149.

Cell-Free Analysis of Membrane Traffic, pages 285–287

ISOLATION OF PLASMA AND VACUOLE MEMBRANES FROM GREEN LEAVES BY PREPARATIVE FREE-FLOW ELECTROPHORESIS

Guy Auderset, A. S. Sandelius, C. Penel, H. Greppin and D. J. Morré

Laboratory of Plant Physiology, University of Geneva, Geneva, Switzerland, Department of Plant Physiology, Botanical Institute, University of Göteborg, S-41319 Göteborg, Sweden and Department of Medicinal Chemistry, Purdue University, West Lafayette, Indiana 47907 USA

The isolation of plasma membrane fractions free from chlorophyll and/or chloroplast fragments from leaves of green plants has been achieved using aqueous two-phase partitioning (Kjellbom and Larsson, 1984). Using preparative free-flow electrophoresis it is also possible to obtain plasma membranes free of contamination (Auderset et al., 1986) from green leaves. Additionally, using this technique it is possible to prepare vacuole membrane (tonoplast) and certain reference fractions from the same homogenates.

In the procedure followed, plants of spinach grown in a growth chamber under short day conditions of illumination (10 mmol $m^{-1}s^{-1}$ of fluorescent light) 8 h daily were homogenized (1 g leaf to 30 ml medium) using a Waring blendor in 0.3 M sucrose, 10 mM KCl, 1 mM $MgCl_2$, 25 mM (Tris/MES, pH 7.5) containing 1 mM ascorbate 1 mg/mm (w/v) polyvinylpyrolidine and 0.1% (w/v) bovine serum albumin. The homogenates were filtered through a single layer of a porous cellulosic fabric (Miracloth, Chicopee Mills, NY) to remove cell wall and debris and the filtrate was centrifuged 10 min at 6,000 g max. The resulting supernatant was centrifuged for 30 min at 60,000 g max and the final resuspended pellet subjected to free-flow electrophoresis as described by Sandelius et al., (1986).

Measurement of absorbance at 280 nm (Fig. 1) with the electrophoretic separations revealed a broad distribution of membranes with two shoulders. The shoulder toward the anode (Fraction A) consisted of thick membranes unreactive with phosphotungstic acid at low pH. These membranes were subsequently identified as tonoplast. Membranes of vesicles of the fraction toward the cathode (Fraction E) consisted of thick membranes nearly all of which stained with phosphotungstic acid at low pH; a test specific for the plasma membranes of plants (Roland et al., 1972).

Different fractions in the midregion of the separation were enriched in mitochondria, peroxisomes, and endoplasmic reticulum fragments, but all fractions of the midregion were green and contained chloroplast and thylakoid membrane fragments.

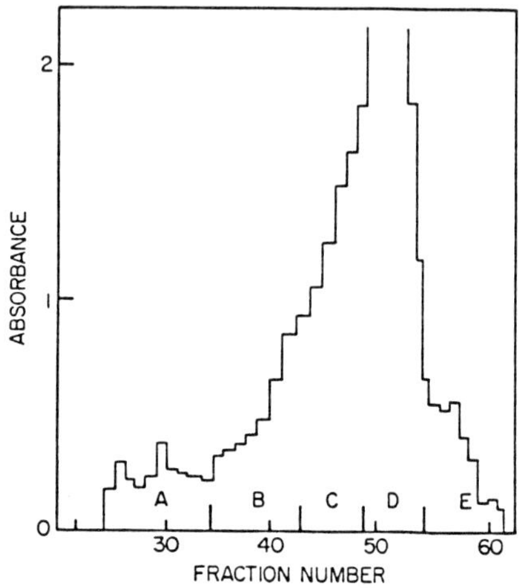

Figure 1. Absorbance at 280 nm of fractions from a free-flow electrophoresis separation of spinach leaves harvested from short-day plants. Letters refer to pooled fractions with A being tonoplast enriched and E being plasma membrane-enriched (From Auderset et al., 1986).

Upon photoinduction of the spinach plants, the electrophoretic separations were broadened (not shown). This broadening was accompanied by a change in the appearance of the plasma membrane vesicles (Fig. 2). Membranes in situ as well as in the isolated vesicles were about 10% thicker in the leaves from photoinduced plants than from plants not photoinduced. The effect was seen primarily on plasma membrane and not on tonoplast and other membranes. It was independent of conditions of fixation and was measured employing several different methods and for species other than spinach.

The findings indicate an alteration of membrane properties upon photoinduction and work is under way to identify the chemical changes responsible for the observed alteration in membrane dimensions. Free-flow electrophoresis provides a convenient procedure for membrane preparation as well as an element of analytical capability sufficient to detect small differences in the membrane properties resulting from developmental change (See also I. Burón, this volume).

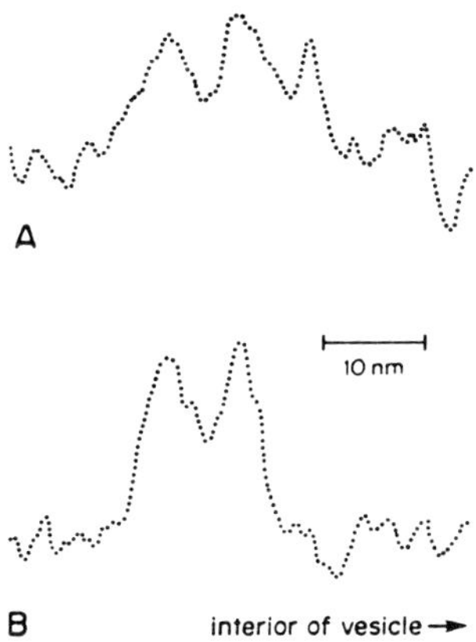

Figure 2. Densitometry traces of plasma membrane from vesicle preparations comparing photoinduced (A) and non-photoinduced (B) spinach leaves. The plasma membrane from photoinduced plants appeared thicker due to a filling in of the shoulder region toward the inner surface of the membrane vesicle (From Auderset et al., 1987).

REFERENCES

Auderset G, Sandelius AS, Penel C, Brightman A, Greppin H, Morré DJ (1986). Isolation of plasma membrane and tonoplast fractions from spinach leaves by preparative free-flow electrophoresis and effect of photo induction. Physiol Plant 68:1-2.

Kjellbom P, Larsson C (1984). Preparation and polypeptide composition of chlorophyll-free plasma membranes from leaves of light-grown spinach and barley. Physiol Plant 62:501-509.

Roland JC, Lembi CA, Morré DJ (1972). Phosphotungstic acid-chromic acid as a selective electron-dense stain for plasma membrane of plant cells. Stain Tech 47:195-200.

Sandelius AS, Penel C, Auderset G, Brightman A, Millard M, Morré DJ (1986). Isolation of highly purified fractions of plasma membrane and tonoplast from the same homogenate of plant cells by free-flow electrophoresis. Plant Physiol 81:177-185.

Cell-Free Analysis of Membrane Traffic, pages 289–290

ASSOCIATION OF A PEROXIDASE TO MEMBRANES OF PHARBITIS COTYLEDONS

Stefanie Kiefer, Hubert Greppin and Claude Penel

Laboratoire de Physiologie végétale, Université de Genève, 3 place de l'Université, 1211 Genève 4, Switzerland

Peroxidase, an enzyme present in plant cell walls and in vacuoles, binds to various membranes upon the addition of calcium or, in some cases, manganese ions (Kiefer et al., 1985). This binding, which only concerns the cathodic isoperoxidase, requires the presence of the carbohydrate moiety of the enzyme and could be necessary for its secretion (Penel et al., 1986). In order to verify the possible involvement of this cation-dependent binding in the routing of the enzyme within plant cells, free-flow electrophoresis (FFE) has been used to obtain pure tonoplast and plasma membranes directly from a crude membrane pellet (Sandelius et al., 1986) for the identification of the membranes able to bind peroxidases and for _in vitro_ studies.

Figure 1a shows the separation profile of membranes from Pharbitis after FFE. As already reported for other plant materials (Sandelius et al., 1986; Auderset et al., 1986) purified plasma membrane identified by naphthylphthalamic acid binding is located farthest from the anode, whereas purified tonoplast identified by ultrastructural criteria (not shown) is present in fractions nearest to the anode. The fractions in between contain the bulk of the other membranes. The membranes present in pooled fractions from a FFE separation were recovered by centrifugation and tested for their capacity to bind the cathodic isoperoxidase in the presence of calcium or manganese ions (Fig. 1b). Expressed as pelleted peroxidase activity per mg membrane protein, the data clearly indicate that tonoplast (fraction A) and plasma membrane (E) have the greatest capacity to bind the enzyme. The bound activity is about 10 times higher than the activity bound to a crude membrane pellet (Kiefer et al., 1985). This observation is consistent with the fact that peroxidases are associated _in situ_ with these 2 cell membranes. It also suggests the existence in plant cells of cation-dependent binding sites for glycoprotein transport.

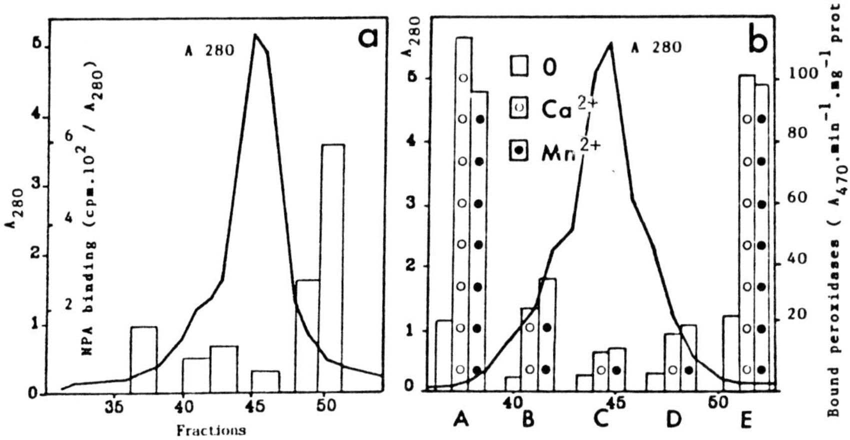

Figure 1. Separation of Pharbitis membranes by FFE (Sandelius et al., 1986) and binding of the cathodic isoperoxidase. a: absorbance at 280 nm and binding of naphthylphthalamic acid (NPA); b: binding of the peroxidase to membranes (Kiefer et al., 1985) from pooled FFE fractions, with or without 10 mM Ca^{2+} or Mn^{2+}.

REFERENCES

Auderset G, Sandelius AS, Penel C, Brightman A., Greppin H, Morré DJ (1986). Isolation of plasma membrane and tonoplast fractions from spinach leaves by preparative free-flow electrophoresis and effect of photoinduction. Physiol Plant 68:1-12.

Kiefer S, Penel C, Greppin H (1985). Ca^{2+}- and Mn^{2+}-mediated binding of the glycoprotein peroxidase to membranes of Pharbitis cotyledons. Plant Sci 39:37-43.

Penel C, Bernardini N, Greppin H (1986). Role du calcium dans le controle de l'activité peroxydasique. Journées Internationales d'Etude du Groupe Polyphénols, Montpellier, in press.

Sandelius AS, Penel C, Auderset G, Brightman A, Millard M, Morré DJ (1986). Isolation of highly purified fractions of plasma membrane and tonoplast from the same homogenate of soybean hypocotyls by free-flow electrophoresis. Plant Physiol 81:177-185.

Cell-Free Analysis of Membrane Traffic, pages 291–295

NBD-PC: A TOOL TO STUDY ENDOCYTOSIS AND PHOSPHOLIPASE ACTIVITY IN PLANT PROTOPLASTS

Lynne A. Dengler, Magaly Rincón, and Wendy F. Boss

Department of Botany, Box 7612
North Carolina State University
Raleigh, NC 27695-7612

INTRODUCTION

There is a controversy as to whether or not endocytosis occurs in plant cells (Morré, 1981). Using electron microscopy, Tanchak et al. (1984) demonstrated rapid (within 10 seconds) uptake of cationized ferritin in protoplasts isolated from soybean cells. Joachim and Robinson (1984) also demonstrated uptake of cationized ferritin by coated vesicles in beans suggesting that the pathway for endocytosis in plants was similar to that for receptor-mediated endocytosis in animal cells. While these EM data suggest endocytosis occurs in plant cells the process has not been documented *in vivo*.

The fluorescent phospholipid, NBD-PC (1-acyl-2(N-4-nitrobenzo-2-oxa-1,3-diazole)-aminocaproyl phosphatidylcholine), has been used to study endocytosis *in vivo* in animal cells (Sleight and Pagano, 1984). In the Chinese hamster V79 lung fibroblasts, the fluorescent phospholipid was associated with the plasma membrane at low temperatures (2°C). As the cells were warmed to 37°C, NBD-PC was internalized in endosomes and became localized in the Golgi apparatus. In these cells, when the NBD-PC was hydrolyzed by the plasma membrane or the lysosomal associated-phospholipases to lysophosphatidylcholine and NBD-caproic acid, NBD-caproic acid did not accumulate within the cells. The NBD-caproic acid was found in the surrounding medium which suggested that it had either diffused or was transported out of the lysosomes and out of the cell.

We have attempted to use NBD-PC to study endocytosis *in vivo* in protoplasts isolated from carrot cells grown in suspension culture. While uptake of the fluorescent phospholipid is rapid, interpretation of the data is complicated by the fact that the NBD-caproic acid formed accumulates in the plant vacuole. In this study we have measured phospholipase A_2 activity and have begun to delineate the uptake of NBD-fluorescence due to phospholipase activity versus a potential endocytic pathway.

MATERIALS AND METHODS

Cell Culture

Fusogenic and nonfusogenic carrot (*Daucus carota* L.) cells were grown in suspension culture as previously described (Boss et al., 1984). Cells were subcultured every 7 days and were used 4 days after transfer. Protoplasts were isolated in 2% Driselase (Plenum Sci. Co.), 0.4 molal sorbitol, 1 mM Mes (pH 4.8) on a rotary shaker (125 rpm) at 25°C for 2 h. Prior to use, the protoplasts were washed twice by centrifugation at 40 g in an osmoticum solution consisting of 0.45 molal sorbitol, and 2 mM Mes (pH 6.0).

Fluorescent Labeling of Protoplasts

C_6-NBD-PC (72 µl of a $CHCl_3$ solution) (Avanti Polar Lipids, Inc., U.S.A.) was dried under N_2 to form a thin film on a glass test tube. A solution of 0.45 molal sorbitol containing 0.68 mg/ml defatted BSA (Sigma Chem. Co.) was added and the mixture was vortexed to obtain a clear, yellow solution. To label the protoplasts, 3 ml of the NBD-PC solution was added to a flask containing 20 ml of osmoticum and approximately 0.2 ml settled volume of protoplasts, and prior to microscopic observation, aliquots of protoplasts were removed and washed twice in the osmoticum. NBD-caproic acid was prepared from NBD-PC in 0.45 molal sorbitol by acid hydrolysis (1 N HCl, pH 1.45) for 1 h. The pH was adjusted to 6.0 with 1 N KOH prior to use. Protoplasts were labeled with NBD-caproic acid in the same manner as with NBD-PC.

Sodium azide and lanthanum chloride were prepared separately in the osmoticum to final concentrations of 30 mM and 1 mM, respectively. The protoplasts were incubated for 20 min with sodium azide or lanthanum chloride prior to the addition of NBD-PC for NBD-caproic acid.

LIPID EXTRACTION

Lipids were extracted according to Folch et al. (1957), dried (in vacuo), and separated on silica gel G (Fisher Sci. Co.) thin-layer plates developed in $CHCl_3$:MeOH:15N NH_4OH (65:35:5).

Fluorescence Microscopy

A Zeiss IM 35 inverted microscope which was equipped with a neofluar 63 x lens for differential interference contrast optics and epifluorescence optics were used. For fluorescence observations, an exciter filter BP450-495, beam splitter FT510, and barrier filter LP520 were used. Photographs were made using Kodak Kodacolor VR ASA 400 film with a Nikon UFX camera.

RESULTS AND DISCUSSION

When NBD-PC was added to isolated fusogenic and nonfusogenic protoplasts, within 10 min there was diffuse yellow-green fluorescence in the cytosol. After 1 to 2 h, a yellow-green fluorescence was detectable in the vacuole of both cells types.

Uptake of the NBD-fluorescence was inhibited by cold temperatures and sodium azide which suggested that uptake of the NBD-PC was energy dependent. Lipid analysis of the protoplasts indicated that whenever NBD-fluorescence was associated with the vacuoles, NBD-caproic acid was present.

When NBD-caproic acid alone was added to the protoplasts it accumulated in the vacuole and this accumulation was inhibited by azide. The question arose as to whether the NBD-caproic acid was formed in an endosome or provacuole and deposited in the vacuole by vesicle fusion, or whether it was formed by phospholipases associated with the plasma membrane and sequestered in the vacuole by diffusion or by a specific transporter.

To determine whether there was phospholipase A_2 activity associated with the plasma membrane, plasma membrane vesicles were isolated by two-phase partitioning (Kjellbom and Larsson, 1984) and lipase activity was assayed according to Moreau (1986) using NBD-PC as a substrate. The isolated plasma membrane from both fusogenic and nonfusogenic protoplasts had active phospholipase A_2 which was inhibited 90% by 1 mM $CaCl_2$ but only approximately 20% by 1 mM $LaCl_3$. Thus, any uptake of NBD-PC fluorescence in the presence of 1 mM $CaCl_2$ should be due to endocytosis.

When 1mM $CaCl_2$ was added, uptake was slower but otherwise very similar to the control. When 1mM $LaCl_3$ was added, there was little to no uptake of fluorescence even though phospholipase A_2 was not dramatically inhibited. In addition, little fluorescence was detectable in the vacuole even after 2 h. These data agree with EM data that endocytosis occurs in protoplasts and suggest that the vacuole may be the site of deposition of the contents of the endocytic vesicle. More experiments are necessary to delineate vacuolar fluorescence due to direct deposition of NBD-PC in the vacuole or diffusion of NBD-caproic acid from an endosome-like compartment into the vacuole. Thus, with plant cells the use of NBD-PC to study endocytosis is confounded by the tendency to sequester compounds in the vacuole. Throughout these studies we found no major differences in plasma membrane phospholipase C in the two cell lines.

REFERENCES

Boss WF, Grimes HD, Brightman AO (1984). Calcium-induced fusion of fusogenic wild carrot protoplasts. Protoplasma 120:209-215.

Folch J, Lees M, Stanley GHS (1957). A simple method for isolation and purification of total lipids from animal tissues. J Biol Chem 226:497-509.

Joachim S, Robinson DG (1984). Endocytosis of cationic ferritin by bean leaf protoplasts. Eur J Cell Biol 34:212-216.

Kjellbom PC, Larsson C (1984). Preparation and polypeptide composition of chlorophyll-free plasma membranes from leaves of light-grown spinach and barley. Physiol Plant 62:501-509.

Moreau, Robert A (1986). Regulation of phospholipase activity in potato leaves by calmodulin and protein phosphorylation-dephosphorylation. Plant Science 47:1-9.

Morré DJ (1981). Exocytosis: Flow routes and kinetics as related to membrane recycling. In Schweiger HG (ed): "InternationalCell Biology 1980-1981," New York: Springer-Verlag, pp 622-632.

Sleight RG, Pagano RE (1984). Transport of a fluorescent phosphatidylcholine analog from the plasma membrane to the Golgi apparatus. J Cell Biol 99:742-751.

Tanchak MA, Griffing LR, Mersey BG, Fowke, LC (1984). Endocytosis of cationized ferritin by coated vesicles of soybean protoplasts. Planta 162:481-486.

Cell-Free Analysis of Membrane Traffic, pages 297-301

EXPERIMENTAL ANALYSIS OF MEMBRANE TRAFFIC DURING SECRETION OF SCALES IN GREEN ALGAE

Michael Melkonian, Dieter Becker, Geoffrey I. McFadden* and Ingo B. Reize

Botanisches Institut, Universität Münster, Schloßgarten 3, D-4400 Münster, Federal Republic of Germany and *Plant Cell Biology Centre, School of Botany, University of Melbourne, Parkville 3052 Victoria, Australia

INTRODUCTION

Many flagellated algae release cell surface-associated structures of discrete size and shape (scales) by exocytosis following assembly of these structures in the Golgi apparatus (GA). Studies on biogenesis and exocytosis of the large cellulosic scales of *Pleurochrysis* showed that these scales are transported through the GA by cisternal progression and are exocytosed still enclosed in a GA-cisterna (Brown 1974). In recent years it has become obvious that the GA-stack of some animal cells consists of at least three functionally and compositionally distinct compartments (cis-, medial-, trans-compartment; e.g. Dunphy and Rothman, 1985). These and other data have led to the now widely accepted hypothesis that secretory proteins are transported through the GA by a shuttling of vesicles between successive GA-cisternae (Rothman and Lenard, 1984) and not by cisternal progression (as proposed by the endomembrane flow hypothesis; Franke et al. 1971). Scale formation in *Pleurochrysis* is then regarded as "a rare formula connected with the unusual geometry and size of the product" (Farquhar and Palade, 1981). We have recently studied scale biogenesis in the green flagellate *Scherffelia dubia* (McFadden and Melkonian, 1986; McFadden et al. 1986; Melkonian et al. 1987). This organism synthesizes numerous small-sized scales (ca. 40 nm diameter) in the GA and should in this respect be more comparable to other secretory systems. Conclusive evidence has been presented that in this scale-secreting system transport of scales through the GA occurs also by cisternal progression and not by shuttle vesicles (summarized in Melkonian et al. 1987).

Here we would like to examine briefly the advantages of the scale-secreting systems Scherffelia and Tetraselmis as model systems for studying exocytosis in plants and for the experimental dissection of membrane traffic.

RESULTS

Scherffelia (freshwater) and Tetraselmis (most species marine) are unicellular algae that can be grown to large cell densities in simple inorganic growth media. In addition the cell cycle can be synchronized by alternating light/dark regimes (Ricketts, 1979; Reize and Melkonian, 1987). In contrast to other scaly algae (e.g. Pleurochrysis) Tetraselmis and Scherffelia synthesize scales only during a limited time period (3 hrs) when cells divide. Scales are of two types: thecal scales (these form the cell wall or theca by extracellular self assembly) and flagellar scales (3 varieties: pentagonal scales, rod-scales, hair-scales). During cell division assembly of thecal and flagellar scales is temporally separated. Scale synthesis can also be induced experimentally by a) flagellar amputation (only flagellar scales are synthesized; McFadden and Melkonian, 1986) or by b) producing protoplasts (only thecal scales are synthesized; Melkonian et al. 1987). Induction of scale synthesis transforms the GA into a scale-synthesizing machinery within 15 min. The precise timing of the induction of scale synthesis allows pulse-chase experiments using radioactive precursors to be performed as well as the application of inhibitors of GA-function. Monensin (100 nM), for example, leads to a dilation of trans-GA cisternae and blocks scale biogenesis and flagellar regeneration in Scherffelia dubia (unpublished observations). Additionally one can trace the signals that are responsible for GA-activation. We know that deflagellation in the absence of external calcium does not lead to GA-activation, so an increase in cytosolic free calcium may be required to activate the GA for scale secretion.

Another advantage of the scale-secreting system Scherffelia/Tetraselmis is that scales are readily visible in the EM and there is no need to use cytochemical and/or immunocytochemical methods to localize the secretory product at the ultrastructural level. It is also easy to quantify the amount of secretory product simply by counting the number of scales present in the cellular compartments and on the cell surface. Since the GA is inactive through most of the cell cycle and scales are always found in regular patterns on the surface of flagella and cell body, it follows that the number of scales present on the cell surfaces represents roughly the number of scales that are synthesized during cell division. When this number is compared with the number of scales present in the intracellular organelles during scale secretion, one can calculate scale production rates for the GA (McFadden and Melkonian,1986; McFadden et al. 1986). Since it has been found that scales are not transported through the GA by shuttle vesicles, but through cisternal progression one can also estimate membrane flow rates through the GA (Melkonian et al. 1987). From these estimations we concluded that a surplus of membrane flow accompanies scale secretion and that a considerable proportion of the transported membrane is retrieved via unknown mechanisms (McFadden and Melkonian,1986; McFadden et al. 1986).

The scale-secreting system Scherffelia/Tetraselmis offers also the possibility to investigate sorting mechanisms of secretory products during secretion. Flagellar scales are of three structurally different types. All types can be found within one GA-cisterna, but they are located in different domains of this cisterna. We found that the scale-type which after exocytosis is attached to the plasmamembrane is exclusively located at the cisternal rims suggesting that the newly synthesized flagellar membrane is also confined to this region of a GA-cisterna (McFadden and Melkonian, 1986). We are now attempting to immuno-localize the prospective flagellar membrane using an antibody against the major flagellar membrane protein (Mr 200000). Interestingly, all three flagellar scale types are abstracted from the trans-side of the GA into one secretory vesicle indicating that in this system sorting of scales does not occur in the GA but later in the secretory pathway. Consequently we have been unable to identify clathrin-coated membranes at the cisternal rims of the trans-GA-region.

The GA in these flagellates consists of 2-4 large Golgi bodies that are closely associated with the nucleus. It should therefore be possible to isolate the Golgi bodies together with the nucleus. If this becomes possible it would pave the way for *in vitro* studies of membrane traffic during scale secretion.

The most perplexing result of the previous investigation is that the transport of secretory products through the GA-stack occurs by cisternal progression. It may be argued that all except the three cis-most cisternae do not belong to the GA, but represent flattened "condensing vacuoles". Since nascent thecal scales can be traced back to the 2nd cis-most cisterna, this interpretation is unlikely. Additionally we note that the secretion-inactive GA consists of even more cisternae (19-21) than the scale-secreting GA. Alternatively it may be possible that the GA of scale-secreting algae is not biochemically compartmentalized into three subcompartments as is the GA of some animal cells. Cytochemical studies to localize marker enzymes for different GA-compartments are underway in this laboratory.

CONCLUSION

Secretion of scales in the green flagellates *Scherffelia* and *Tetraselmis* represents a unique model for studying membrane traffic in lower plants. The system is amenable to experimental manipulation and in vitro studies should be possible. Since secretory products are transported through the GA-stack by cisternal progression this system could represent an alternative type of GA-membrane traffic operative mainly in plant cells.

REFERENCES

Brown RM (1974). The Golgi apparatus and endomembrane system: its role in the biosynthesis, transport, and secretion of cell wall constituents in *Pleurochrysis*. Port Acta Biol 14: 369-384.

Dunphy WG, Rothman JG (1985). Compartmental organization of the golgi stack. Cell 42: 13-21.

Farquhar MG, Palade GE (1981). The Golgi apparatus (complex) - (1954-1981) - from artifact to center stage. J Cell Biol 91: 77s-103s.

Franke WW, Morré DJ, Deumling B, Cheetham RD, Kartenbeck J, Jarasch ED, Zentgraf HW (1971). Synthesis and turnover of membrane protein in rat liver: an examination of the membrane flow hypothesis. Z Naturforsch 26b: 1031-1039.

McFadden GI, Melkonian M (1986). Golgi apparatus activity and membrane flow during scale biogenesis in the green flagellate *Scherffelia dubia* (Prasinophyceae). I. Flagellar

regeneration. Protoplasma 130: 186-198.
McFadden GI, Preisig HR, Melkonian M (1986). Golgi apparatus activity and membrane flow during scale biogenesis in the green flagellate Scherffelia dubia (Prasinophyceae). II. Cell wall secretion and assembly. Protoplasma 131: 174-184.
Melkonian M, McFadden GI, Reize IB, Becker D (1987). Secretion of organic scales in green algae: secretory products are transported through the Golgi apparatus by cisternal progression. Ber Deutsch Bot Ges 100 (in press).
Reize IB, Melkonian M (1987). Flagellar regeneration in the scaly green flagellate Tetraselmis striata (Prasinophyceae). I. Regeneration kinetics and effect of inhibitors. Helgol Meeresunters (in press).
Ricketts TR (1979). The induction of synchronous cell division in Platymonas striata Butcher (Prasinophyceae). Br Phycol J 14: 219-223.
Rothman JE, Lenard J (1984). Membrane traffic in animal cells. Trends Biochem Sci 9: 176-178.

Cell-Free Analysis of Membrane Traffic, pages 303–304

A METHOD TO STUDY THE IN VIVO INTERMEMBRANE TRANSFER OF LIPIDS AND FATTY ACIDS TO THE PLASMA MEMBRANE IN HIGHER PLANTS

Patrick Moreau, Hélène Juguelin, René Lessire and Claude Cassagne

IBCN-CNRS, 1 rue Camille Saint-Saëns, 33077 Bordeaux cedex, France

Compared to proteins, little is known about the intracellular movement of lipids during plasma membrane biogenesis. This is particularly true for higher plants. To investigate intermembrane transfer events of lipids and fatty acids to the plasma membrane _in vivo_, the plant material used must have high and reproducible levels of _in vivo_ biosynthesis and transfer of lipids and fatty acids, and the material must allow the preparation of subcellular membrane fractions. As well, the methodology used must discriminate between actual transfer of lipids and fatty acids and de novo synthesis or turnover.

A suitable plant material is seven-day-old etiolated leek seedlings. Seedlings grown in the dark favour the incorporation of labeled acetate into the lipids (phospholipids, neutral lipids, very long chain fatty acids (VLCFA)) of the endomembranes and the plasma membrane, rather than those (galactolipids) of plastid membranes (Moreau et al., 1984). The procedure for the study of the intermembrane transfer of lipids is summarized in Figure 1.

Pulse chase experiments are required to follow the redistributions of the label between the various lipids and the transfer of labeled lipids (including fatty acids) between membranes. In such experiments, the efficiency of the chase is the crucial point. For this, _in vivo_ incorporations of labeled acetate should be done at pH 6.5 and chased by unlabeled acetate at pH 5.5.

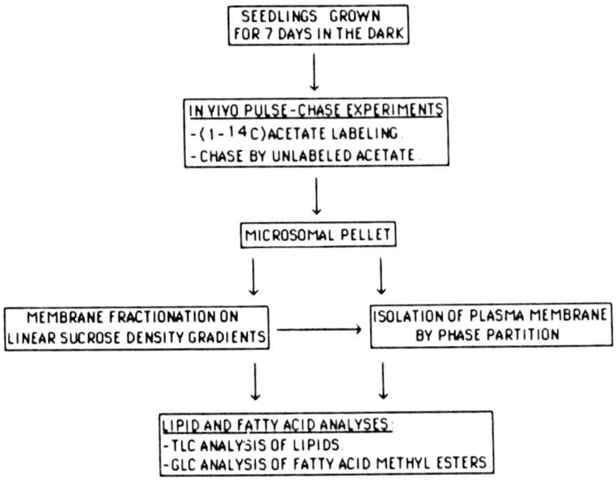

FIGURE 1: EXPERIMENTAL STEPS REQUIRED FOR THE ANALYSIS OF LIPID TRANSPORT TO THE PLASMA MEMBRANE OF ETIOLATED LEEK SEEDLINGS.

Membrane subfractionation on a linear density sucrose gradient, followed by the analysis of the labeled lipids showed intermembrane transfers of lipids and VLCFA between a light membrane fraction (1.08-1.09 g/cm^3) and a heavier one (1.16-1.18 g/cm^3). However, the heavy fraction, which contains the plasma membrane, is heterogeneous as shown by marker enzymes (Moreau et al., 1986). Aqueous phase partitioning applied to this heavy fraction yields a highly purified plasma membrane fraction. Using this modified procedure, we detected the transfer of PC, PE, neutral lipids and VLCFA to the plasma membrane *in vivo*.

This procedure combined with perturbants of intracellular membrane flow, such as ionophores, will allow the elucidation of the path, or paths, involved in the transport of lipids to the plasma membrane of higher plant cells.

REFERENCES

Moreau P, Juguelin H, Lessire R, Cassagne C (1984). In vivo incorporation of acetate into the acyl moieties of polar lipids from etiolated leek seedlings. Phytochem 23: 67-71.

Moreau P, Juguelin H, Lessire R, Cassagne C (1986). Intermembrane transfer of long chain fatty acids synthesized by etiolated leek seedlings. Phytochem 25: 387-391.

Section III
RECONSTITUTION OF MEMBRANE TRAFFIC IN CELL-FREE SYSTEMS

Cell-Free Analysis of Membrane Traffic, pages 307–309

Reconstitution of Membrane Traffic in Cell-Free Systems

Cell-free systems, as they increase in versatility and degree of sophistication, are expected to play an increasingly more important role to help unravel the details of membrane traffic. To introduce the section, Rothman emphasizes the need to reduce complex cell biological processes to forms that are amenable to the tools of biochemistry. Inevitably, this requires development of appropriate cell-free assays. To reconstitute intercompartmental transport under suitably controlled conditions, isolated subcellular compartments prepared from homogenates provide the starting material. To date, all such experiments have taken the form of 'complementation' assays where a 'donor' fraction provides the transferred protein (e.g. VSV-encoded G-glycoprotein) that interacts with an 'acceptor' fraction containing a protein or enzyme that modifies the transferred substrate in an acceptor or destination compartment. This strategy was first developed by Rothman and his colleagues to study transport of material from one Golgi apparatus compartment to another . The transport of VSV-encoded G protein from the cis to the medial compartment of the Golgi stacks was determined using donor Golgi stacks from CHO cells which lacked UDP-GlcNAc transferase I. Acceptor stacks were prepared from wild type CHO cells which possessed the enzyme. Transfer of proteins from donor to acceptor was assayed by the incorporation of GlcNAc. Transport between Golgi compartments required absolutely the presence of a cytosol fraction, ATP and proteins on the surface of the Golgi membranes. In discussing the function of cytosolic factors, Rothman points out that cytosol from yeast and from plants can substitute for animal cell cytosol and suggests that the ability to use yeast materials opens up the prospect of using genetic methods to speed the isolation and identification of cytosolic components for which assays are in hand.

Gruenberg and Howell describe the need to establish a cell-free system that reconstitutes the stages of the endocytic pathways in the cell. By implanting the G-glycoprotein of vesicular stomatitis virus into the plasma membrane of BHK-cells in its membrane-spanning conformation, these authors have immuno-isolated plasma membranes and endosomes as time-defined fractions using antibodies to the exoplasmic and cytoplasmic domain of the G-protein. Results to date indicate that internalized membrane proteins are present in an early sub-compartment from where recycling to the plasma membrane or routing to lysosomes for degradation occurs. Biochemical analysis may now be combined with assays of cell-free fusion to characterize the molecular details of the fusion process.

Balch et al. report on the transport of G-glycoprotein from the rough endoplasmic reticulum to the cis compartment of the Golgi apparatus. In vitro, the transport of G-glycoproteins from the endoplasmic reticulum to Golgi results in the processing of the Man_{8-9} $GlcNAc_2$ residues of the glycoprotein to the Man_5 $GlcNAc_2$ form, and ongoing efforts are to develop an in vitro system that faithfully reproduces these events. A protocol is described for preparing large quantities of extracts from mitotic cells. During cell homogenization, the endoplasmic reticulum normally is sheared into small resealed vesicles. These vesicles did not support trimming of G protein. However, the problem was overcome by homogenizing cells undergoing mitosis, where both the nuclear envelope and the Golgi complex are disassembled. In contrast to interphase ER microsomes, the physiologically disassembled organelles of mitotic cells were found to be functionally stable to the homogenization conditions and active in transport of G protein from the ER to the Golgi.

The section concludes with a description of cell-free systems by Edwardson and Woodman and by Hurtley and Warren for the insertion of a newly synthesized glycoprotein into the plasma membrane and the endocytic fusion event, respectively. Both make use of a 'complementation' assay in which fusion is monitored by the release of free sialic acid, as a result of the action of neuraminidase brought into contact with an appropriate substrate by one of the compartments upon fusion. Edwardson and Woodman use a donor vesicle prepared from BHK-21 cells infected with influenza virus. These vesicles contain a viral neuraminidase on the inner surface of the vesicle membrane. The acceptor vesicles were prepared from cells that had Semliki Forest virus (SFV) envelope glycoproteins containing [^{3}H] N-acetyl-neuraminic acid bound to their membranes at 4°C. Since no endocytosis occurs at 4°C, homogenates of these cells will carry tritium only in their plasma membranes. Hurtley and Warren similarly bound SFV carrying tritiated sialic acid residues to BHK cells. They then allowed internalization to proceed at 20°C for 2 hours and prepared a labeled endosome fraction from the cells. In this assay, the donor membranes were prepared from MDCK cells to which fowl plague virus, which possesses a neuraminidase as one of 5 envelope proteins, had been bound at 4°C and then internalized at 37°C for 20 minutes. Both these assays allow for a detailed study of the properties of vesicle fusion events in endocytosis and the factors required.

Cell-Free Analysis of Membrane Traffic, pages 311–316

RECONSTRUCTION OF ORGANELLE FUNCTION IN CELL-FREE SYSTEMS: INTRODUCTION

James E. Rothman

Department of Biochemistry, Stanford University, Stanford, California 94305

A great deal has been learned in recent years about the overall organization of membrane traffic, and about the flow of proteins and of information between organelles. Essentially, we have a road map that tells us which proteins follow which routes of vesicular transport between which membrane-bound compartments. Of course, the resolution of this map needs to, and will, improve, but it is already sufficient for use to begin to focus attention on mechanistic as well as geographical issues. Virtually nothing is known of how transport vesicles bud, how they select (or possibly reject) content, or of how the selective fusion of transport vesicles is programmed and catalyzed. An understanding of membrane traffic consists of the resolution of these issues for all transport steps that occur.

The mechanisms underlying vesicular transport must ultimately consist of a series of molecular interactions and the ensuing chemical events. To elucidate these mechanisms requires that this complex process of cell biology be reduced to a form that is amenable to the tools of biochemistry. That is, each of the components needed for vesicle production and metabolism will have to be isolated and its role established in relation to the others. The essential step to permit this is the reconstitution of membrane traffic in cell-free systems. This provides functional assays with which to discover and study the mechanism of action of needed components. Genetic and other methods can greatly facilitate this process, but it is only the resolution and reconstitution approach that will someday enable us to discuss membrane traffic in the same terms in which we can now describe other aspects of macromolecular metabolism, such as protein, nucleic acid, and polysaccharide synthesis.

The ensuing papers in this volume will offer evidence for the reconstitution of transport from ER to Golgi (Balch and colleagues), transport to plasma membrane (Edwardson and Woodman), steps in receptor-mediated endocytosis (Howell and colleagues), and in assembly and disassembly of the nuclear envelope (Burke and colleagues). The major challenge is how to exploit these and related kinds of assays to isolate the components. A combination of these biochemical with genetic approaches holds the promise of revealing the responsible machinery.

To date, the designs of reconstitutions of intercompartmental transport involve "complementation" assays that employ a "donor" fraction that provides, as a substrate in the starting compartment, the transferred protein or enzyme, and an "acceptor" fraction that contains a protein or enzyme that modifies the transferred substrate at the destination compartment. This strategy was first developed in the reconstitution of vesicular transport between successive cisternae of the Golgi stack. An analysis of this system has offered some initial insights into the mechanisms of vesicle budding, targeting, and fusion.

The transport of the VSV-encoded G protein, from the cis compartment of "donor" CHO Golgi stacks that lack UDP-GlcNAc Transferase I to the medial compartment of "acceptor" wild-type CHO Golgi stacks, could be detected by monitoring the incorporation of GlcNAc as a consequence of transport in a "complementation" assay (Fries and Rothman, 1980; Balch et al., 1984). The unique localization of UDP-GlcNAc Transferase I to the medial Golgi (Dunphy et al., 1985) permitted biochemical detection of the arrival of G protein into this compartment, and a variety of controls ruled out the possibilities of nonspecific fusion of entire Golgi stacks, or delivery of the transferase to mutant Golgi, rather than G protein to wild-type medial Golgi (Balch et al., 1984; Braell et al., 1984). Transport absolutely required the presence of a crude cytosol fraction, ATP, and proteins on the surface of the Golgi membranes (Fries and Rothman, 1980; Balch et al., 1984; Balch and Rothman, 1985).

Electron microscopy revealed that when the isolated Golgi stacks were incubated with cytosol and ATP to initiate transport, coated vesicles of a non-clathrin variety form from their rims (Orci et al., 1986; Balch et al., 1984). These vesicles contain G protein (Orci et al., 1986) and are thought to be responsible for the transport between the cisternae in the stacks of this cell-free system, resulting in the maturation of glycoprotein as measured in the biochemical assay.

Intercompartmental transport of G protein occurs via a series of sequential intermediates that most likely reflect stages in the budding and fusion of these transport vesicles (Balch et al., 1984; Wattenberg et al., 1986). ATP is required at every stage, and cytosol at all but the last step, which seems to involve the actual fusion of the transport vesicle with the target membrane. The process of targeting a vesicle is probably quite complex, since a series of prefusion complexes is involved.

The budding of transport vesicles requires both ATP and a mixture of proteins from the cytosol. This establishes that vesicle budding is an active process that does not occur in the fashion of a self-assembly. Therefore, even though a coat protein like clathrin can self-assemble to yield empty coats, the budding off of a transport vesicle is apparently a much more complex process. Clathrin represents only the outermost shell of certain vesicle coats (Pearse, 1982; Pfeffer et al., 1983; Zaremba and Keen, 1983). Therefore, several other proteins are likely to act together with clathrin and clathrin homologues to contribute a substantial portion of the driving force for budding, determine the size of the vesicle, and in the case of signal-mediated transport, possibly provide a scaffold for content selection. Once a coated vesicle is formed, the coat must be removed to permit the contact between bilayers needed for fusion with the target membrane. Uncoating is an energy-dependent process (Patzer, et al., 1982). A cytosolic uncoating enzyme that couples ATP hydrolysis to the removal of the clathrin type of coat has been identified and studied extensively (see Rothman and Schmid, 1986, for review). The uncoating enzyme for non-clathrin coats in the Golgi has yet to be found.

A cytosolic factor that utilizes fatty acyl-Coenzyme A as if a cofactor has recently been identified, raising the possibility that a cycle of acylation and deacylation may promote rounds of transport. The sensitivity of donor and acceptor membranes to N-ethylmaleimide (NEM) treatment (Balch and Rothman, 1985) led to the identification of this NEM-Sensitive Factor (NSF) that is essential for intra-Golgi transport (Glick and Rothman, 1987). NSF is present in both cytosol and in Golgi membranes. NSF activity, as measured by the ability of a given fraction to restore activity to NEM-inactivated donor and acceptor Golgi fractions, is abolished by protease, and is about 400,000 daltons native MW as judged by filtration. NSF is present on the surfaces of CHO Golgi membranes and can be extracted from the membranes by high salt concentrations and ATP. The ability of fatty acyl Coenzyme A to greatly stimulate NSF activity has led to the hypothesis that acyl CoA acts as an acyl chain donor to activate a recipient molecule for transport. Deacylation might recycle this component for another round of transport. The G-protein to be transported is not the acyl acceptor in this cycle, since fatty acyl CoA stimulation of an NSF-dependent transport reaction could be demonstrated using a type of G protein that cannot be acylated.

What might be the function of cytosolic factors that facilitate intra-cellular transport reactions? In the case of intra-Golgi transport, the reactions that must be reconstituted include collection of transported proteins into a vesicle bud followed by budding, targeting, and fusion of the transport vesicle with the appropriate target membrane. One can envision a series of proteins that act at each of these distinct steps. Remarkably, the cytosol from yeast (Dunphy, et al., 1986) and from plants (Paquet et al., 1986) will substitute for animal cell cytosol in promoting transport in the Golgi stack and in forming the coated vesicles. This interchangeability means that the transport machinery is extremely similar, even in detail, in all eukaryotes. The ability of yeast cytosol to replace mammalian cytosol opens the door to genetics, and with this offers the prospect that genetic methods can now be employed to speed the isolation of components for which assays are at hand.

REFERENCES

Balch WE, Dunphy WG, Braell WA, Rothman JE (1984). Reconstitution of the transport of protein between successive compartments of the Golgi measured by the coupled incorporation of N-acetylglucosamine. Cell 39:405-416.

Balch WE, Glick BS, Rothman JE (1984). Sequential intermediates in the pathway of intercompartmental transport in a cell-free system. Cell 39:525-536.

Balch WE, Rothman JE (1985). Characterization of protein transport between successive compartments of the Golgi apparatus: Asymmetric properties of donor and acceptor activities in a cell-free system. Arch Biochem Biophys 240:413-425.

Braell WA, Balch WE, Dobbertin DC, Rothman JE (1984). The glycoprotein that is transported between successive compartments of the Golgi in a cell-free system resides in stacks of cisternae. Cell 39:511-524.

Dunphy WG, Brands R, Rothman JE (1985). Attachment of terminal N-acetylglucosamine to asparagine-linked oligosaccharides occurs in central cisternae of the Golgi stack. Cell 40:463-472.

Dunphy WG, Pfeffer SR, Clary DO, Wattenberg BW, Glick BS, Rothman JE (1986). Yeast and mammals utilize similar cytosolic components to drive protein transport through the Golgi complex. Proc Natl Acad Sci USA 83:1622-1626.

Fries E, Rothman JE (1980). Transport of vesicular stomatitis virus glycoprotein in a cell-free extract. Proc Natl Acad Sci USA 77:3870-3874.

Orci L, Glick BS, Rothman JE (1986). A new type of coated vesicular carrier that appears not to contain clathrin: Its possible role in protein transport within the Golgi stack. Cell 46:171-184.

Paquet MR, Pfeffer SR, Burczak JD, Glick BS, Rothman JE (1986). Components responsible for transport between successive Golgi cisternae are highly concerned in evolution. J Biol Chem, 261:4367-4370.

Patzer EJ, Schlossman DJ, Rothman JE (1982). Release of clathrin from coated vesicles dependent upon a nucleoside triphosphate and a cytosol fraction. J Cell Biol 93:230-236.

Pearse BMF (1982). Coated vesicles from human placenta carry ferritin, transferrin, and immunoglobulin G. Proc Natl Acad Sci USA 79:451-455.

Pfeffer SR, Drubin DG, Kelly RB (1983). Identification of three coated vesicle components as α- and β- tubulin

linked to a phosphorylated 50,000-dalton polypeptide. J Cell Biol 97:40-47.
Rothman JE, Schmid SL (1986). Enzymatic recycling of clathrin from coated vesicles. Cell 46:5-9.
Wattenberg BW, Balch WE, Rothman JE (1986). A novel prefusion complex formed during protein transport between Golgi cisternae in a cell-free system. J Biol Chem 261:2202-2207.
Zaremba S, Keen JH (1983). Assembly polypeptides from coated vesicles mediate reassembly of unique clathrin coats. J Cell Biol 97:1339-1347.

Cell-Free Analysis of Membrane Traffic, pages 317–331

FUSION IN THE ENDOCYTIC PATHWAY RECONSTITUTED IN A CELL-FREE SYSTEM USING IMMUNO-ISOLATED FRACTIONS

Jean Gruenberg and Kathryn E. Howell

European Molecular Biology Laboratory
Postfach 10.2209, 6900 Heidelberg FRG

INTRODUCTION

Endosomes consist of morphologically distinct elements where internalized molecules, e.g. receptors and their ligands, are sorted (reviews: Steinman et al, 1983; Goldstein et al,1985; Wileman et al, 1985). Specific properties of the different endosomal sub-compartments have not been characterized. It is also unclear which endosomal sub-compartments are connected by vesicular traffic or possibly interact with elements of the exocytic pathway (Helenius et al, 1983; Farquhar, 1985). To unravel the functional organization of the endosomal sub-compartments new methodological approaches are required. These include both a means to fractionate different sub-compartments of the pathway and biochemical assays to study endosomal functions in vitro.

We have established a cell-free system that reconstitutes the stages of the endocytic pathway connected by vesicular traffic in the cell (Gruenberg and Howell, 1986 & 1987). For these studies a foreign antigen, the trans-membrane glycoprotein G of vesicular stomatitis virus (VSV), is implanted into the plasma membrane of BHK cells and subsequently endocytosed in a single and synchronous wave. The compartments of the pathway through which G is transported from the plasma membrane to the lysosomes are then immuno-isolated with specific antibodies (time-defined fractions). This time-controlled distribution of the antigen in the compartments of the endocytic pathway is possible because the implanted G-protein is restricted to the plasma membrane at the beginning of the experiment. The procedure we have used is summarized below (Fig 1).

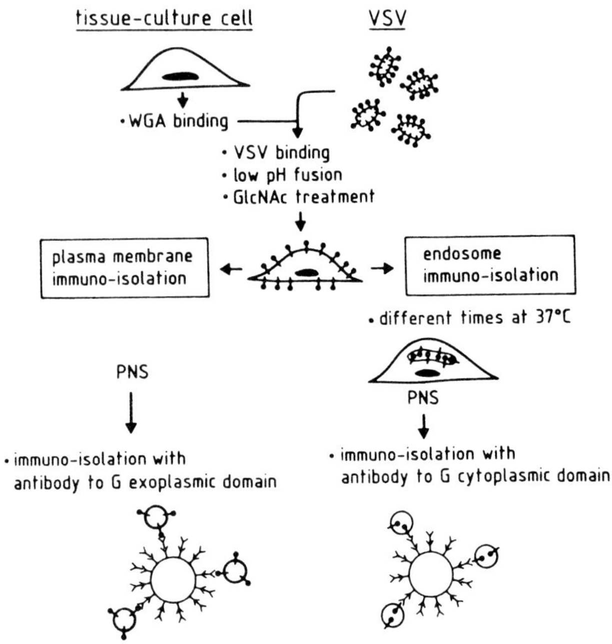

Figure 1: Outline of G implantation and immuno-isolation.

IMMUNO-ISOLATION OF TIME-DEFINED ENDOSOMAL FRACTIONS

The VSV is first bound to the cell surface via wheat germ agglutinin to render binding of the virions independent of potential cell surface receptors. The envelope of the bound virions is then fused at low pH with the plasma membrane and the unfused viral particles are removed with the lectin competing sugar, N-acetylglycosamine (White et al, 1980; Gruenberg and Howell, 1985). With this protocol the G-protein is implanted into the plasma membrane in its membrane-spanning conformation. Mis-implantation of the G, resulting in hairpin insertion, would expose the cytoplasmic domain on the cell surface. The cytoplasmic domain of the G-protein was not detected on the cell surface over a wide range of G densities using a time-resolved immuno-florometric assay (Fig 2). The cells at 4°C are reacted with a monoclonal antibody either against the endo- or the ecto-domain of the G-protein and then with a second antibody labelled with Eu, an element of the lanthanide series. The delayed fluorescence emitted by Eu is now used for clinical immunoassays because of an increased sensitivity and a dynamic range wider than any other labelling technique (Soini & Hemmilä, 1979; Soini & Kojola, 1983; Hemmilä et al, 1984). We have extended this technique to the in vivo quantitation of cell surface antigens.

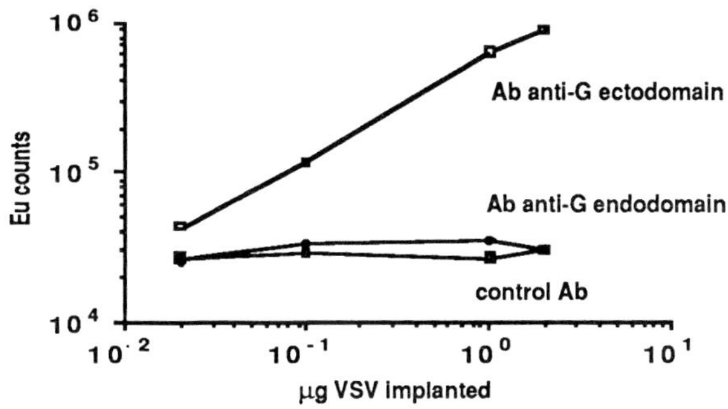

Figure 2. Orientation of the implanted G-protein. Increasing amount of G-protein was implanted in the plasma membrane of BHK cells, corresponding to 0.02 to 2 µg VSV fused per 2 X 10^6 cells. The orientation of the implanted G was monitored with two specific monoclonal antibodies respectively against the ecto- (17-2-21-4; Simons, Europ. Mol. Biol. Lab., Heidelberg, FRG) and the endo-domain (P5D4; Kreis, 1986) as well as a control mouse IgG. The monoclonal antibodies were bound for 1 hr at 4°C. After washing the cells, the cells with bound antibody were incubated for 1 hr at 4°C with a second antibody against the Fc domain of mouse IgG labelled with Eu (I. Hemmilä, Wallac/LKB, Turku, Finland). The cells were washed and then treated directly with Enhancement solution (Wallac/LKB) to release the bound Eu atoms. The Eu was counted with a DELFIA flurometer (Wallac/LKB). The Eu counts correspond to the actual measurements of the delayed fluorescence on a log scale.

Endocytosis of the implanted G-protein is initiated by allowing membrane traffic to resume at 37°C and stopped after defined times of internalization from 5 to 30 min by returning the cells to ice-temperature. The kinetics of G transport were determined by assaying the amount of G present in the 3 compartments of the endocytic pathway: plasma membrane, endosomes and lysosomes (Gruenberg and Howell, 1987). The implanted G-protein is quickly internalized into an early endosome (60% in 5min, Fig 3). From this compartment the G is transported along the 2 major routes followed by membrane proteins during endocytosis: approximately 50% is recycled back to the plasma membrane and 50% is routed to the lysosomes via later endosomes (TABLE I).

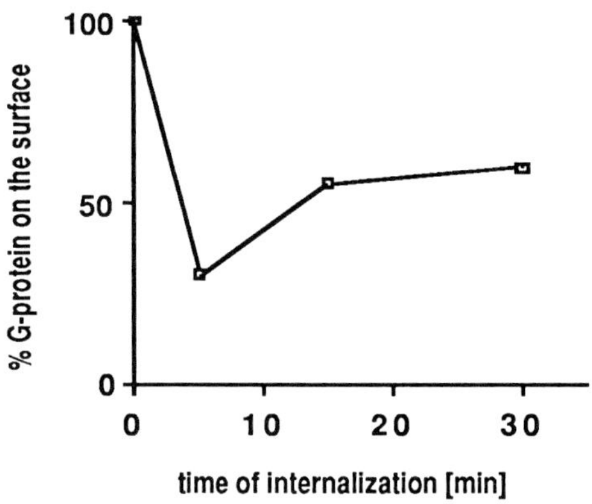

Figure 3. G-protein endocytosis. Implantation of the G-protein was carried out using 4 µg VSV for 1.5 X 10^6 cells, resulting in 0.8 µg VSV fused and a density of 70 G/μm^2 membrane surface area. Following implantation, the G-protein was internalized for the indicated time. Detection of the G-protein on the cell surface at each time-point was by time-resolved immuno-fluorimetry using the monoclonal antibody against G ecto-domain, as described in the legend of Figure 2.

TABLE I Endocytosis of G-protein and HRP:

	0 min	5 min	15 min	30 min
G-protein*				
Cell surface:	100 (100)	40 (40)	50 (20)	60 (2)
Endosomes:	0 (0)	60 (40)	50 (75)	30 (70)
Lysosomes:	0 (0)	0 (0)	0 (0)	12 (30)
HRP **				
Intact cells:	0	20	50	100
Fractions:	0	20	50	100

* The amount of G-protein in each compartment is expressed as the percent of the amount originally present on the cell surface (see Fig 3). The numbers in parenthesis correspond to values obtained in experiments carried out after cross-linking the G molecules with a polyclonal antibody at the cell surface prior to internalization.
-The amount of G present at the cell surface is taken from the data in Fig 1.
-The amount of G-protein present in endosomes is obtained by immuno-isolation of time-defined endosomal fractions. The VSV is metabolically labelled and the immuno-isolated ^{35}S-labelled G is quantitated as described (Gruenberg and Howell, 1987).
-The amount of G-protein degraded in lysosomes is determined as described (Gruenberg and Howell, 1987) using metabolically labelled VSV. Complete degradation of the endocytosed G-protein occured after 60 min incubation.
** The amount of HRP internalized in the cells and immuno-isolated in the temporal fractions was determined as described (Gruenberg and Howell, 1986) and is expressed in percent of the total amount endocytosed after 30 min. We estimated the volume of the endosomal compartment to 0.6 % of the cell volume after 15 min HRP internalization at 37°C. Under the conditions of these experiments 70% of the HRP present in the post-nuclear supernatant is immuno-isolated (yield = 32% of the total cell homogenate).

Sorting of the G-protein to both recycling and degradation may be a consequence of interactions between the luminal portions of the G molecules in the acidic environment of endosomes. This lateral association may direct to degradation G molecules which would otherwise recycle. This is supported by our other observations that the percentage of G molecules transported to the lysosomes is increased as the density of G in the endosomal membrane increases (Gruenberg and Howell, 1987). Lateral association of the G molecules can be mimicked by cross-linking the G with a polyclonal antibody at the cell surface prior to the internalization step. This abolishes recycling and directs all of the G molecules to degradation (Table 1), confirming the observation of Mellman and Plutner (1984) and Ukkonen et al (1986) with the Fc receptor in macrophages. In the exocytic pathway, the role of oligomerization as a signal for intracellular transport has been documented with the G-protein of VSV (Kreis and Lodish, 1986) as well as the spike protein of influenza virus (Copeland et al, 1986; Doyle et al, 1986; Schuy et al, 1986).

The cytoplasmic domain of the G-protein is exposed on the surface of endosomal vesicles and provides the antigen for the immuno-isolation of the time-defined fractions. We are using new magnetic solid supports described by Howell et al (1987 and this book) which facilitate the isolation, since the magnetic properties of the beads are used to retrieve the immuno-isolated fraction and to remove the unbound material. The immuno-adsorbent (ImAd) consists of the solid support with covalently attached linker antibody raised in sheep against the Fc domain of mouse IgG and bound to it the specific monoclonal antibody against G cytoplasmic domain. For the immuno-isolation of the plasma membrane, cells with implanted G are used and the ImAd is prepared with the monoclonal antibody against G exoplasmic domain. The isolation is carried out by mixing the ImAd for 2 hr at 4°C with a post-nuclear supernatant prepared from cells with either implanted or internalized G.

The internalized G-protein is quantitatively immuno-isolated in each time-defined fraction (Table I). When horseradish peroxidase (HRP) is co-internalized with the G-protein as an independent marker of the endosomal fluid phase, the amount of HRP in the temporal fractions increases with the time of incubation as in the intact cells (Table I). As shown in Fig 4, this increase reflects the distribution of HRP in morphologically distinct endosomal sub-compartments. After 5 min internalization, vesicular and tubular elements at the cell periphery are labelled. Multivesicular endosomes which are positive after 15 min internalization become heavily labelled after 30 min.

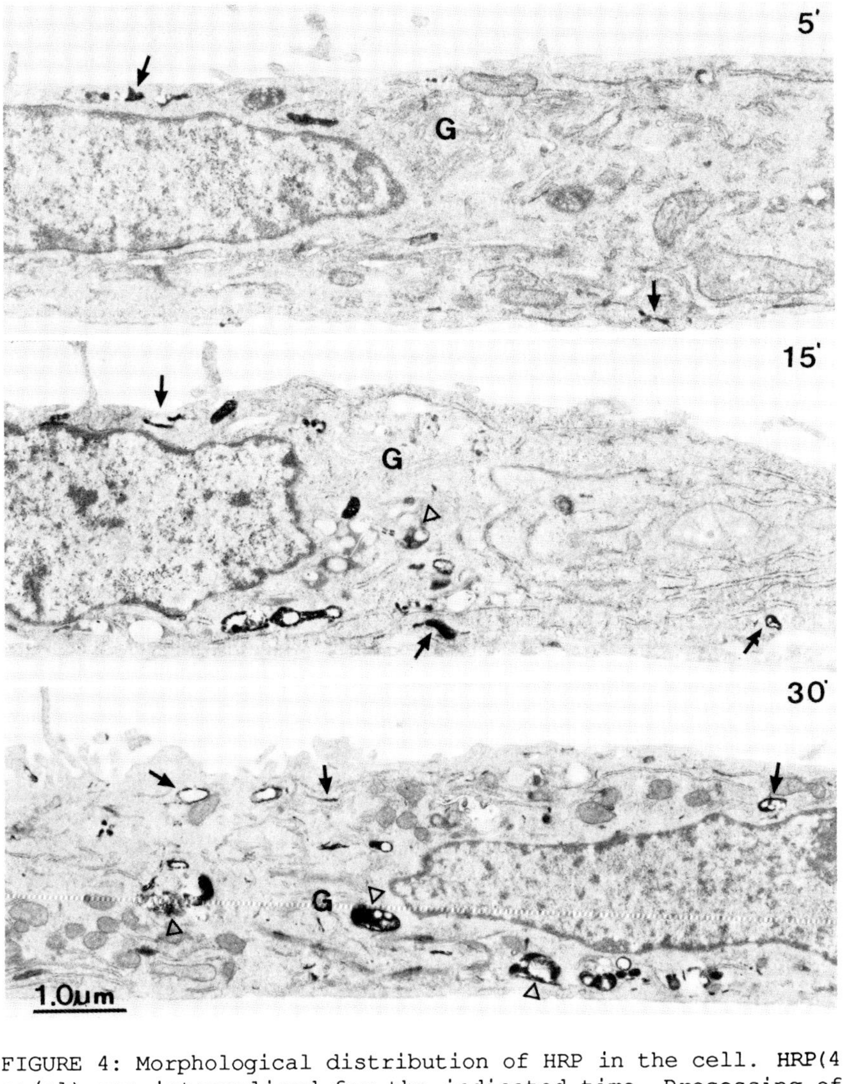

FIGURE 4: Morphological distribution of HRP in the cell. HRP(4 mg/ml) was internalized for the indicated time. Processing of the cells for electron microscopy was as described (Gruenberg and Howell, 1986). Arrows point at tubular and vesicular elements already labelled after 5 min and arrowheads at multivesicular endosomes labelled after 15 and 30 min internalization. G indicates the position of the Golgi complex. The bar is 1.0μm

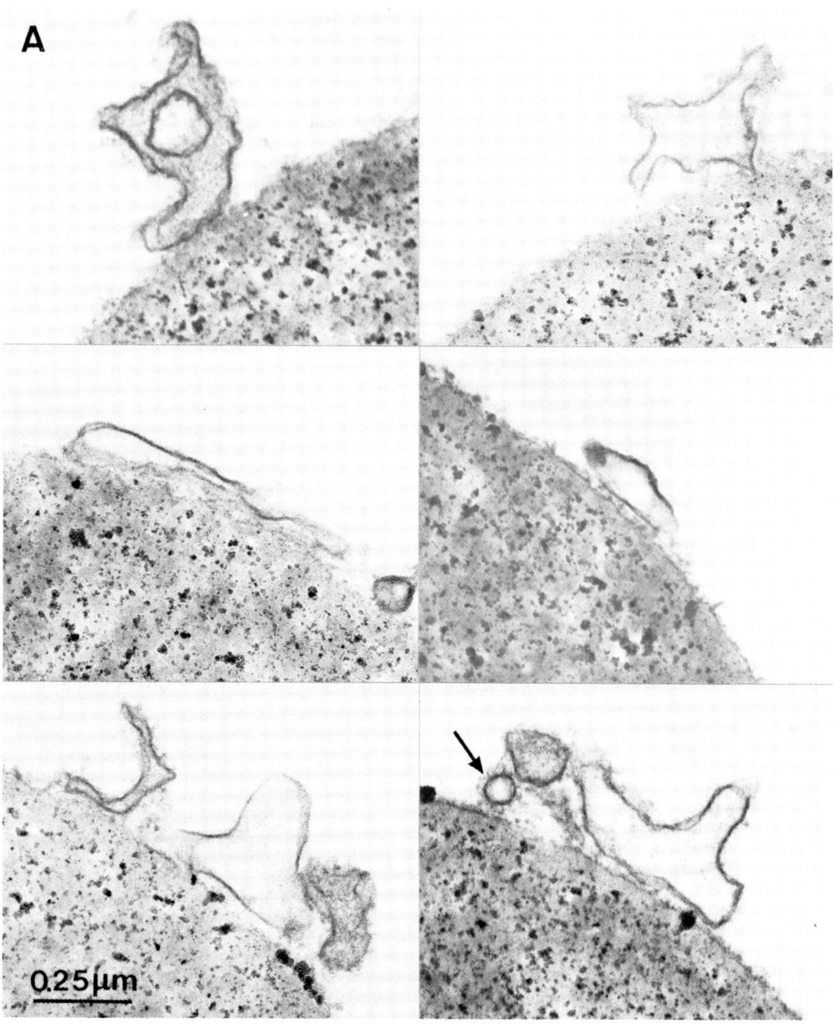

FIGURE 5: Morphology of immuno-isolated endosomal fractions. Time-defined endosomal fractions immuno-isolated after 5 (A) and 15 min (B) G endocytosis, were processed for electron microscopy as described (Gruenberg and Howell, 1986). In Fig. 5A, the arrow points to an immuno-isolated coated vesivcle. The magnetite present in the solid support is easily seen as dark precipitates. The bar is 0.25µm.

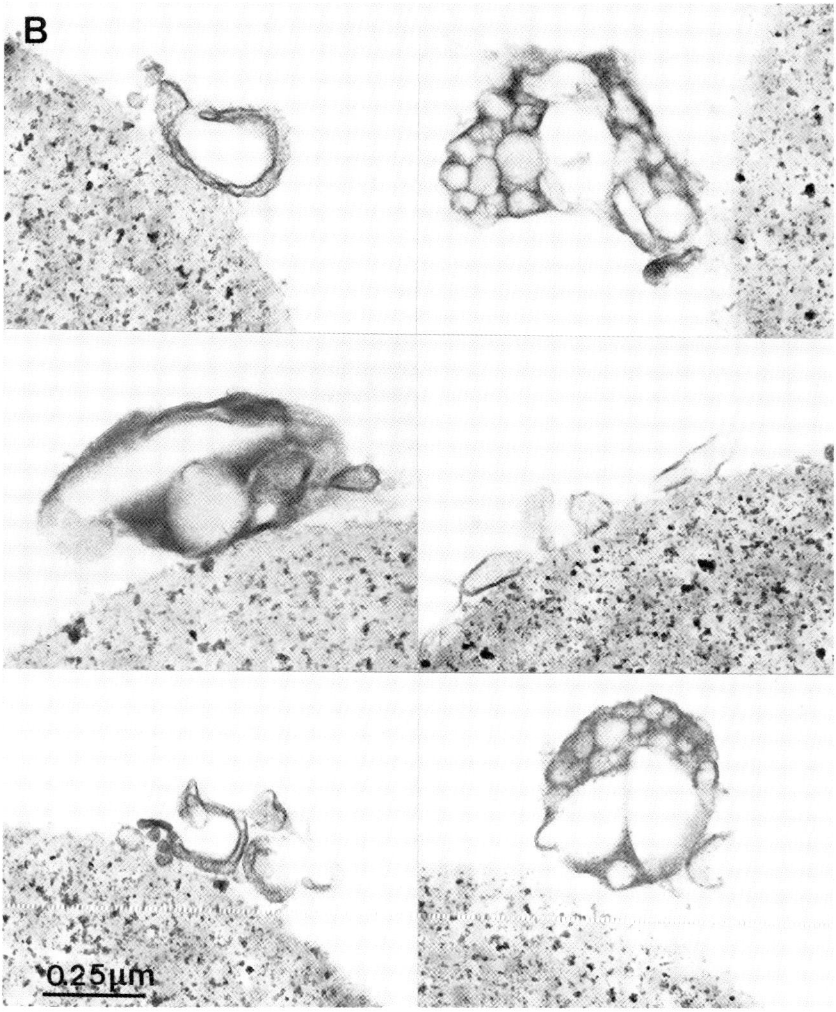

Figure 5B

Morphological analysis of the fraction immuno-isolated 5 min after G internalization show lucent vesicles and cisternae ranging from 0.1 to 0.5 µm in diameter (Fig 5). After 15 min internalization dense multi-vesicular endosomes (diameter: 0.3-0.6 µm) are also immuno-isolated. The immuno-isolated structures correspond to the distinct elements visualized in the cells both after 5 and 15 min of internalization.

We chose to monitor the maintenance of an acidic luminal pH in the time-defined fractions to determine if functional integrity is retained. We used the weak base ^{14}C-labelled methylamine (Fig 6) as a probe for the vesicle pH (Russel and Holz, 1981; Forgac et al, 1983; Schneider, 1983). The methylamine is accumulated in the fractions and is sensitive to the protonophore FCCP. The experimental conditions do not permit an accurate calculation of the endosomal pH. However, an estimate ranging from pH 5.2 to 6.0 can be made using internalized HRP to determine the volume of the immuno-isolated fractions. This range agrees well with in vivo determinations (Maxfield, 1982). These observations extend our earlier finding using acridine orange (Gruenberg and Howell, 1986) and demonstrate that the immuno-isolated vesicles are sealed and have a low luminal pH, retaining the major function reported for the endosomal compartment (review: Mellman et al, 1986).

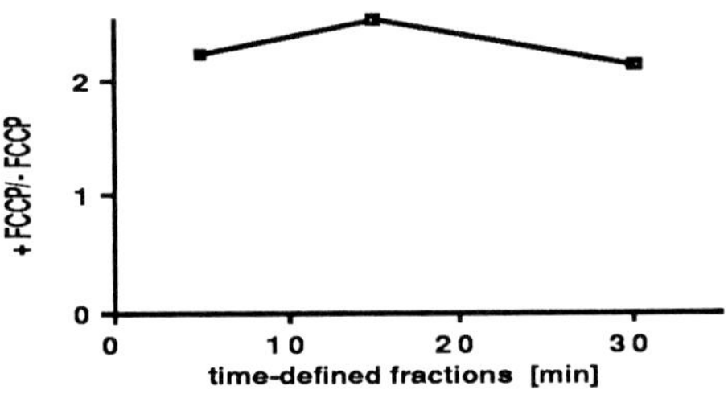

Figure 6. Methylamine accumulation by time-defined endosomal fractions. The immuno-isolated fractions were incubated at room temperature for 30 min in the presence of 2.5 mM ATP and 0.083 mM ^{14}C-methylamine (58 mCi/mmol) with or without 10 µg/ml FCCP. The immuno-isolated fractions were then washed quickly and the label associated with the fraction was counted. The FCCP-sensitive accumulation of ^{14}C-methylamine is plotted as the ratio of the label accumulated in the absence and in the presence of FCCP. For each fraction the amount of immuno isolated HRP was used to normalize the methylamine taken up +/- FCCP.

CELL-FREE RECONSTITUTION OF ENDOCYTIC VESICLE FUSION

The cell-free system for the reconstitution of vesicle fusions occurring during endocytosis is described in (Gruenberg and Howell, 1986 & 1987) and outlined below. The reconstitution is efficient because the time-defined endosomal fractions are immobilized on the solid support and can be easily introduced into and retrieved from the different reaction mixtures. The reaction is carried at 37°C in the presence of ATP by mixing the time-defined endosomal fraction immuno-isolated via the G-protein (acceptor) with a post-nuclear supernatant prepared from other cells that have internalized lactoperoxidase (LPO) for 30 min at 37°C (donor). The immuno-isolated fraction is then retrieved to remove the excess donor and an iodination reaction is carried out without adding LPO. The excess ^{125}I is removed with a second round of retrieval. The occurrence of vesicle fusion is monitored by the iodination of the G-protein, a unique marker of the acceptor, with LPO delivered by the donor. This assay is specific because LPO is an enzymatic activity exogenous to the cell and G is an integral constituent of endosomal acceptor membranes.

Detection of fusion events in a cell-free system is limited by the assay selected. We used two biochemical and a morphological detection system for the cell-free fusion assay and obtained consistent results. For biochemical analysis the G-protein present in the immuno-isolated fraction is first detected after gel electrophoresis and transfer to nitrocellulose by Western blotting using an antibody against G. Iodination of G is then analysed by autoradiography of the nitrocellulose sheet. This protocol provides a direct comparision of the amount of G present and the degree of iodination (Gruenberg and Howell, 1986). Alternatively, after reconstitution the G-protein is immuno-precipitated and the extent of iodination is analysed by gel electrophoresis, autoradiography and direct counting (Gruenberg and Howell, 1987). For a morphological detection of fusion, it is necessary to provide a marker of the acceptor fraction that is abundant enough to be detected in every vesicle in a thin section. We chose an antibody that recognizes BHK cell surface antigens. The antibody is bound to the cell surface at 4°C after G implantation, labelled with colloidal gold-Protein A and cointernalized with the G protein. Fusion is then visualized by the presence of the reaction product of the peroxidase delivered by the donor in the lumen of immuno-isolated acceptor vesicles labelled with gold (Gruenberg and Howell, 1986). These studies show that the fusion of endocytic vesicles is ATP-dependent, confirming the finding of Davey et al (1985). Maximal fusion is detected with an early endosomal fraction prepared 5 min after G internalization at 37°C. With this fraction, morphological studies showed that peroxidase is delivered to approximately 50% of the acceptor vesicles labelled with gold. As shown in Fig 9, the fusion competence decreases exponentially with later time-defined fractions, exhibiting a $t_{1/2} < 5$ min.

No fusion is detected with a fraction prepared 30 min after G internalization. This loss of fusion competence does not reflect differences in the amount of G present in the fractions, as seen by immuno-blotting of the fractions and by direct quantitation of the amount of immuno-isolated G-protein (Table I). This profile of fusion competence shows that the cell-free fusion of endocytic vesicles is specific. As an additional control, we have also shown that no iodination of the G-protein occurs when an immuno-isolated plasma membrane fraction is used as acceptor. These results demonstrate that we have immuno-isolated two functionally distinct endosomal sub-compartments, an early fusion competent sub-compartment of the pathway and a later non-fusogenic sub-compartment.

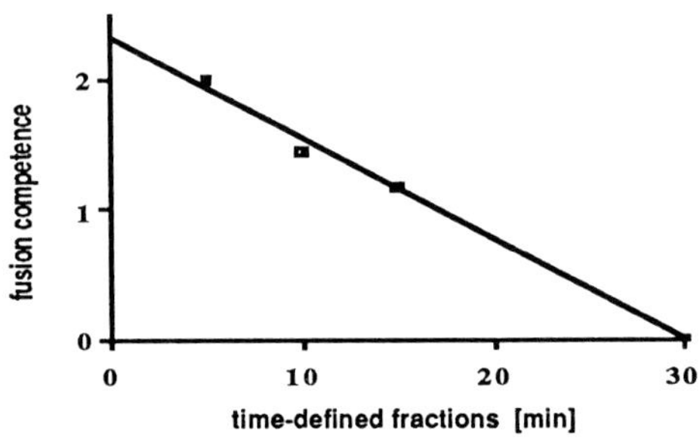

Figure 8: Fusion competence of the time-defined endosomal fractions. Following the cell-free assay, immuno-precipitation of the G-protein and SDS-PAGE, the band corresponding to the G-protein is cut and the amount of $[^{125}I]G$ counted. Iodination of the G with each acceptor is expressed in percent of the amount iodinated at 5 min. The exponential decrease of the fusion competence is shown on a semi-log scale (regression coefficient = 0.9966) and the $t_{1/2}$ of the process is 3.1 ± 0.6 min. This data is taken from Gruenberg and Howell, 1987.

CONCLUSION

After 5 min internalization, the wave of G-protein is present in an early sub-compartment from where recycling to the plasma membrane and routing to degradation occurs (Table I). This early endosome represents the fusion-competent sub-compartment, and is likely to be the endosomal sub-compartment where internalized membrane proteins are sorted. A high fusion activity of these endosomal elements is expected to account for the continuous fusion of vesicles originating from the plasma membrane with endosomes. We have not identified the donor vesicle in our fusion reaction. Likely candidates are either incoming plasma membrane vesicles delivering their cargo to endosomes by membrane fusion or the early endosomal elements of the donor fraction. From the early fusion competent sub-compartment the G molecules are transported to later elements of the pathway which do not undergo fusion.

These results provide evidence that internalized membrane proteins are delivered by vesicle fusion to an early sub-compartment of the endocytic pathway and that the $t_{1/2}$ of this process is < 5 min. This number agrees well with the $t_{1/2}$ reported for the recycling of both the fluid phase (Besterman et al, 1981) and the plasma membrane glycoproteins (Burgert and Thilo, 1983; Thilo, this volume). This also implies that sorting to recycling or degradation occurs in this early endosomal sub-compartment shortly after internalization. The slower rate of transport to the lysosomes (≈30 min for the G-protein as well as other proteins in different systems) (Oka and Weigel, 1983; Mellman and Plutner, 1984; Duprez and Dautry-Varsat, 1986) when compared to recycling ($t_{1/2}$ = 7.5 min for the G-protein and the transferrin receptor) (Hopkins and Trowbridge, 1983) may reflect the movement of later endosomes from the periphery of the cell to the lysosomes.

We will now combine biochemical analysis of the time-defined fractions with the assay of cell-free fusion to characterize the physiological partners of the fusion and to identify the factors involved at the molecular level.

REFERENCES:

Besterman JM, Airhart JA, Woodworth RC, Low RB (1981). Exocytosis of pinocytosed fluid in cultured cells: kinetic evidence for rapid turnover and compartmentation. J Cell Biol 91: 716-727.

Burgert HG, Thilo L (1983). Internalization and recycling of plasma membrane glycoconjugates during pinocytosis in the macrophage cell line, P388D1. Exp Cell Res 144: 127-142.

Copeland CS, Doms RW, Bolzau EM, Webster RG and Helenius A (1986). Assembly of influenza hemagglutinin trimers and its role in intracellular transport. J Cell Biol 103: 1179-1191.

Davey J, Hurtley SM, Warren G (1985). Reconstitution of an endocytic fusion event in a cell-free system. Cell 43: 643-652.

Doyle C, Sambrook J, Gething M-J (1986). Analysis of progressive deletions of the transmembrane and cytoplasmic domains of influenza hemagglutinin. J Cell Biol 103: 1193-1204.

Duprez V, Dautry-Varsat A (1986). Receptor-mediated endocytosis of interleukin 2 in a human tumor T cell line. J Biol Chem 261: 15450-15454.

Farquhar MG (1985). Progress in unraveling pathways of Golgi traffic. Ann Rev Cell Biol 1: 447-488.

Forgac M, Cantley L, Wiedenmann B, Altstiel L, Branton D (1983). Clathrin-coated vesicles contain an ATP-dependent proton pump. Proc Natl Acad Sci (USA) 80: 1300-1303.

Goldstein JL, Brown MS, Anderson RGW, Russel DW, Schneider WJ (1985). Receptor-mediated endocytosis: concepts emerging from the LDL receptor system. Ann Rev Cell Biol 1: 1-39.

Gruenberg J, Howell KE (1985). Immuno-isolation of vesicles using antigenic sites either located on the cytoplasmic or the exoplasmic domain of an implanted viral protein. A quantitative analysis. Europ J Cell Biol 38: 312-321.

Gruenberg J, Howell KE (1986). Reconstitution of vesicle fusions occurring in endocytosis with a cell-free system EMBO J 5: 3091-3101.

Gruenberg J, Howell KE (1987). An internalized trans-membrane protein resides in a fusion competent endosome for less than 5 min. Proc Natl Acad Sci (USA) 84: 5758-5762.

Helenius A, Mellman I, Wall D, Hubbard A (1983). Endosomes Trends Biochem Sci 8: 245-250.

Hemmilä I, Dakubu S, Mukkala V-M, Siitari H, Lövgren T (1984). Europium as a label in time-resolved immunofluorometric assays. Analyt Biochem 137: 335-343.

Hopkins CR, Trowbridge IS (1983).. Internalization and processing of transferrin and the transferrin receptor in human carcinoma A431 cells. J Cell Biol 97: 508-521.

Howell KE, Ansorge W, Gruenberg J (1987) Magnetic free flow immuno-isolation system designed for sub-cellular fractionation. In "Microspheres: medical and biological applications," Rembaum A, Tokes Z, (eds): Boca Raton, Florida. CRC Press, (in Press).

Kreis T. (1986) Microinjected antibodies against the cytoplasmic domain of vesicular stomatitis virus glycoprotein block its transport to the cell surface. EMBO J 5: 931-941.

Kreis T, Lodish HF (1986) Oligomerization is essential for transport of vesicular stomatitis viral glycoprotein to the cell surface. Cell 46: 929-937.

Maxfield FR (1982). Weak bases and ionophores rapidly and reversibly raise the pH of endocytic vesicles in culture mouse fibroblasts. J Cell Biol 95: 676-681.

Mellman I, Plutner H (1984) Internalization and degradation of macrophage Fc receptors bound to polyvalent immune complexes. J Cell Biol 98: 1170-1177.

Mellman I, Fuchs R, Helenius A (1986) Acidification of the endocytic and exocytic pathways. Ann Rev Biochem 55: 663-700.

Oka JA, Weigel PH (1983). Recycling of the asialoglycoprotein receptor in isolated rat hepatocytes. J Biol Chem 258: 10253-10262.

Russel JT, Holz RW (1981). Measurement of ΔpH and membrane potential in isolated neurosecretory vesicles from bovine neurohypophyses. J Biol Chem 256: 5950-5953.

Schneider DL (1983). ATP-dependent acidification of membrane vesicles isolated from purified rat liver lysosomes. Acidification activity requires phosphate. J Biol Chem 258: 1833-1838.

Schuy W, Will C, Kuroda K, Scholtissek C, Garten W, Klenk HD (1986). Mutations blocking the transport of the influenza hemagglutinin between the rough endoplasmic reticulum and the plasma membrane. EMBO J 5: 2831-2836.

Soini E, Hemmilä I (1979). Fluoroimmunoassays: present status and key problems. Clin Chem 25/3: 353-361.

Soini E, Kojola H (1983). Time-resolved fluorometer for lanthanide chelates -A new generation of nonisotopic immunoassays. Clin Chem 29/1: 65-68.

Steinmam RM, Mellman IS, Muller WA, Cohn ZA (1983). Endocytosis and the recycling of plasma membrane. J Cell Biol 96: 1-27.

Ukkonen P, Lewis V, Marsh M, Helenius A, Mellman I (1986). Transport of macrophage Fc receptors and Fc receptor-bound ligands to the lysosomes. J Exp Med 163: 952-971.

White J, Kartenbeck J, Helenius A (1980). Fusion of Semliki Forest virus with the plasma membrane can be induced by low pH. J Cell Biol 87: 264-272.

Wileman T, Harding C, Stahl P (1985). Receptor-mediated endocytosis. Biochem J 232: 1-14.

Cell-Free Analysis of Membrane Traffic, pages 333–342

RECONSTITUTION OF PROTEIN TRANSPORT FROM THE ENDOPLASMIC RETICULUM TO THE GOLGI USING A CELL FREE SYSTEM

William E. Balch, Con J.M. Beckers, and David S. Keller

Department of Molecular Biophysics and Biochemistry, Yale University, New Haven, Connecticut 06510

INTRODUCTION

An understanding of the biochemical mechanisms which control the vesicular transport of protein between compartments of the secretory pathway of eucaryotic cells will require successful reconstitution of individual segments of the pathway using cell-free assays. Transport of membrane glycoprotein between the *cis* and *medial* compartments of the Golgi complex has been successfully reconstituted *in vitro* (Fries and Rothman, 1981; Balch et al., 1984a; Balch et al., 1984b; Braell et al., 1984; Orci et al., 1986; Wattenberg and Rothman, 1986; Wattenberg et al., 1986). Other stages in the exocytic pathway remain to be reconstituted *in vitro* with high efficiency. An important step in the delivery of protein to the cell surface is the selective export of proteins from their site of synthesis in the cellular endoplasmic reticulum (ER). While the biochemical basis for synthesis (Walters et al., 1984), insertion (Wickner and Lodish, 1985), and core-glycosylation (Robbins et al., 1977) of protein in the rough endoplasmic reticulum (RER) is the focus of many studies, little is known of the biochemical events which regulate export of protein from the ER to subsequent compartments of the exocytic pathway.

To understand the enzymology of the first stage of transport, export of newly synthesized protein from the ER, we have recently developed a model system based on the G protein of vesicular stomatitis virus (VSV)

(Balch et al., 1986; Balch et al., 1987). The synthesis (Lenard, 1978), post-translational processing (Tabas et al., 1977; Li et al., 1978; Kornfeld and Kornfeld, 1985), and transport of G protein from the ER to the cell surface (Bergmann et al., 1981; Bergmann and Singer, 1983) has been studied extensively *in vivo*. It is indistinguishable from normal cellular membrane-bound glycoproteins. During synthesis G protein acquires two high mannose core-oligosaccharides (Li et al., 1978). Transfer of G protein from the ER to the *cis* compartment of the Golgi *in vivo* results in the G protein oligosaccharides being trimmed from the $man_{8-9}GlcNAc_2$ forms to the $man_5GlcNAc_2$ species as a result of the early Golgi oligosaccharide processing enzyme α-mannosidase I (Tabas and Kornfeld, 1979; Dunphy and Rothman, 1985; Kornfeld and Kornfeld, 1985; Balch and Keller, 1986; Balch et al., 1986). Trimming of G protein can be measured directly in the mutant Chinese hamster ovary (CHO) cell line clone 15B which is missing the *medial* Golgi processing enzyme N-acetylglucosamine transferase I (Gottlieb et al., 1975; Dunphy et al., 1985). In this mutant further processing of the $man_5GlcNAc_2$ oligosaccharide form to the complex structure by sequentially acting α-mannosidase II and glycosyltransferases in later compartments of the Golgi is blocked (Gottlieb et al., 1975; Kornfeld and Kornfeld, 1985). However, G protein is delivered to the cell surface in the $man_5GlcNAc_2$ form since processing, not transport, is the only known defect in clone 15B (Schlessinger et al., 1976).

Export of G protein from the ER *in vivo* can be distinguished from later transport steps between Golgi compartments through use of the temperature-sensitive VSV mutant strain ts045. The mutant G protein of ts045 is found exclusively in the ER when cells are incubated at the restrictive temperature (39.5°C) during infection (Lafay, 1974; Knipe et al., 1977; Zilberstein et al., 1980; Lodish and Kong, 1983; Balch et al., 1986). Shift of cells from the restrictive to the permissive temperature (32°C) *in vivo* results in the synchronous release of G protein to the Golgi complex (Bergmann and Singer, 1983; Bergmann et al., 1981; Balch et al., 1986). Transport of ts045 G protein between the ER and the α-mannosidase I containing *cis* Golgi compartment *in vivo* has been carefully characterized and shown to occur in two biochemically distinct steps. An early temperature-sensitive, ATP-dependent export (budding) step is followed by a temperature and ATP-insensitive step providing for delivery (fusion) of G protein to the *cis* Golgi compartment containing α-mannosidase I (Balch and Keller, 1986; Balch et al., 1986).

In order to study transport of G protein between the ER and *cis* Golgi compartment using a cell-free system, authentic transfer *in vitro* should faithfully reproduce the events observed *in vivo* resulting in processing of G protein from the $man_{8-9}GlcNAc_2$ form to the $man_5GlcNAc_2$ form. When a cell homogenate was prepared from a VSV infected 15B interphase cell population in which ts045 G protein was localized to the ER by infection at the restrictive temperature (39.5^{o}C) prior to homogenization, no trimming of G protein *in vitro* was observed using incubation conditions which support inter-Golgi transport *in vitro* (Balch et al., 1987). These results raised the possibility that the transport function of the ER compartment was disrupted during cell homogenization. The ER is a single, reticular network of tubules and cisternae distributed throughout the cytoplasm, and found in continuity with the nuclear envelope. During cell homogenization the ER is sheared into small resealed vesicles (microsomes). To circumvent the problem of vesiculation during homogenization, one approach is to obtain homogenates from cells undergoing mitosis. In this stage of the cell cycle both the nuclear envelope (14), ER (W.E. Balch and A. Hermanowsky, unpublished results), and the Golgi complex (54) are disassembled. In contrast to interphase ER microsomes, the physiologically disassembled organelles found in mitotic cells are functionally stable to the homogenization conditions, and active in transport of G protein from the ER to the Golgi (Balch et al., 1987). To demonstrate this result a post-chromosomal supernatant (PCS) was prepared from mitotic clone 15B cells infected with the ts045 virus at the restrictive temperature. When the PCS was incubated at 32°C (permissive temperature) for 60 min in the presence of ATP, a form of G protein with increased electrophoretic mobility comparable to that observed for the $man_5GlcNAc_2$ species formed *in vivo* was observed upon SDS-gel electrophoresis of the incubated fraction (Fig. 1)(Balch et al., 1987).

The preparation of cell-free homogenates, and incubation conditions necessary to achieve transport of G protein between the ER and Golgi of mitotic 15B cells are outlined below.

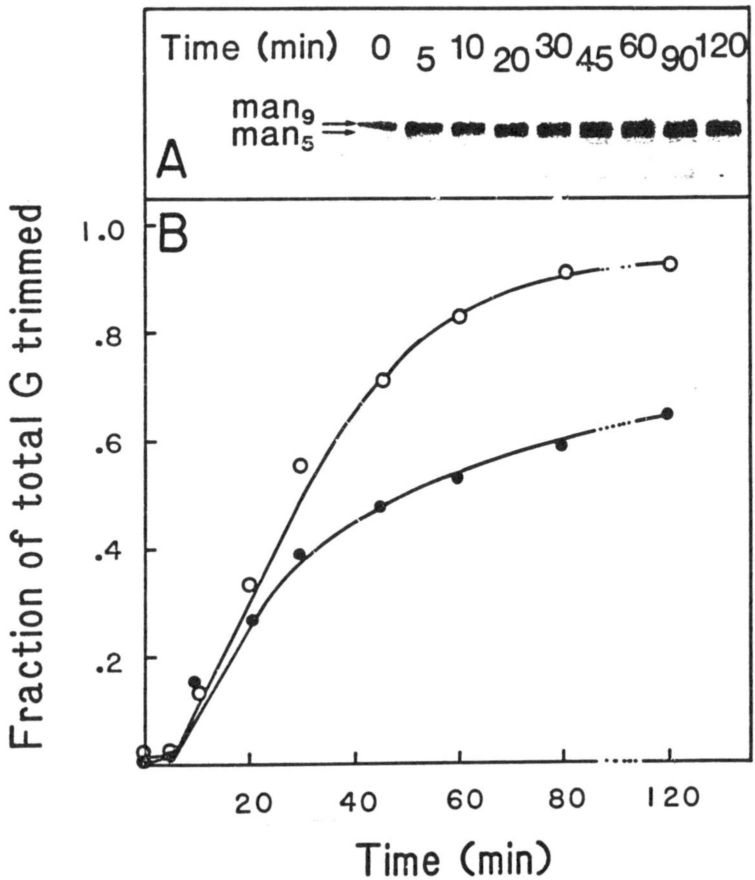

Figure 1. Kinetics of processing of G protein *in vitro* (reproduced with permission from Rockefeller University Press (see Balch et al., 1987). A 15B mitotic cell homogenate containing [^{35}S]methionine labeled ts045 G protein was incubated *in vitro* for the indicated time at 32°C, processed for SDS-gel electrophoresis, and the amount of G protein trimmed to the man$_5$GlcNAc$_2$ form was quantitated as described in the methods section. The autoradiogram (panel A) was used to quantitate the amount of G protein trimmed at each time point *in vitro* as shown in panel B (closed circles). A typical time course for processing of G protein *in vivo* is shown (panel B,open circles)(Balch et al., 1986).

METHODOLOGY

Growth and synchronization of cells. The Chinese hamster ovary cell lines were grown in monolayer (10 cm dishes) in minimal essential medium (α-mem) (Gibco, Grand Island, NY) supplemented with 10% fetal calf serum.

For preparation of synchronized mitotic cell populations of clone 15B the following protocol was used. Confluent cells (1×10^7 cells per 10 cm LUX^R tissue culture dish (Miles Laboratories)) were trypsin released and seeded into new tissue culture dishes at a cell density of 2.5×10^6 cells per 10 cm dish. After 24 hr, the medium was replaced with fresh medium and the cells were incubated for an additional 24 hr. At this time, closely packed cells (2×10^7 cells per 10 cm dish) were trypsin released and seeded into new dishes at a cell density of 6.7×10^6 cells per 10 cm dish in the presence of 2.5 mM thymidine. After 14 hours, cells were washed twice with 2 ml of TD buffer (138 mM NaCl, 5 mM KCl, 25 mM TRIS base, 0.4 mM Na_2HPO_4; pH 7.4), and reincubated in the presence of 5 ml of α-mem for 5.5 hr. At 5.5 hr a stock solution of nocodazole dissolved at 50 μg/ml in DMSO was added to each dish with mixing to a final concentration of 0.1 μg/ml. After an additional 3 hour incubation cells were harvested by the mitotic shake procedure (Taylor, 1965). Using this procedure 50 to 70% of the total cell population in each dish was arrested in mitosis.

Infection of mitotic cells with VSV. Protein synthesis is required just prior to entry of cells into mitosis (Tobey et al., 1966). Since virus infection rapidly inhibits host protein synthesis, the timing of virus infection is important to insure a maximal yield of infected, mitotic cells. For infection of clone 15B with tsO45, one-third of the total synchronized cells in each experiment were infected after a 5.5 hr period of incubation in the absence of thymidine. Each dish was incubated with 10-20 plaque forming units of virus per cell in 1 ml of an infection medium containing α-mem, 25 mM HEPES (pH 7.4), 0.1 μg/ml nocodazole per 10 cm dish for 30 min at 32^oC with gentle rocking. Each dish was subsequently supplemented with an additional 4 ml of α-mem containing 10% fetal calf serum and 0.1 μg/ml nocodazole, and incubated at 37^oC for 2.0 hr prior to harvest of cells by mitotic shake. Using this procedure 15-30% of the total cell population in each dish was arrested in mitosis. The mitotic index of harvested cells ranged from 60 to 80% for different preparations.

Labeling of cells. Mitotic cell populations infected with tsO45 were washed 3 times with 5 ml portions of a labeling medium (Joklik's minimal essential medium (JMEM) minus methionine containing non-essential amino acids (Gibco), 0.1 μg/ml of nocodazole, and 20 mM Hepes-NaOH, pH 7.4) at 39.5°C. Cells were incubated for 5 min at 39.5°C (the restrictive temperature) prior to the addition of [^{35}S]methionine (25-50 μCi/ml). After a 10 min pulse, unlabeled methionine was added to a final concentration of 2.5 mM, and the incubation continued for 3 min at 39.5°C prior to transfer of cells to a 10X volume excess of ice-cold labeling medium to rapidly chill cells and inhibit transport. The infected, labeled mitotic cells were pooled with the uninfected mitotic cells harvested from the remaining two-thirds of the cells used in each experiment.

Preparation of cell homogenates. Cells in ice-cold labeling medium were pelleted for 3 min at 800 X g, and washed 1X with a TEA buffer containing 10 mM triethanolamine and 150 mM KCl (pH 7.4). Cells were pelleted and resuspended in a H/KCl buffer containing 50 mM HEPES, 90 mM KCl (pH 7.2) at a ratio of 100 - 200 μl pelleted cells per 1 ml of H/KCl buffer, and homogenized using a stainless steel ball-bearing homogenizer (Balch and Rothman, 1985). A post-chromosomal supernatant (PCS) was prepared where indicated by centrifugation of the homogenate for 5 min at 800 X g. Cell homogenates (5-15 mg/ml protein with ~40% of total protein being membrane protein) were used immediately for assay, or transferred in 125 μl portions to 500 μl microcentrifuge tubes and rapidly frozen in liquid nitrogen and stored at -80°C. For assay, frozen fractions are thawed by a brief incubation at 37°C, and transferred to ice.

Incubation conditions to achieve transport in vitro. In addition to cell homogenate (10 to 50 μg per incubation), each assay contained in a total volume of 50 μl (final concentrations): 50 mM HEPES-KOH (pH 7.2), 90 mM KCl, 2.5 mM magnesium chloride, 1 mM ATP, 5 mM creatine phosphate (CP), and 0.2 IU of rabbit muscle creatine phosphokinase (CPK). The ATP regenerating system was made fresh daily by mixing 20 μl CPK (2000 IU/ml,stored at -80°C), 100 μl of 200 mM CP, 100 μl of 40 mM ATP (Na form, neutralized with NaOH); and added as a concentrate to the assay to obtain the indicated final concentrations. Additional modifications of the standard incubation conditions are indicated in the Results. Each assay was initiated by incubation for indicated time and temperature as described in the Results. Transport was terminated by transfer to ice, followed by the addition of 15 μl of a 5X concentrate of a gel sample buffer (Laemmli, 1970) containing 12.5 ml of a 1 M TRIS (pH 6.8) buffer, 20 ml glycerol, 0.5 g DTT, 4 g sodium dodecyl sulfate (SDS), 4 mg bromophenol blue, in a final volume of 40 ml H_2O.

Quantitation of transport. Samples were processed for SDS-gel electrophoresis using 10% polyacrylamide gels (Laemmli, 1970), treated for autoradiographic enhancement by incubation for 30 min in 100 ml of a 30% methanol solution containing 0.125 M salicylic acid (Na form; pH 7.0), dried, and autoradiographed for 1-3 days (XAR-5 film, Kodak). The fraction of G protein processed to the $man_5GlcNAc_2$ form was determined by densitometry of the exposed autoradiogram using a GS 300 transmission scanning densitometer (Hoefer Instru.) connected to an IBM-XT^R with the GS350 integrating software (Hoefer Inst.). The quantitative relationship between appearance of the faster migrating band using SDS-gel electrophoresis, and processing to the $man_5GlcNAc_2$ oligosaccharide form has been established previously (Balch et al., 1986).

DISCUSSION

The methodology used to reconstitute transport of G protein from the ER to the Golgi relies heavily upon the ability to routinely prepared large quantities of mitotic extracts for biochemical analysis. The protocol presented in the Methods has been optimized to yield with high efficiency mitotic extracts for use in the transport assay. Several criteria have been used to establish the validity of the events observed *in vitro* to those occurring *in vivo* (Balch et al., 1987). The cellular location of the transported protein prior to incubation and the processing activity marking the acceptor compartment (*cis* Golgi compartment containing mannosidase I) are rigorously defined (Bergmann and Singer, 1981; Balch and Keller, 1986). These data identify the segment of the total cellular pathway being studied *in vitro*. In addition, we have demonstrated that the events occurring *in vitro* are a consequence of transfer between sealed membrane-bound compartments as is known to occur *in vivo*, transport *in vitro* requires physiological incubations conditions, and trimming occurs with the efficiency, specificity and kinetics approaching those observed *in vivo* (Balch et al., 1987).

What is the role of the mitotic ER in this cell-free assay? It was proposed that prior physiological disassembly of the ER in mitotic cells prior to cell homogenization is essential for recovery of activity. It is likely that these mitotic structures are not susceptible to the shear and extensive vesiculation that an intact interphase ER structure encounters during cell homogenization. In addition, a second possibility consistent with our results is that the mitotic homogenate promotes reassembly or "repair" of organelles *in vitro*. It has recently been established that in addition to organelle disassembly, all vesicular trafficking is arrested in mitotic cells (Featherstone et al., 1985). G protein accumulates in the ER in the $man_{8-9}GlcNAc_2$ oligosaccharide forms in mitotic cells arrested at metaphase with nocodazole (Featherstone et al., 1985). Three lines of evidence support the interpretation that factors required for maintenance of the mitotic state are inactivated during cell homogenization. In addition to appearance of the trimmed form of G protein *in vitro* reflecting reinitiation of transport, interphase cytosol was found to substitute for mitotic cytosol (Balch et al., 1987), and we have observed that both the nuclear envelope and the RER are reassembled during incubation of the donor and acceptor crude homogenates *in vitro* (A. Hermanowsky and W.E. Balch, unpublished observations). The appearance of reassembled nuclei *in vitro* from CHO homogenates using similar incubation conditions has been well documented (Burke and Gerace, 1986). In addition, nuclei reassembled *in vitro* have been shown to provide a useful model system to study import of protein into the nucleus in a cell-free system (Newmeyer et al., 1986A; Newmeyer et al., 1986B), attesting to the facility of reassembled organelles to authentically reconstitute functions observed *in vivo*.

If the efficiency of G protein export observed here is a consequence of reassembly of donor ER and Golgi acceptor membranes into functional organelles, then the present assay conditions will allow us to explore in addition to the enzymological basis of transport between the ER and Golgi, the biochemical mechanisms leading to reassembly of the ER subsequent to cell division.

REFERENCES

Balch WE, Keller DS (1986). ATP-coupled transport of vesicular stomatitis virus G protein: functional boundaries of secretory compartments. J Biol Chem 261: 14681-14689.

Balch WE, Rothman JE (1985). Characterization of protein transport between successive compartments of the Golgi apparatus: asymmetric properties of donor and acceptor activities in a cell-free system. Arch Biochem Biophys 240: 413-425.

Balch WE, Dunphy WG, Braell WA, Rothman JE (1984). Reconstitution of the transport of protein between successive compartments of the Golgi measured by the coupled incorporation of N-acetylglucosamine. Cell 39: 405-416.

Balch WE, Elliott MM, Keller, DS (1986). ATP-coupled transport of vesicular stomatitis virus G protein between the endoplasmic reticulum and the Golgi. J Biol Chem 261: 14690-14696.

Balch WE, Glick BS, Rothman JE. (1984) Sequential intermediates in the pathway of intercompartmental transport in a cell-free system. Cell 39: 525-536.

Balch, WE, Wagner, KR, Keller, DS (1987). Reconstitution of transport of vesicular stomatitis virus G protein from the endoplasmic reticulum to

the Golgi complex using a cell-free system. J Cell Biol 104: 749-760.
Bergmann JE, Singer SJ (1983). Immunoelectron microscopic studies of the intracellular transport of the membrane glycoprotein (G) of vesicular stomatitis virus in infected Chinese hamster ovary cells. J Cell Biol 97: 1777-1787.
Bergmann JE, Tokuyasu KT, Singer SJ (1981). Passage of an integral membrane protein, the vesicular stomatitis virus glycoprotein, through the Golgi apparatus en route to the plasma membrane. Proc Natl Acad Sci USA 76: 4350-4354.
Braell WA, Balch WE, Dobbertin DC, Rothman JE (1984). The glycoprotein that is transported between successive compartments of the Golgi in a cell-free system resides in stacks of cisternae. Cell 39: 511-524.
Burke B, and Gerace L (1986). A cell free system to study reassembly of the nuclear envelope at the end of mitosis. Cell 44: 639-652.
Dunphy WG, Brands R, Rothman JE (1985). Attachment of terminal N-acetylglucosamine to asparagine-linked oligosaccharides occurs in central cisternae of the Golgi stack. Cell 40: 463-472.
Dunphy WG, Rothman JE (1985). Compartmental organization of the Golgi stack. Cell 42: 13-21.
Featherstone C, Griffiths G, Warren G (1985). Newly synthesized G protein of vesicular stomatitis virus is not transported to the Golgi complex in mitotic cells. J Cell Biol 101: 2036-2046.
Fries E, Rothman JE (1981). Transient activity of Golgi-like membranes as donors of vesicular stomatitis viral glycoprotein *in vitro*. J Cell Biol 90: 697-704.
Gottlieb C, Baenziger J, Kornfeld S (1975). Deficient uridine diphosphate-N-acetylglucosamine: glycoprotein N-acetylglucosaminetransferase activity in a clone of Chinese hamster ovary cells with altered surface glycoproteins. J Biol Chem 250: 3303-3309.
Knipe DM, Baltimore D, Lodish HF (1977). Maturation of viral proteins in cells infected with temperature-sensitive mutants of vesicular stomatitis virus. J Virol 21: 1149-1158.
Kornfeld R, Kornfeld S (1985). Assembly of asparagine-linked oligosaccharides. Ann Rev Biochem 54: 631-664.
Laemmli UK (1970). Cleavage of structural proteins during the assembly of the head of bacteriophage T4. Nature (London) 227: 680-685.
Lafay F (1974). Envelope proteins of vesicular stomatitis virus: effect of temperature-sensitive mutations in complementation groups III and V. J Virol 14: 1220-1228.
Lenard J (1978). Virus envelopes and plasma membranes. Ann Rev Biophys Bioeng 7: 139-165.
Li E, Tabas I, Kornfeld S (1978). The synthesis of complex oligosaccharides. I. Structure of the lipid-linked oligosaccharide precursor of the complex-type oligosaccharides of the vesicular stomatitis virus G

protein. J Biol Chem 253: 7762-7770.
Lodish HF, Kong N (1983). Reversible block in intracellular transport and budding of mutant vesicular stomatitis virus glycoproteins. Virology 125: 335-348.
Newmeyer DD, Finlay DR, Forbes DJ (1986A). In vitro transport of a fluorescent nuclear protein and exclusion of non-nuclear protein. J Cell Biol 103: 2091-2102.
Newmeyer DD, Lucocq JM, Burglin TR, De Robertis EM (1986B). Assembly *in vitro* of nuclei active in nuclear protein transport: ATP is required for nucleoplasmin accumulation. EMBO J 5: 501-510.
Orci L, Glick BS, Rothman JE (1986). A new type of coated vesicular carrier that appears not to contain clathrin: its possible role in protein transport within the Golgi stack. Cell 46: 171-184.
Robbins P, Hubbard SC, Turco SJ, Wirth DF (1977). Proposal for a common oligosaccharide intermediate in synthesis of membrane glycoproteins. Cell 12: 893-900.
Schlessinger S, Gottlieb C, Feil P, Gelb N, Kornfeld S (1976). Growth of enveloped RNA viruses in a line of chinese hamster ovary cells with deficient N-acetyl-glucosaminyltransferase activity. J Virol 17: 239-246.
Tabas I, Kornfeld S (1979). Purification and characterization of rat liver Golgi α-mannosidase capable of processing asparagine-linked oligosaccharides. J Biol Chem 254: 11655-11663.
Tabas I, Schlessinger S, Kornfeld, S (1977). Processing of high mannose oligosaccharides to form complex type oligosaccharides on the newly synthesized polypeptides of the vesicular stomatitis virus G protein and the IgG heavy chain. J Biol Chem 253: 716-722.
Taylor EWJ (1965). The mechanism of colchicine inhibition of mitosis. I. Kinetics of inhibition and binding of ^{3}H-colchicine. J Cell Biol 25: 145-150.
Tobey RA, Anderson EC, Peterson DF (1966). RNA stability and protein synthesis in relation to the division of mammalian cells. Proc Natl Acad Sci USA 56: 1520-1527.
Walters P, Gilmore R, Blobel G (1984). Protein translocation across the endoplasmic reticulum. Cell 38: 5-8.
Wattenberg BW, Rothman JE (1986). Multiple cytosolic components promote intra-Golgi protein transport. Resolution of a protein acting at a late stage, prior to membrane fusion. J Biol Chem 261: 2208-2221.
Wattenberg BW, Balch WE, Rothman JE (1986). A novel prefusion complex formed during protein transport between Golgi cisternae in a cell-free system. J Biol Chem 261: 2202-2207.
Wickner WT, Lodish HF (1985). Multiple mechanisms of insertion into and across membranes. Science 230: 400-407.
Zilberstein A, Snider MD, Porter M, Lodish HF (1980). Mutants of vesicular stomatitis virus blocked at different stages in maturation of the viral glycoprotein. Cell 21: 417-427.

Cell-Free Analysis of Membrane Traffic, pages 343–354

CELL-FREE RECONSTITUTION OF THE INSERTION OF A NEWLY SYNTHESISED GLYCOPROTEIN INTO THE PLASMA MEMBRANE

J. Michael Edwardson and Philip G. Woodman

Department of Pharmacology, University of Cambridge, Hills Road, Cambridge CB2 2QD, U.K.

INTRODUCTION

The membranes of the organelles present within eukaryotic cells contain characteristic sets of proteins. Consequently, newly synthesised membrane proteins must be segregated during their intracellular transport for delivery to these different destinations. Although little definite information is available at present, it seems reasonable to propose that at each step in the transport pathway, populations of vesicles are able to select proteins bound for a common destination, to which the vesicles are then targetted. The elucidation of the molecular details of this process will require the reconstitution of the various transport steps in cell-free systems. We set out to develop a cell-free assay for the final step in the intracellular transport pathway of a viral glycoprotein, that is, its insertion into the plasma membrane. The assay exploits the enzyme activity of influenza virus neuraminidase. In cells infected with influenza virus, this envelope glycoprotein is synthesised in large amounts and is transported through the cells along the route taken by host cell proteins (Fuller et al., 1985).

METHODS

Basic Design of the Assay

Two types of post-nuclear supernatant, which act as donor and acceptor of neuraminidase, are prepared from baby hamster kidney (BHK-21) cells. Donors are prepared from cells infected with influenza virus and containing neuraminidase en route to the plasma membrane. Acceptors are prepared from cells that have had Semliki Forest virus (SFV), with envelope glycoproteins containing [^{3}H]N-acetylneuraminic acid ([^{3}H] NANA), bound to their plasma membranes at 4°C. Since no endocytosis occurs at this temperature (Fries and Helenius, 1979), homogenates of these cells will contain [^{3}H]SFV bound only to the plasma membrane. [^{3}H]SFV is prepared by harvesting virus from BHK-21 cells that have been infected with SFV in the presence of [^{3}H]N-acetyl-D-mannosamine (Monaco and Robbins, 1973). Fusion between vesicles from the two preparations permits access of enzyme to substrate and results in the release of free [^{3}H]NANA. This release is detected through the transfer of radioactivity from a trichloroacetic acid (TCA)-insoluble to a TCA-soluble fraction. The cell-free assay is illustrated schematically in Fig. 1. The NANA analogue 2,3-dehydro-2-desoxy-NANA (DD-NANA), an inhibitor of neuraminidase (Meindl et al., 1974), is included routinely in all assays to prevent release of [^{3}H]NANA as a consequence of contact between vesicles bearing outward facing enzyme and substrate.

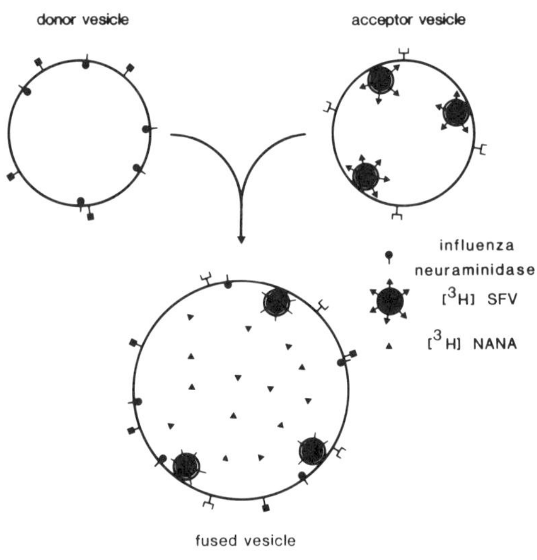

Figure 1. Basic design of the cell-free assay. From Woodman and Edwardson (1986).

Determination of Acceptor Latency

The percentage of the [^{3}H]SFV in an acceptor that is located within sealed vesicles (acceptor latency) can be determined by comparing the ability of neuraminidase (from Clostridium perfringens) to cleave [^{3}H]NANA from [^{3}H]SFV and from samples of acceptor. In a typical experiment, neuraminidase cleaved maximally 79% of the total radioactivity from [^{3}H]SFV and 68% from a sample of acceptor. The percentage of the total [^{3}H]SFV that is exposed to enzyme is, therefore, 100 x 68/79, or 86%. Hence, 14% is protected inside sealed vesicles. The mean latency value (± SEM) obtained was 16±4% (n=12). Neuraminidase in the form of influenza virions cleaved maximally 61% of [^{3}H]NANA from [^{3}H]SFV. Since acceptor latency is 16%, therefore, the maximum possible fusion-dependent release of [^{3}H]NANA will be approximately 10% of the total.

RESULTS

A Typical Experiment

Since vesicle-mediated intracellular transport is known to require energy (Palade, 1975), we expected that the cell-free fusion event that we were looking for would also be energy-dependent. Accordingly, donors and acceptors were incubated together at 37°C (usually for 1h) in the presence of either an ATP-depleting cocktail (25IU/ml hexokinase/5mM glucose) or an ATP-generating cocktail (1mM ATP/8mM creatine phosphate/7IU/ml creatine phosphokinase). A blank donor, prepared from uninfected cells, was also included to control for any influenza neuraminidase-independent release of [^{3}H]NANA. Table 1 shows the results obtained in a typical experiment. Donor gave both ATP-dependent and ATP-independent components of release of [^{3}H]NANA. Blank donor gave a small and statistically non-significant ATP-dependent component of release and no ATP-independent component. The ATP-independent component of release given by the donor appears to be a result of incomplete inhibition of outward-facing neuraminidase by the DD-NANA, since this component fell progressively when the concentration of DD-NANA in the incubation mix was increased (not shown). Of more interest was the ATP-dependent component of release given by the donor, which was obtained reproducibly with different donors and acceptors.

TABLE 1. Results from a typical cell-free assay.

donor	ATP	release of [^{3}H]NANA (%)	ATP-dependent release of [^{3}H]NANA (%)	ATP-dependent, influenza neuraminidase-dependent release of [^{3}H]NANA (%)
blank	-	0.0±0.5	-	-
	+	0.8±0.4	0.8±0.6	-
donor	-	6.9±0.6	-	-
	+	13.6±0.9	6.7±1.1	5.9±1.3

Values are means ± SEM of triplicate determinations.

ATP-dependent Release of [^{3}H]NANA Given by Donor is a Consequence of Vesicle Fusion.

If the ATP-dependent, influenza neuraminidase-dependent component of release of [^{3}H]NANA is a consequence of vesicle fusion, it should be abolished by the inclusion in the incubation mix of the detergent Triton X-100, at a concentration (0.1% w/v) that should lyse vesicles. Table 2 shows that this is indeed the case. No significant ATP-dependent, influenza neuraminidase-dependent release of [^{3}H]NANA was obtained in the presence of Triton X-100. In addition, no significant ATP-dependent, influenza neuraminidase-dependent release was obtained when blank donors and acceptors were used in combination with added influenza virus or [^{3}H]SFV (not shown). Hence, this component of release is only produced when both enzyme and substrate are within sealed vesicles.

Specificity of Fusion

Since the intracellular transport of proteins is known to be a highly specific process, it is reasonable to expect that any significant cell-free fusion event would also show specificity. Hence, in the case of the present cell-free assay, only post-Golgi vesicles from the donor should fuse with plasma membrane vesicles from the acceptor. The design

TABLE 2. Effect of Triton X-100 on the cell-free assay.

donor	TX-100	ATP-dependent release of [^{3}H]NANA (%)	ATP-dependent, influenza neuraminidase-dependent release of [^{3}H]NANA (%)
blank	-	1.2±0.4	-
donor	-	4.5±0.9	3.3±1.0
blank	+	0.9±1.0	-
donor	+	1.3±0.6	0.4±1.2

Values are means ± SEM from five separate experiments.

of the assay ensures that all of the [^{3}H]SFV is bound to vesicles derived from the plasma membrane. However, there is no simple way of preparing donors containing neuraminidase only in post-Golgi vesicles. Standard donors prepared from cells maintained at 37°C will contain neuraminidase throughout the transport pathway. It has been shown (Matlin and Simons, 1983) that incubation of virally-infected cells at 20°C causes an arrest in the intracellular transport of the viral envelope glycoproteins in a compartment on the _trans_ side of the Golgi complex that has been termed the _trans_ Golgi network (TGN). In addition, it has been found that rewarming of the cells to 37°C causes the rapid (5-30min) appearance of the blocked glycoproteins at the cell surface (Griffiths et al., 1985). We have exploited these observations in designing an experiment to test the specificity of the cell-free fusion event. The aim of the experiment was to block the transport of neuraminidase in the TGN, and then to release the block, causing a wave of enzyme to move through to the plasma membrane. If the fusion event is indeed specific, one would expect to chase the neuraminidase into and out of a fusion-competent compartment using this protocol.

It was first necessary to establish whether the 20°C block technique was applicable to influenza neuraminidase in BHK-21 cells. The appearance of neuraminidase at the cell surface was detected by assaying enzyme activity. At various times post-infection, infected cells were cooled to 4°C and a suspension of [^{3}H]SFV (approximately 20,000 dpm) was added above the monolayer. Cells and virus were incubated together at 4°C for 30min, with gentle rocking, and the transfer of radioactivity from a TCA pellet to a TCA supernatant was determined. Itwas shown in a preliminary experiment that release of [^{3}H]NANA was linear over the 30min incubation period. Hence, release will be proportional to the amount of neuraminidase at the cell surface. Cells were either incubated at 37°C throughout, or cooled to 20°C at 3.5h post-infection. Cooled cells were either incubated at 20°C for 2h or for 1h, followed by a 1h period at 37°C. The results obtained are shown in Fig. 2. It is clear that reduction of the temperature to 20°C causes an almost complete block in the transport of neuraminidase to the cell surface, and that rewarming to 37°C causes, after a delay of approximately 5min, the rapid appearance of neuraminidase at the cell surface.

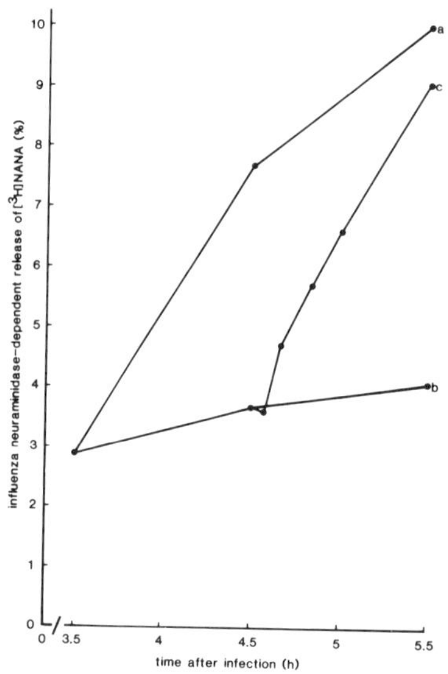

Figure 2. Appearance of neuraminidase at the cell surface. (a) Incubation at 37°C throughout. (b) Incubation at 20°C throughout the 2h period shown. (c) Incubation at 20°C for 1h and then at 37°C for a further 1h.

Donors were prepared from cells incubated at 37°C for 4h post-infection, followed by a 1h period at 20°C. Cells were either homogenised immediately or incubated at 37°C for a further 10 or 25min. Donors were centrifuged through 5% (w/v) metrizamide onto a 35% metrizamide cushion (2h at 100,000g), to prepare membrane fractions. Neuraminidase activities of the different donors were determined and equal amounts of activity were used in separate cell-free assays with a similarly fractionated acceptor. Membrane protein concentrations were equalised by addition of blank membranes, and equal amounts of a cytosol preparation (high-speed supernatant) were added. As a negative control, a 'plasma membrane' donor was included. This was prepared by binding influenza virus to BHK-21 cells at 4°C. This donor was processed in the same way. The results obtained are shown in Fig. 3. The most active donor was the one that had been prepared from cells warmed to 37°C for 10min, at which time a wave of neuraminidase should be moving from the TGN to the plasma membrane. After a 25min warm-up a considerable amount of the blocked neuraminidase should have reached the plasma membrane. At

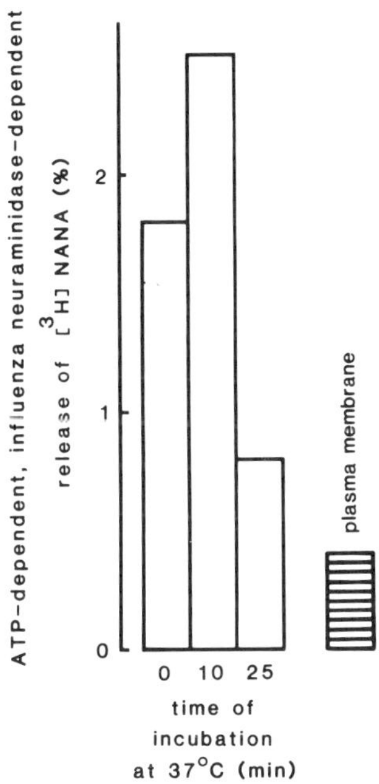

Figure 3. Specificity of fusion.

this time the activity of the donor had fallen considerably, but was still higher than that of the plasma membrane donor, which had the lowest activity. This experiment shows that donor activity can be chased into and out of a compartment between the TGN and the plasma membrane, as would be expected of a cell-free assay for the insertion of a protein into the plasma membrane.

Time-course of [^{3}H]NANA Release

The time-course of the ATP-dependent, influenza neuraminidase-dependent component of release of [^{3}H]NANA is shown in Fig. 4. The time-course has a distinct sigmoid shape, which may reflect a requirement for time for vesicles from the two preparations to fuse together and deliver enzyme to substrate. The time-course of the action of enzyme on substrate almost certainly contributes to the time-course of release, which will not, therefore, represent the time-course of vesicle fusion.

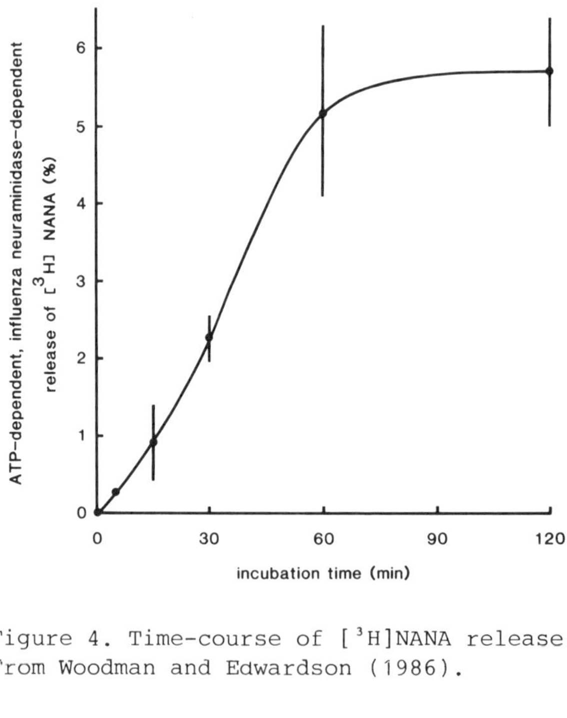

Figure 4. Time-course of [^{3}H]NANA release. From Woodman and Edwardson (1986).

Efficiency of Fusion

After a 2h incubation of donor with acceptor, the ATP-dependent, influenza neuraminidase-dependent release of [^{3}H]NANA produced is 5.7% of the total (Fig. 4). Since the latent releasable [^{3}H]NANA accounts for 10% of the total, about 57% of the [^{3}H]SFV present within sealed vesicles is participating in the fusion reaction.

Protease-sensitivity of Fusion

Since post-Golgi vesicles fuse specifically with the plasma membrane in the intact cell, it is reasonable to propose that they recognise their target through an interaction between marker proteins on the cytoplasmic surface of the vesicles and complementary receptors on the cytoplasmic surface of the plasma membrane. It is also likely that cytosolic proteins are involved in the fusion reaction. Hence, it was expected that the cell-free fusion reaction would be sensitive to pre-treatment of the preparations with trypsin. Mixed donor and acceptor preparations were treated with trypsin-TPCK (100µg/ml) at 4°C for 1h. Soybean trypsin inhibitor (100µg/ml) was then added, and the preparations were incubated at 37°C for 1h. The results obtained are shown in Table 3. After trypsin treatment, the ATP-dependent release

TABLE 3. Effect of trypsin on fusion.

donor	trypsin	ATP-dependent release of [^{3}H]NANA (%)	ATP-dependent, influenza neuraminidase-dependent release of [^{3}H]NANA (%)
blank	-	1.0±0.2	-
donor	-	5.9±0.4	4.9±0.4
blank	+	1.0±0.6	-
donor	+	1.7±0.3	0.7±0.7

Values are means ± SEM of triplicate determinations.

given by donor was not significantly different from that given by blank donor. This experiment indicates that proteins are involved in the fusion reaction. However, since crude post-nuclear supernatants were used, it is not possible to say whether the crucial proteins are present on the surfaces of the vesicles or in the cytosol.

DISCUSSION

We have devised a cell-free assay for the insertion of a model glycoprotein, influenza neuraminidase, into the plasma membrane. The assay involves placing neuraminidase and an appropriate substrate, [^{3}H]NANA-labelled SFV, into two populations of vesicles and monitoring fusion through the appearance of product, free [^{3}H]NANA. An ATP-dependent component of release is obtained that appears to be a consequence of vesicle fusion, and that is maximised by concentrating neuraminidase in a cellular compartment between the TGN and the plasma membrane. We believe that the assay represents a useful starting point for an approach to the problem of the biogenesis of the plasma membrane.

A more detailed study of the properties of the fusion reaction is now required. For example, we must determine the time-course of the fusion reaction itself, the nature of the reconstituted event, and whether it can be divided into substages, the roles of proteins on the cytoplasmic surfaces of the vesicles and in the cytosol, and the characteristics of these proteins. If this information can be obtained, it will begin to shed light on the mechanisms involved in the assembly of the plasma membrane.

The two envelope glycoproteins of the influenza virus, neuraminidase and haemagglutinin, are delivered exclusively to the apical surface of monolayers of cultured epithelial cells, such as Madin-Darby canine kidney (MDCK) cells (Rodriguez-Boulan and Sabatini, 1978; Rodriguez-Boulan and Pendergast, 1980). It has been shown that these glycoproteins are segregated from basolaterally directed glycoproteins, such as glycoprotein G of vesicular stomatitis virus (VSV), after they have passed through the Golgi complex (Rindler et al., 1984; Fuller et al., 1985) but before they have reached the plasma membrane (Matlin and Simons, 1984; Misek et al., 1984). In fact, the most likely sorting site is the TGN (Griffiths and Simons, 1986). It seems likely, therefore, that the two sets of glycoproteins are packaged into separate populations of vesicles budding from the TGN, and that these two populations of vesicles are then targetted to the two domains of the plasma membrane. This raises the possibility that a modified version of the present cell-free assay may contribute to the elucidation of the mechanisms underlying the asymmetric delivery of proteins to the cell surface in epithelia. However, before this possibility can be realised, it will be necessary to devise a method for binding a target [^{3}H]NANA-labelled glycoprotein separately to both surfaces of an epithelial monolayer.

We thank the Wellcome Trust and the National Kidney Research Fund for financial support. PGW is a Science and Engineering Research Council Scholar.

REFERENCES

Fries E, Helenius A (1979). Binding of Semliki Forest virus and its isolated glycoproteins to cells. Eur J Biochem 97: 213-220.

Fuller SD, Bravo R, Simons K (1985). An enzymatic assay reveals that proteins destined for the apical or basolateral domains of an epithelial cell line share the same late Golgi compartments. EMBO J 4: 297-307.

Griffiths G, Pfeiffer S, Simons K, Matlin K (1985). Exit of newly synthesised membrane proteins from the trans cisterna of the Golgi complex to the plasma membrane. J Cell Biol 101: 949-964.

Griffiths G, Simons K (1986). The trans Golgi network : sorting at the exit site of the Golgi complex. Science 234: 438-443.

Matlin KS, Simons K (1983). Reduced temperature prevents transfer of a membrane glycoprotein to the cell surface but does not prevent terminal glycosylation. Cell 34: 233-243.

Matlin KS, Simons K (1984). Sorting of an apical plasma membrane glycoprotein occurs before it reaches the cell surface in cultured epithelial cells. J Cell Biol 99: 2131-2139.

Meindl P, Bodo G, Palese P, Schulman J, Tuppy H (1974). Inhibition of neuraminidase activity by derivatives of 2-deoxy-2,3-dehydro-N-acetylneuraminic acid. Virology 58: 457-463.

Misek DE, Bard E, Rodriguez-Boulan E (1984). Biogenesis of epithelial polarity : intracellular sorting and vectorial exocytosis of an apical plasma membrane glycoprotein. Cell 39: 537-546.

Monaco F, Robbins J (1973). Incorporation of N-acetylmannosamine and N-acetylglucosamine into thyroglobulin in rat thyroid in vitro. J Biol Chem 248: 2072-2077.

Palade G (1975). Intracellular aspects of the process of protein secretion. Science 189: 347-358.

Rindler MJ, Ivanov IE, Plesken H, Rodriguez-Boulan E, Sabatini DD (1984). Viral glycoproteins destined for apical and basolateral domains traverse the same Golgi apparatus during their intracellular transport in doubly infected Madin-Darby canine kidney cells. J Cell Biol 98: 1304-1319.

Rodriguez-Boulan E, Pendergast M (1980). Polarized distribution of viral envelope proteins in the plasma membrane of infected epithelial cells. Cell 20: 45-54.

Rodriguez-Boulan E, Sabatini DD (1978). Asymmetric budding of viruses in epithelial monolayers : a model system for study of epithelial polarity. Proc Natl Acad Sci USA 75: 5071-5075.

Woodman PG, Edwardson JM (1986). A cell-free assay for the insertion of a viral glycoprotein into the plasma membrane. J Cell Biol 103: 1829-1835.

Cell-Free Analysis of Membrane Traffic, pages 355–360

RECONSTITUTION OF AN ENDOCYTIC FUSION EVENT IN A CELL-FREE SYSTEM: SOME CHARACTERISTICS OF FUSION *IN VITRO*

Stella Hurtley and Graham Warren*.

European Molecular Biology Laboratory, Heidelberg, Germany and *Dept. Biochemistry, University of Dundee, Scotland, U.K.

INTRODUCTION

During receptor-mediated endocytosis particles bind to the cell surface and are internalised by the formation of clathrin-coated pits which pinch off into the cytosol to form coated vesicles. The loss of coat from the vesicle presumably then allows it to fuse with an endosome. Subsequently, the endosomal contents are delivered to the lysosome and the membrane is recycled to the cell surface (for review see Goldstein et al., 1979).

In order to study the mechanisms involved in the endocytic pathway we attempted to reconstitute stages of this process in a cell-free system. An assay was designed to monitor the fusion of vesicles derived from the endocytic pathway (Davey et al., 1985).

Briefly, for the assay, we produce two components. First we bind ^{3}H-sialic acid labeled-Semliki Forest virus (^{3}H-SA-SFV) to the surface of baby hamster kidney (BHK) cells at 4°C for 2 hours, and then the virus is internalised by the cells during an incubation at 20°C for 2 hours; surface virus is removed with protease and a post nuclear supernatant (PNS) is prepared. This PNS containing ^{3}H-SA-SFV accumulated in "endosomes" is termed acceptor membranes (AM). To produce the second component Fowl Plague virus (FPV) is bound to the surface of Madin-Darby canine kidney (MDCK) cells for 2 hours at 4°C and then internalisation is triggered by incubation at 37°C for 20 minutes; again a PNS is prepared. These are termed donor membranes (DM) and contain FPV, which has a neuraminidase as one of its envelope proteins, in endocytic vesicles. With appropriate controls, fusion is monitored by the release of ^{3}H-SA.

The assay revealed an energy-dependent, time-dependent cell-free fusion reaction between vesicle preparations. Fusion was shown to require intact membranes. Only vesicles derived from certain stages of the endocytic pathway (after internalisation, before lysosomes) were capable of participating in the reaction.

Here we present data from the further characterisation of the assay. Efficiency of fusion in the assay did not depend upon the low pH found inside endocytic vesicles and was partially inhibited by the presence of μM free Ca^{2+}. Fusion only occurred at temperatures greater than 20°C. Fusion in the cell-free assay showed an absolute requirement for soluble cytoplasmic factors, and these factors could be inactivated by brief exposure to high temperatures.

RESULTS AND DISCUSSION

One characteristic of endocytic vesicles is their low intra-lumenal pH (Galloway et al., 1983). Figure 1 demonstrates that chloroquine, a weak base which can raise the pH inside vesicles to that of the surrounding buffer, had no inhibitory effect on the energy-dependent, Fowl Plague neuraminidase-dependent release of ^{3}H-SA in the cell-free fusion assay. This suggests that the low pH found inside endocytic vesicles is not important for their fusion *in vitro*. This is not surprising since *in vivo* endocytosed ligands do efficiently enter cells as far as lysosomes in the presence of chloroquine (Marsh and Helenius, 1980).

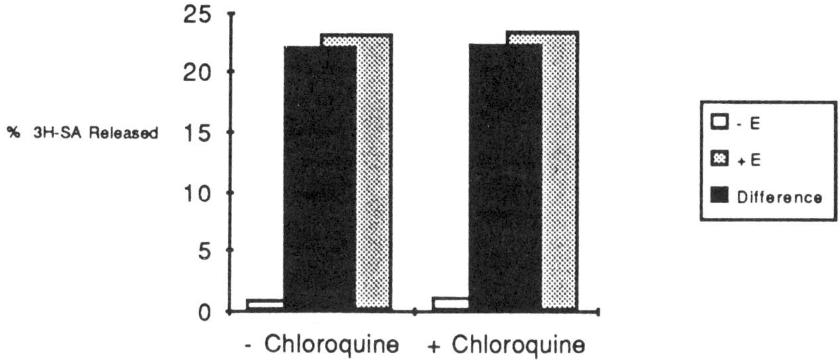

Figure 1. Effect of chloroquine.
AM and DM were incubated in a standard assay mixture in the presence or absence of energy (E) and in the presence or absence of 0.1mM chloroquine. All incubations are for 2 hours at 37°C, unless otherwise stated. Releases produced by mock DM (DM with no added FPV) have been subtracted. Standard assays, in a total volume of 250μl, contained: 100μl each AM and DM, 25mM Hepes pH 7.0, 25mM KCl, 1.5mM Mg acetate, 1mM DTT, 1mM EGTA pH 7.0, 10μg/ml each of chymostatin, leupeptin, antipain and pepstatin, 1mM DDNA; in tubes - energy: 0.2mg/ml hexokinase, 50mM D-glucose; in tubes + energy: 1mM ATP, 8mM creatine phosphate and 0.04 mg/ml creatine phosphokinase.

Ca^{2+} has often been implicated in fusion reactions, but its involvement in the regulation of endocytosis has not been clearly demonstrated, and no involvement could be demonstrated by Iacopetta et al., 1986. In Figure 2 the effect of titrating in increasing amounts of free Ca^{2+} into the cell-free assay is illustrated. Generally the assay is performed in the presence of 1mM EGTA with no added Ca^{2+}. Addition of only μM free Ca^{2+} actually inhibits the cell-free fusion of endocytic vesicles by 30%. The reason for this inhibition is not clear, but it is clear that levels of free Ca^{2+} in the μM range or higher do not play a role in mediating the fusion of vesicles derived from the endocytic pathway in this cell-free system.

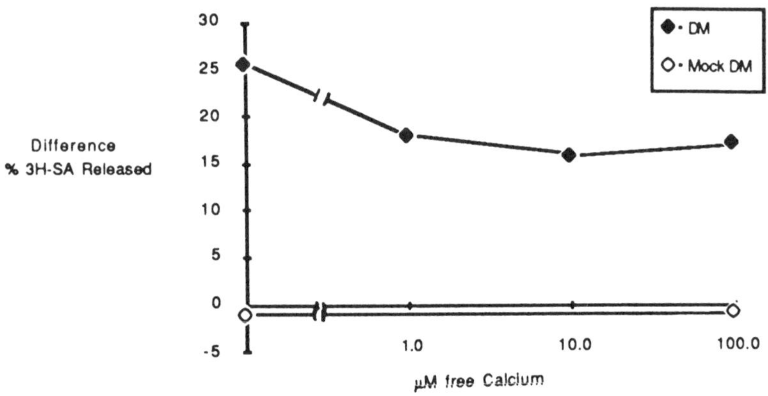

Figure 2. Calcium titration.
A standard assay was performed in the presence of increasing concentrations of added Ca^{2+}

Fusion is most efficient in the cell-free assay at 37°C, the temperature at which its physiological counterpart takes place in intact cells (Figure 3). At 20°C no efficient fusion can be monitored and at 45°C fusion activity in the assay is abolished. At all the temperatures studied the FPV neuraminidase in the DM was working at between 65% and 120% of its activity at 37°C.

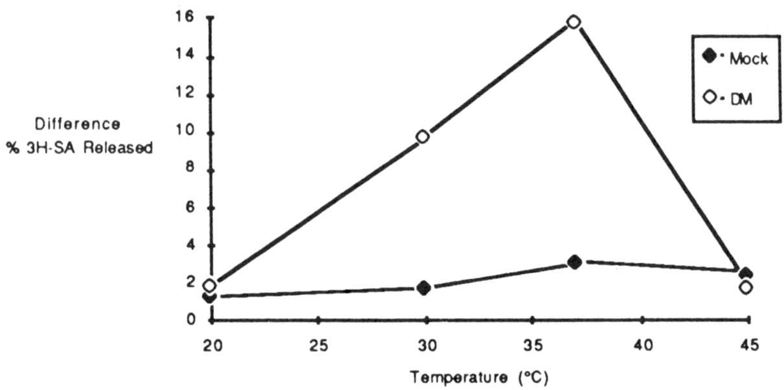

Figure 3. Temperature dependence.
Standard assays were performed for 2 hours at the indicated temperatures.

The membranes in the AM and DM preparations were sedimented and resuspended either in cytosol or in homogenisation buffer containing BSA to normalise protein concentration. Figure 4 demonstrates that there is an absolute requirement for soluble cytoplasmic factors to mediate the fusion of endocytic vesicles *in vitro*. In the absence of added cytosol there is absolutely no energy-dependent, Fowl Plague-dependent release of ^{3}H-SA.

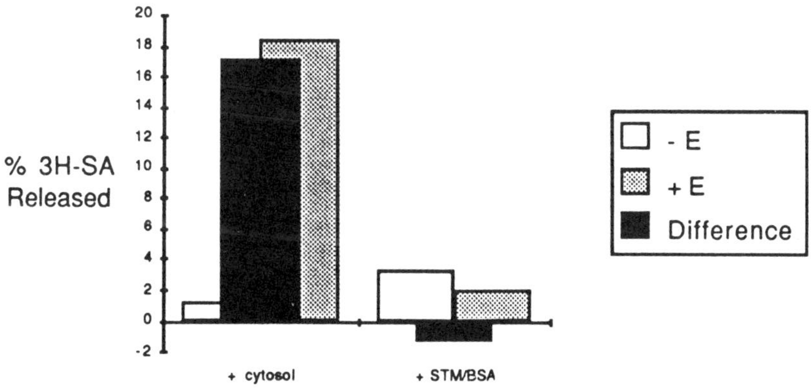

Figure 4. Requirement for cytosol.
AM and DM preparations were sedimented at 100,000g for 5 minutes in a Beckman airfuge and the membrane pellets resuspended in either a cytosol fraction (100,000g cell supernatant) or in 5mg/ml BSA in STM (0.25M Sucrose, 10mM Tris-Cl pH 7.5, 1mM Magnesium acetate), the resuspended preparations were then assayed as usual for 2 hours at 37°C.

Furthermore, as shown in Figure 5, when the cytosol used to resuspend the AM and DM is incubated briefly at elevated temperatures, it loses its ability to support the fusion event in the cell-free assay.

These findings can now be exploited in order to identify the factors required to facilitate efficient, accurate fusion of endocytic vesicles in the cell-free system.

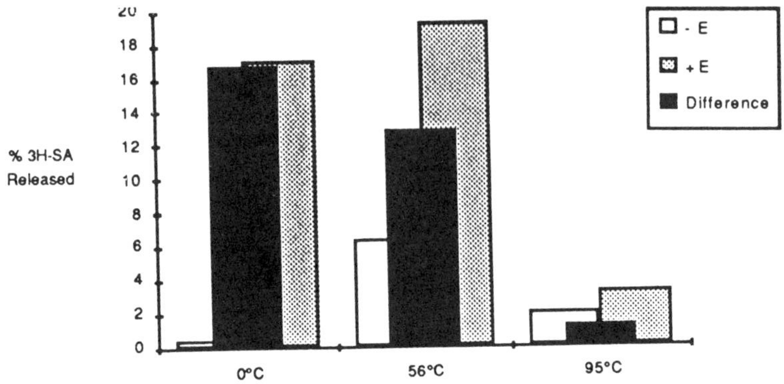

Figure 5. Heat treatment of cytosol.
AM and DM preparations were sedimented as before and resuspended in cytosol fractions which had been treated for 5 minutes at the indicated temperatures.

REFERENCES

Davey J, Hurtley SM, Warren G (1985). Reconstitution of an endocytic fusion even in a cell-free system. Cell 43:643-652.

Galloway CJ, Dean GE, Marsh M, Rudnick G, Mellman I (1983). Acidification of macrophage and fibroblast endocytic vesicles *in vitro*. Proc Natl Acad Sci USA 80:3334-3338.

Goldstein JL, Anderson RGW, Brown MS (1979). Coated pits, coated vesicles and receptor-mediated endocytosis. Nature 279:679-685.

Iacopetta B, Carpentier J-L, Pozzan T, Lew DP, Gorden P, Orci L (1986). Role of intracellular calcium and Protein Kinase C in the endocytosis of transferrin and insulin by HL60 cells. J Cell Biol 103:851-856

Marsh M, Helenius A (1980). Adsorptive endocytosis of Semliki Forest virus. J Mol Biol 142:439-454.

Section IV
ANALYSES OF MEMBRANE TRAFFIC AND APPLICATIONS

Cell-Free Analysis of Membrane Traffic, pages 363–364

Analyses of Membrane Traffic and Applications

The final section of the book summarizes a range of new and developing approaches to the study of membrane traffic. Topics include a detailed consideration of the mannose-6-phosphate receptor and the transport of lysosomal hydrolases, the development of a covalent membrane marker to quantify membrane internalization and recycling, some functional characteristics of the rat liver endosomal apparatus, an *in vitro* system for studying microlipid droplet fusion, transferrin-iron uptake by the liver, the identification of endosome specific antigens, the mitotic form of the Golgi apparatus and the application of endocytosis research to drug delivery.

Pfeffer et al., discuss the targeting of proteins to lysosomes, and show by endosome-specific radiolabeling, that mannose-6-phosphate (cation independent) receptor appears to cycle constitutively between intracellular compartments independently of the presence of newly synthesized ligands. Minnifield et al. use free-flow electrophoresis (see also Part IA) to separate a fraction enriched in transitional elements and small vesicles from rough endoplasmic reticulum which contain the phosphomannosyl receptor. This finding is of interest since endoplasmic reticulum elements are implicated in the transfer of lysosomal enzymes and phosphomannosyl receptor from the sites of translation on polysomes attached to rough endoplasmic reticulum to sites of concentration in the cis elements of the Golgi apparatus.

Thilo shows that the terminal galactosylation of cell surface glycoconjugates is a useful procedure for quantifying membrane internalization. By clearing the surface of remaining label by -galactosidase treatment, under conditions where membrane flow is arrested, it is possible to study membrane recycling as previously internalized label again becomes accessible to release when it is recycled to the cell surface under appropriate conditions.

Bergeron et al. examine the physiological significance of the internalization of the peptide hormones insulin and prolactin, and epidermal growth factor in concert with their receptors. It appears that peptide hormone and growth factor internalization may be related to ligand action since transmembrane signalling implicating the insulin and EGF receptors has been demonstrated with isolated endosomes. Valivullah et al. introduce an *in vitro* system to study microlipid droplet fusion and explore the various factors and conditions controlling this event. Fusion was facilitated both by calcium and a cytosol fraction. Goldenberg et al. describe transferrin uptake by livers of rats *in situ* where a specific transferrin receptor mediated uptake and recycling mechanism is operative. Ahluwalia et al. suggest that the availability of endosome-specific antibodies would be useful not only for dissecting the endocytic pathway, but would also help to establish endosomes as distinct entities. Antisera raised to Na_2CO_3 washed 'late' endosomal membranes are described that recognize an immunologically distinct set of proteins of the parent endosome fraction.

Little is known about the detailed mechanisms of fragmentation and re-assembly of the Golgi apparatus of cells undergoing mitosis. Lucocq et al. report on their immunoelectron microscopic studies on the localization in mitotic cells of galactosyltransferase, a resident protein of Golgi apparatus membranes. It appears that in such cells galactosyltransferase is found almost exclusively in distinct globular clusters of vesicles and tubules which these authors term 'Golgi clusters.' As yet, it is neither known whether all Golgi membrane proteins are present in 'Golgi clusters' nor what is the role of these clusters in division. However, from a methodological point of view, this chapter highlights the utility of immunocytochemical techniques to follow the fate of an organelle undergoing major changes in morphology.

In the final chapter, Wilson et al. review how recent progress in understanding the morphological and biochemical details of endocytic compartments will aid in the design of drug delivery systems. For example, the use of replication-defective retroviral envelopes for the delivery of replacement genes to cells *ex vivo* is close to human clinical trials, and these and related developments may have future applications for the delivery of drugs *in vivo*.

Cell-Free Analysis of Membrane Traffic, pages 365–375

INTRACELLULAR TRANSPORT OF THE MANNOSE-6-PHOSPHATE RECEPTOR

Suzanne R. Pfeffer, Yukiko Goda and Nancy W. Chege

Department of Biochemistry, Stanford University School of Medicine, Stanford, California 94305

INTRODUCTION

Proteins destined for the plasma membrane, secretory vesicles and lysosomes traverse a common pathway of intracellular transport until they reach the Golgi apparatus, where they are somehow sorted to their correct destinations. The targetting of proteins to lysosomes is presently understood in the greatest detail (see Kornfeld, 1986, von Figura and Hasilik, 1986, for review). Upon arrival in the Golgi apparatus, newly synthesized lysosomal hydrolases are recognized by an N-acetylglucosaminyl phosphotransferase that adds GlcNAc-P to terminal α-1,2 mannose residues. A specific phosphodiesterase then removes the GlcNAc to generate the Man-6-P residues that enable lysosomal enzymes to bind to Man-6-P receptors in the Golgi apparatus. It is likely that receptor-bound hydrolases are then transported to an acidic prelysosomal compartment, within which the ligand is released for transport to lysosomes, and the free receptor recycles back to the Golgi apparatus.

Two types of Man-6-P receptors have been identified. One has been purified from a number of sources, and is a single polypeptide of 215,000 daltons (Sahagian et al., 1981). However, the 215kD receptor is not present in several cultured cell lines that adequately target their hydrolases to lysosomes (Gabel et al., 1983). Such cells were recently found to contain a second Man-6-P receptor which appears to be a glycoprotein comprised of three, 46kD subunits (Hoflack and Kornfeld, 1985a,b). This receptor requires divalent cations, and has thus been termed Man-6-P receptor "CD" (cation dependent) to distinguish it from the 215kD cation-independent, Man-6-P receptor "CI".

At steady state, the Man-6-P receptor^{CI} is located in the Golgi apparatus and in endosomes; a small fraction is present on the cell surface (Willingham et al., 1983; Geuze et al, 1984a; Brown and Farquhar, 1984a). Recent experiments suggest that the Man-6-P receptor^{CI} equilibrates rapidly between these intracellular locations, since at least 90% of the 215kD Man-6-P receptors contact extracellularly administered receptor antibodies within 90 minutes (Sahagian, 1984; von Figura et al., 1984). However, several reports have indicated that the receptor may accumulate in the cis Golgi under conditions in which synthesis of lysosomal hydrolases or the Man-6-P recognition marker is blocked (Brown and Farquhar, 1984b; Brown et al., 1984). These findings could only be explained if the Man-6-P receptor^{CI} was selectively retained in the Golgi complex and only exported after ligand binding.

In order to study the intracellular transport of the Man-6-P receptor^{CI}, we have employed an endosome-specific radiolabelling procedure to biochemically detect the 215kD Man-6-P receptor when present in endosomes. To our surprise, we found that the steady state concentration of the Man-6-P receptor^{CI} in endosomes of HepG2 cells was quantitatively unaltered under conditions in which lysosomal hydrolase synthesis was completely blocked. Thus, it appears that the 215kD Man-6-P receptor cycles constitutively between intracellular compartments, independent of the presence of newly synthesized ligands.

EXPERIMENTAL PROCEDURE

Preparation of conjugate: Asialoorosomucoid (ASOR) coupled to lactoperoxidase (ASOR-LPO) was prepared exactly as described by Watts (1984). The product corresponding to one molecule of lactoperoxidase conjugated to one or two molecules ASOR was purified by chromatography on a 250 ml Sephacryl S-300 column run at

16 ml/h.

Endosome labelling: The procedure of Watts (1984) was employed using 10 cm dishes per reaction. Monolayers were incubated with conjugate for 2h in a 23°C waterbath to "load" endosomes; iodinations were carried out at 4°C for 2 hours. Culture dishes were placed onto ice-cooled steel blocks covered with water-soaked paper towels to maintain an iodination reaction temperature of $\leq$4°C. After terminating the reaction, cells were harvested directly into 1 ml, 0.15M NaCl, 20mM Na phosphate, 1% Triton X-100, 0.1% deoxycholate, 5mM KI and frozen in liquid nitrogen for subsequent analyses.

Affinity chromatography: Pentamannosyl-6-phosphate fragments of yeast phosphomannan were derivatized with p-(aminophenyl)ethylamine and coupled to CNBr-activated Sepharose CL-4B as described (Brown and Farquhar, 1984a). Labelled extracts (0.4 ml) were chromatographed at 4°C on 0.5 ml columns in 50 mM imidazole, pH 7, 150 mM NaCl, 5 mM β-glycerophosphate, 0.05% Triton X-100 according to the procedure of Hoflack and Kornfeld (1985b).

Other techniques: Immunoprecipitations were carried out as described (Louvard et al., 1982) except that *Staphylococcus aureus* cells were substituted for protein A-Sepharose. Densitometry was carried out using a Quick Scan JR scanning densitometer. Care was taken to ensure linearity of the response.

RESULTS

One can easily imagine a model in which the Man-6-P receptor, upon ligand binding, undergoes a conformational change that triggers its vesicular transport from the Golgi to pre-lysosomes. Alternatively, the receptor may shuttle between the Golgi and pre-lysosomes, whether or not it is occupied. Since the Man-6-P receptor is in rapid equilibrium between cellular compartments, a strong prediction of the first model is that in the absence of new ligand synthesis, the Man-6-P receptor should disappear from pre-lysosomes and accumulate in the Golgi complex. We have employed an endosome-labelling procedure in an attempt to distinguish between these two models for Man-6-P receptor trafficking.

Mellman and colleagues have shown that

lactoperoxidase internalized by fluid phase endocytosis can be used to selectively label endocytic vesicle proteins from within intact cells (Mellman et al, 1980; Mellman and Galloway, 1983). By conjugating lactoperoxidase to a specific ligand, Watts has adapted this procedure to make use of the advantages of receptor-mediated endocytosis (Watts, 1984). Unlike fluid phase uptake, ligands internalized by receptor-mediated endocytosis can be concentrated within cells by virtue of their high affinity association with specific cell surface receptors. The receptor-mediated labelling procedure devised by Watts (1984) was employed to iodinate proteins within the endosome compartment of human hepatoma (HepG2) cells. Cells were incubated with a conjugate of asialoorosomucoid coupled to lactoperoxidase (ASOR-LPO). As shown previously (Watts, 1984), the conjugate bound to cell surface asialoglycoprotein receptors, since after binding at 4°C, ASOR-LPO could catalyze a cell surface iodination reaction in the presence of ^{125}I, glucose oxidase and glucose that depended entirely upon the presence of the conjugate, was greatly diminished by addition of excess ASOR and was eliminated after washing the cells with 5 mM EDTA, a treatment that interferes with receptor-ligand association. If cells were allowed to endocytose ASOR-LPO at 23°C and were then washed with EDTA to remove cell-surface conjugate, endosome-specific iodination could be achieved after incubating intact cells with ^{125}I, glucose oxidase, and glucose. Under these conditions, most of the ASOR-LPO is not transported beyond endosomes to lysosomes (Dunn et al., 1980), and the ^{125}I and extracellularly generated H_2O_2 must permeate endosomal and plasma membranes for intra-organellar iodination (Mellman et al., 1980).

Watts has shown that the transferrin receptor is among the major polypeptides labelled within endosomes of HepG2 cells using this procedure. To determine if the Man-6-P receptor^{CI} could also be detected in endosomes, HepG2 cells were processed for endosome labelling, and the resulting extract subjected to affinity chromatography on a pentamannosyl-6-phosphate-Sepharose column. The Man-6-P receptor^{CI} should bind selectively to this resin in the absence of divalent cations. As can be seen in Fig. 1, a peak of radioactive material was eluted from this column after addition of Man-6-P but not glucose-6-P, characteristic of this receptor. Analysis of the fractions by SDS-polyacrylamide gel electrophoresis and autoradiography confirmed that the eluted material corresponded to the 215kD Man-6-P receptor. Therefore, the endosome-radiolabelling protocol led to iodination of the Man-6-P $receptor^{CI}$ in HepG2 cell endosomes.

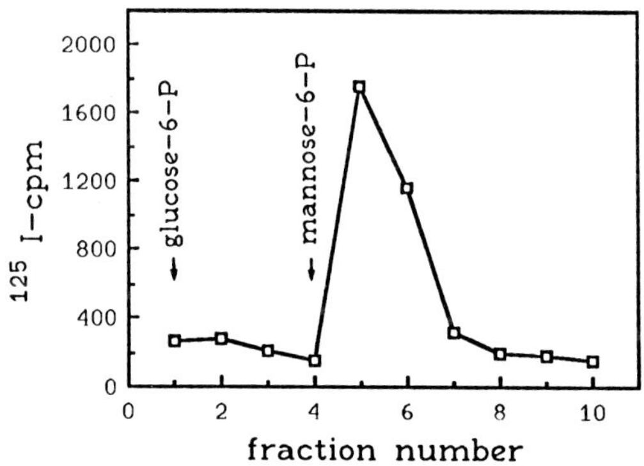

Figure 1. Pentamannosyl-6-phosphate-Sepharose chromatography of endosomal proteins.

If the 215kD Man-6-P receptor accumulates in the Golgi apparatus in the absence of lysosomal hydrolase synthesis, it should be correspondingly depleted from endosomes after a period of cycloheximide treatment. To test this, cells were treated with cycloheximide (100 µg/ml) for up to six hours to permit complete transit of newly synthesized lysosomal hydrolases to lysosomes. A variety of hydrolases require only 1-3 hours for complete lysosome delivery. The conditions employed blocked at least 98% of protein synthesis as measured by [^{35}S]-methionine incorporation, and did not effect Man-6-P receptor turnover (data not shown).

Cells treated with cycloheximide for various times were incubated with ASOR-LPO for 2 hours at 23° and then processed as before for endosomal radiolabelling. Individual samples were then analyzed for both the Man-6-P $receptor^{CI}$ (Fig. 2a) and the transferrin receptor (Fig. 2b). The level of the transferrin receptor in endosomes should not be affected by cycloheximide treatment since this protein has been shown to cycle through the endocytic pathway at the same rate in the presence or absence of bound transferrin (Watts, 1985), and cycloheximide has no effect on the recycling of this receptor in the same cell line under virtually identical conditions (Ciechanover et al., 1983). Therefore, the transferrin receptor serves as an internal standard for endosomal labelling.

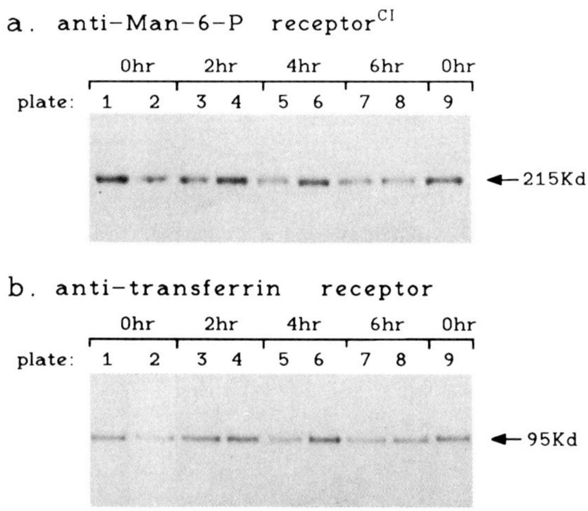

Figure 2. Effect of cycloheximide on endosomal Man-6-P and transferrin receptors. Individual plates of HepG2 cells were pretreated for the indicated times with 100µg/ml cycloheximide and then processed for endosomal protein iodination. Panel a, of Man-6-P receptor^{CI}; panel b, transferrin receptor. The plate samples used for immunoprecipitation in panel a. are identical with those used in panel b. Plate 9 was incubated with ASOR-LPO in the absence of cycloheximide; all other ASOR-LPO incubations were carried out in the continued presence of cycloheximide.

Despite some variability, the ratio of labelled Man-6-P receptor to transferrin receptor was constant (Figs. 2,3). The observed variability was due entirely to differences in cell number between plates, since the radioactive incorporation per mg protein was comparable for all samples (data not shown). Quantitation of these data by densitometric analysis showed that the relative endosomal concentrations of the Man-6-P receptor^{CI} and the transferrin receptor were unaffected by cycloheximide treatment (Fig. 3).

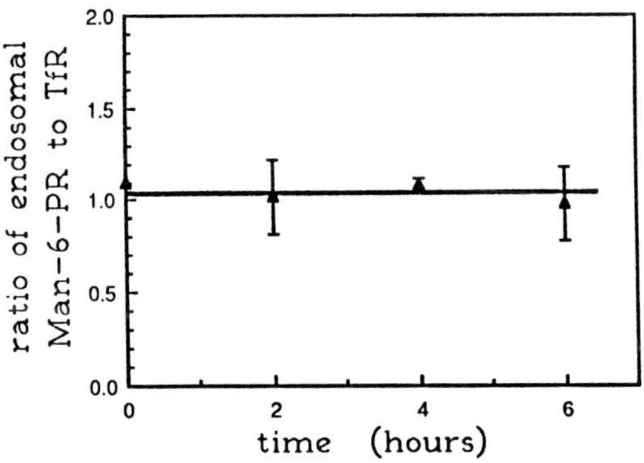

Figure 3. Quantitation of the effect of cycloheximide on endosomal Man-6-P and transferrin receptors. Data presented in Fig. 2, panels a and b were quantitated by densitometric scanning. The relative amounts of the two labelled receptors in a single sample after various times of treatment are shown in arbitrary units; triangles represent the average of duplicate samples.

To investigate the sensitivity of the endosome labelling procedure, antibody treatment was used to deplete cells of a large fraction of their total cellular Man-6-P receptors (von Figura et al., 1984), and the level of Man-6-P receptors remaining in endosomes was then determined. Under conditions that deplete cells of two-thirds of their Man-6-P receptors, the level of Man-6-P receptors detected by the endosome radiolabelling procedure also decreased by two-thirds, while the level of transferrin receptors was unchanged. This experiment demonstrates that the endosome-specific radiolabelling procedure can detect a change in endosomal receptor content. Furthermore, the correspondence between the loss of total cellular

Man-6-P receptors (66%) and the loss of endosomal receptors (67%) indicates that the pool of Man-6-P receptors labelled in this experiment is in equilibrium with the total cellular Man-6-P receptor pool.

The experiments described above demonstrate that the level of Man-6-P receptors present in endosomes does not change in the absence of new ligand synthesis. These findings are consistent with the notion that the Man-6-P receptor cycles continuously between cellular compartments, whether or not it is occupied. This conclusion would only be valid if the receptor remained in equilibrium with other cellular compartments in the absence of ligand synthesis. The ability of a singificant fraction of intracellular Man-6-P receptors to contract extracellular antibodies was used as a measure of the cycling of this receptor between cellular compartments, and no difference was detected in the presence or absence of cycloheximide (data not shown).

DISCUSSION

Immunocytochemical experiments indicated that the Man-6-P receptor^{CI} accumulates in the Golgi apparatus under conditions in which synthesis of lysosomal hydrolases or the Man-6-P recognition marker is blocked (Brown and Farquhar, 1984; Brown et al., 1984; but see also Geuze et al., 1984b). This would require that the departure of the Man-6-P receptor from the Golgi is coupled to ligand binding. In an attempt to confirm these observations using biochemical methods, we have used an endosome-specific labelling procedure to detect and quantify the Man-6-phosphate receptor^{CI} as well as the transferrin receptor in HepG2 cell endosomes. If the receptor accumulates in the Golgi, it should be concomitantly depleted from endosomes. Under conditions in which 98% of protein synthesis was blocked, and after incubation periods that should have been far beyond what was necessary to transfer most newly synthesized lysosomal hydrolases to lysosomes, we were unable to detect any alteration in the steady state levels of either receptor in endosomes. Analogous results were obtained using tunicamycin to block N-linked oligosaccharide addition and thus attachment of the Man-6-P recognition marker (data not shown). We cannot explain the discrepancy between these biochemical data and the immunocytochemical experiments of Brown and Farquhar (1984), (Brown et al., 1984).

The steady state measurements reported here indicate that the cellular distribution of the Man-6-P receptor is not affected by the presence or absence of transiting lysosomal enzymes. Since Man-6-P receptors in the Golgi, endosomes, and at the cell surface are in rapid equilibrium, even in the absence of protein synthesis, it is highly unlikely that the endosomal Man-6-P receptor measured in these experiments represents a separate pool of receptor that cycles independently. Taken together, these findings strongly suggest that the Man6P receptor cycles continuously between intracellular compartments with or without bound ligand. A major challenge for the future will be to identify the signals used to target Man-6-P receptors in the Golgi apparatus to a prelysosomal compartment, and then return them to the Golgi for another round of lysosomal hydrolase transport.

REFERENCES

Brown WJ, Farquhar MG (1984a). The mannose-6-phosphate receptor for lysosomal enzymes is concentrated in cis Golgi cisternae. Cell 36: 295-307.

Brown WJ, Farquhar MG (1984b). Accumulation of coated vesicles bearing mannose 6-phosphate receptors for lysosomal enzymes in the Golgi region of I-cell fibroblasts. Proc Natl Acad Sci USA 81: 5135-5139.

Brown WJ, Constantinescu E, Farquhar MG (1984). Redistribution of mannose-6-phosphate specific receptors induced by tunicamycin and chloroquine. J Cell Biol 99: 320-326.

Ciechanover A, Schwartz AL, Dautry-Varsat A, Lodish, HF (1983). Kinetics of internalization of transferrin and the transferrin receptor in a human hepatoma cell line. Effect of lysosomotropic agents. J Biol Chem 258: 9681-9689.

Dunn WA, Hubbard AL, Aronson NN (1980). Low temperature selectively inhibits fusion between pinocytic vesicles and lysosomes during heterophagy of ^{125}I-asialofetuin by the perfused rat liver. J Biol Chem 255: 5971-5978.

Gabel CA, Goldberg DE, Kornfeld S (1983). Identification and characterization of cells deficient in the mannose-6-phosphate receptor: Evidence for an alternate pathway for lysosomal enzyme targeting. Proc Natl Acad Sci USA 80: 775-779.

Geuze HJ, Slot JW, Strous GJAM, Hasilik A, von Figura K (1984a). Ultrastructural localization of the mannose-6-phosphate receptor in rat liver. J Cell Biol 98: 2047-2054.

Geuze HJ, Slot JW, Strous GJ, Luzio JP, Schwartz AL (1984b). A cycloheximide-resistant pool of receptors for asialoglycoproteins and mannose 6-phosphate residues in the Golgi complex of hepatocytes. EMBO J 3: 2677-2685.

Hoflack B, Kornfeld S (1985a). Lysosomal enzyme binding to mouse P388D$_1$ macrophage membranes lacking the 215-kDa mannose-6-phosphate receptor: Evidence for the existence of a second mannose 6-phosphate receptor. Proc Natl Acad Sci USA 82: 4428-4432.

Hoflack B, Kornfeld S (1985b). Purification and characterization of a cation-dependent mannose-6-phosphate receptor from murine P388D$_1$ macrophages and bovine liver. J Biol Chem 260: 12008-12014.

Kornfeld S (1986). Trafficking of lysosomal enzymes in normal and disease states. J Clin Invest 77: 1-6.

Louvard D, Reggio H, Warren G (1982). Antibodies to the Golgi complex and the rough endoplasmic reticulum. J Cell Biol 92: 92-107.

Mellman IS, Steinman RM, Unkeless JC, Cohn, ZA (1980). Selective iodination and polypeptide composition of pinocytic vesicles. J Cell Biol 86: 712-722.

Mellman IS, Galloway CJ (1983). Selective labeling and quantitative analysis of internalized plasma membrane. Meth Enzym 98: 545-555.

Sahagian GG, Distler J, Jourdian GW (1981). Characterization of a membrane associated receptor from bovine liver that binds phosphomannosyl residues of bovine testicular β-galactosidase. Proc Natl Acad Sci USA 78: 4289-4293.

Sahagian GG (1984). The mannose 6-phosphate receptor: Function, biosynthesis and translocation. Biol Cell 51: 207-214.

von Figura K, Gieselmann V, Hasilik A (1984). Antibody to mannose-6-phosphate specific receptors induces receptor deficiency in human fibroblasts. EMBO J 3: 1281-1286.

von Figura K, and Hasilik A (1986). Lysosomal enzymes and their receptors. Ann Rev Biochem 55: 167-193.
Watts C (1984). *In situ* ^{125}I-labelling of endosome proteins with lactoperoxidase conjugates. EMBO J 3: 1965-1970.
Watts C (1985). Rapid endocytosis of the transferrin receptor in the absence of bound transferrin. J Cell Biol 100: 633-637.
Willingham MC, Pastan IH, Sahagian, GG (1983). Ultrastructural immunocytochemical localization of the phosphomannosyl receptor in chinese hamster ovary cells. J Histochem Cytochem 31: 1-11.

ACKNOWLEDGEMENTS

I thank Judith Townsend for expert preparation of this manuscript.

This research was funded by grant DK37332 from the National Institutes of Health and an Institutional Research grant from the American Cancer Society.

Cell-Free Analysis of Membrane Traffic, pages 377–390

KINETICS OF INTERNALIZATION AND RECYCLING OF CELL-SURFACE GLYCOPROTEINS

Lutz Thilo

Centre d'Immunologie INSERM-CNRS de Marseille-Luminy, Case 906, 13288 Marseille Cedex 9, France

The membrane events associated with endocytosis (review: Steinman et al., 1983) have previously been studied by either following the intracellular fate of non-specific endocytic contents markers, or by observing the behaviour of cell surface receptors via their specific ligands (review: Goldstein et al., 1985). After internalization, these markers dissociate rapidly from the membrane they tag, and as a result very limited quantitative information has been obtained for intracellular membrane traffic during endocytosis (review: Thilo, 1985b). In order to allow a more quantitative approach, a technique was developed to covalently label a large population of cell surface glycoproteins (Thilo, 1983). This made it possible to analyse internalization and recycling in a quantitative way, simultaneously for a large number of different membrane constituents, and in different cell types. The results as presented below could accommodate earlier observations obtained for endocytic content markers and for cell-surface receptors. In addition, by remaining membrane bound for extended periods of endocytosis, the marker also provided quantitative information on recycling pathways and on membrane sorting events.

THE LABELLING SYSTEM

The principle of the labelling technique is outlined in Figs. 1 and 2 (Thilo, 1983). Residues of ^{3}H-galactose (or ^{14}C-galactose) are covalently introduced on terminal N-acetylglucosamine (GlcNAc) moieties of membrane glycoproteins at ice-temperature for 20 min with milk galactosyl transferase (Fig. 3). A pretreatment with neuraminidase and β-galactosidase will increase the amount of exposed GlcNAc residues and, hence, increase the labelling efficiency. Membrane traffic is allowed to resume by raising the temperature. At selected time points, internalized label is quantified by its inaccessibility to enzymatic removal using β-galactosidase (Fig. 4). In order to demonstrate recycling of internalized label, cells are returned to ice temperature, treated with β-galactosidase to remove label on the cell surface, and the temperature is raised again to initiate recycling of previously internalized label. The amount of labelled glycoproteins that has been transported to the cell surface during this time is measured from the radioactivity released in a second round of β-galactosidase treatment.

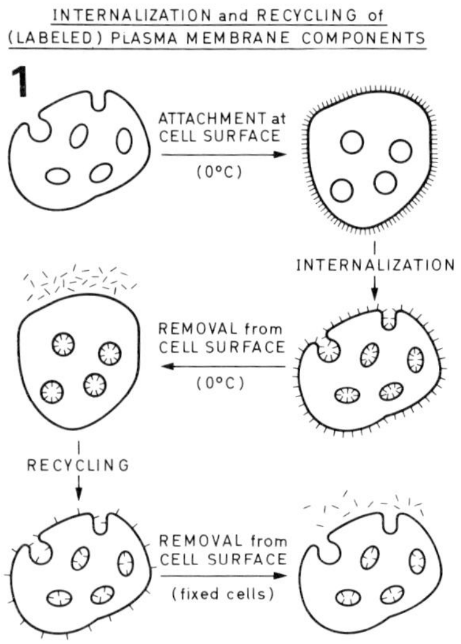

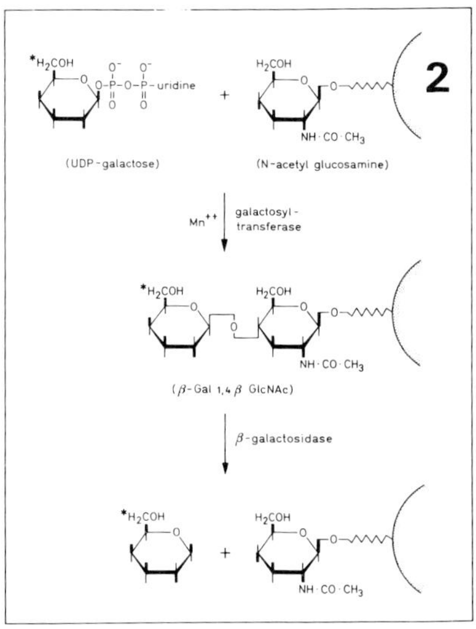

Figure 1. Internalized label is no longer accessible to removal and can therefore be distinguished quantitatively from label remaining on the cell surface. Previously internalized label again becomes accessible to release when it is recycled to the cell surface.

Figure 2. Reversible labelling of membrane glycoconjugates. ^3H or ^{14}C-labelled galactose can be used as a marker (Thilo, 1983; reprinted with the permission of Academic Press, Inc.).

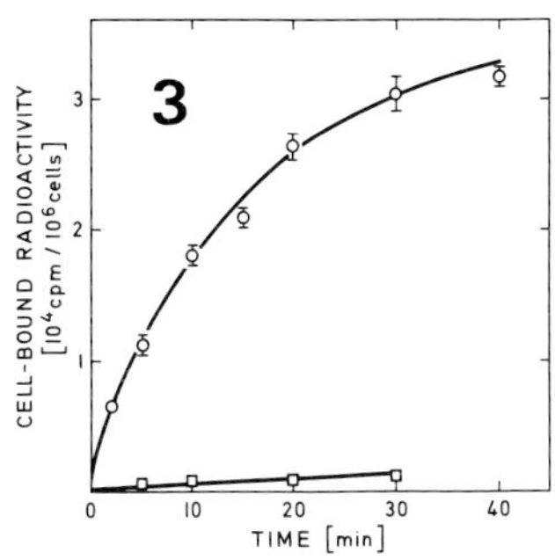

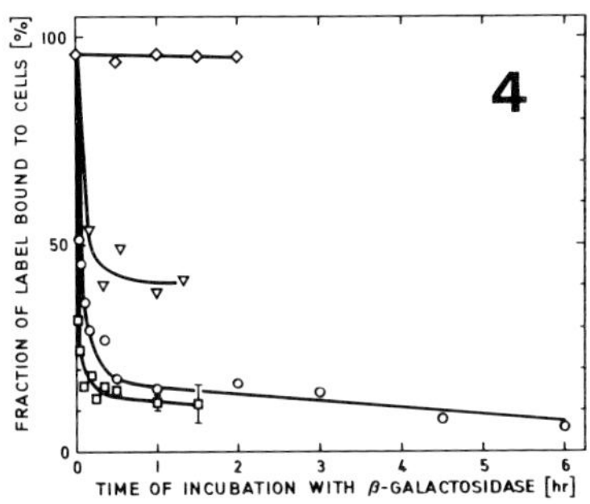

Figure 3. Kinetics of enzymatic transfer of ^{3}H-galactose (16 Ci/mmol)to the cell surface of P388D1, macrophage cells, in the presence (○) or absence (□) of galactosyltransferase (0.5 U/ml). ^{3}H counting efficiency about 20% (Burgert and Thilo, 1983; reprinted with the permission of Academic Press, Inc.).

Figure 4. Kinetics of hydrolytic release of ^{3}H-galactose from the cell surface of P388D1 cells on ice (□,▽,◇) or, when fixed with 2.5% glutaraldehyde, at 20°C (○), in the presence of β-galactosidase (0.5 U/ml, from *Streptococcus pneumoniae*;□,○), in the absence of enzyme (◇), or in the presence of pronase (5 U/ml,▽) (Burgert and Thilo, 1983; reprinted with the permission of Academic Press, Inc.).

The label provides a useful membrane marker according to the following criteria (Burgert and Thilo, 1983; Schwarz and Thilo, 1983; Thilo and Vogel, 1980; Thilo, 1985a): (i) more than about 20 different molecular size species are labelled at the cell surface (cf. Mellman et al., 1980, for cell surface iodination); (ii) during fluid-phase pinocytosis all labelled species are internalized to the same relative extent (this is also observed for iodinated cell-surface molecules, Mellman et al., 1980); (iii) the label is stable during many rounds of membrane recycling, being lost at a rate compatible with the normal turnover rate of plasma membrane proteins (T 1/2 = 10-100 h, Morré et al., 1979); (iv) the label remains membrane bound in a way characteristic for integral membrane proteins; (v) the label remains unmodified in terms of its susceptibility to hydrolysis by β-galactosidase and neuraminidase.

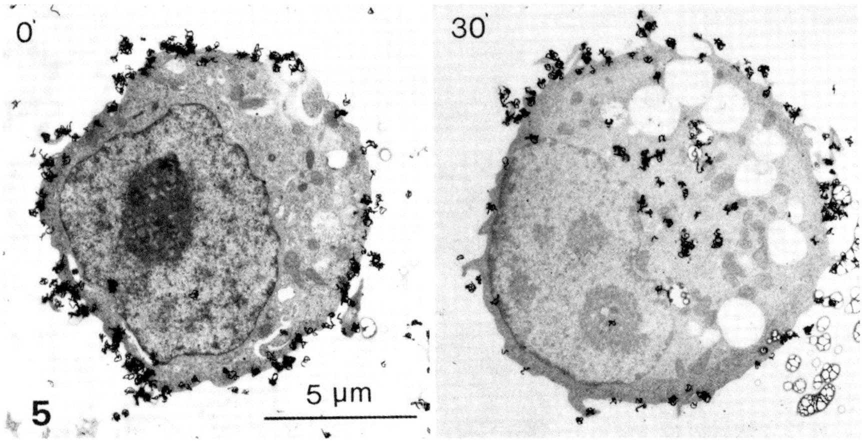

Figure 5. Autoradiographs prepared from P388D1 cells either immediately after labelling (0') or after 30 min incubation at 37°C (30'). Cells were labelled to about 0.5 dpm/cell and exposure was for about 6 weeks (courtesy, Heinz Schwarz, Tubingen).

Direct observation by means of electron microscopic autoradiography shows that the label is indeed internalized (Fig. 5). When membrane flux is measured directly by autoradiography, the results are quantitatively the same as those obtained by the faster, but indirect, biochemical method which analyses the redistribution kinetics of label (Schwarz and Thilo, 1983; De Chastellier et al., 1983, 1987).

KINETICS OF MEMBRANE TRAFFIC IN ENDOCYTOSIS

The time-dependence of internalization of the glycoproteins originally labelled at the cell surface is shown in Fig. 6, as observed for cells of the macrophage cell line, P388D1. After ~120 min, an equilibrium is reached where the rate of internalization (Fig. 6) is compensated by recycling of the internalized molecules to the cell surface (Fig. 7). In the steady-state condition 17% of the label is intracellular at any time point. To guarantee a quantitative comparison between the overall rate of endocytosis and the rate of recycling, the volume of fluid-phase uptake was measured in the two experiments (using FITC-dextran) and remained similar.

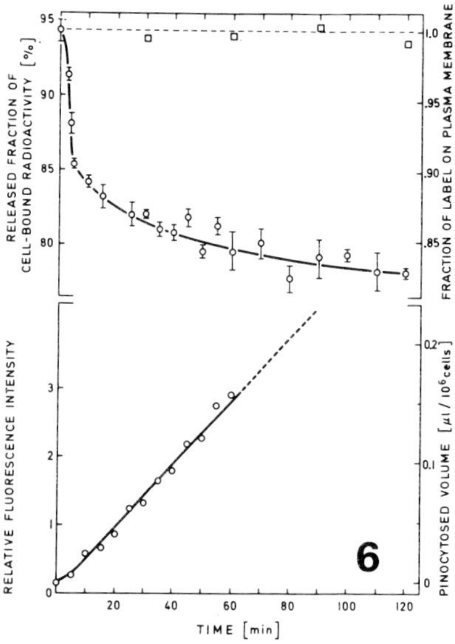

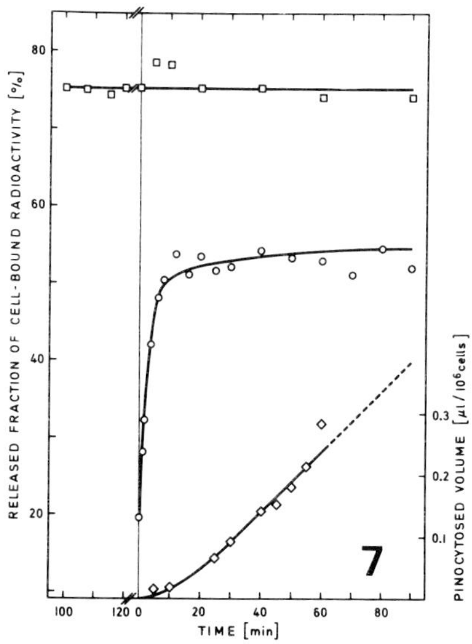

Figure 6. Internalization of plasma membrane glycoconjugates, labelled with ^{3}H-galactose. Cells were labelled at 0°C and resuspended in medium containing FITC-dextran as a fluid-phase marker. At 0 min (before warming), and at the indicated times after warming to 37°C, samples were taken to measure internalization of label (upper part) and fluid- phase uptake (lower part). Upper part: the fraction of label remaining on the cell surface after various times of internalization was determined by fixing cells in 2% glutaraldehyde, incubating the cells with β-galactosidase for 8 h at 20°C, and measuring the fraction of label hydrolysed off the cell surface. From the cells taken at 0 min about 94% of the label was released (left ordinate). This was taken to represent 100% of label on the plasma membrane (right ordinate). All other values were normalized accordingly:○, Internalization at 37°C; □, at 0°C. Lower part: the cell-associated fluorescence intensity was measured (left ordinate) and by comparison with a standard curve was converted to represent the endocytosed volume (right ordinate). (Burgert and Thilo, 1983; reprinted with the permission of Academic Press, Inc.).

Figure 7. See legend on page 382.

Figure 7. Recycling of previously internalized label. After labelling, cells were incubated in medium at 37°C for 120 min until a steady-state distribution of label was indicated (left part) with about 17% of the label internalized (cf. Fig. 6). Cells were cooled and 85% of the label remaining on the cell surface was removed during an intermediate treatment with β-galactosidase (60 min, 0°C; control was incubated without enzyme). Cells were resuspended in medium containing FITC-dextran as fluid-phase marker, warmed to 37°C and samples were taken to measure the endocytosed volume (◇), and the fraction of label that could be released from the cell surface (○; control, □) by subsequent treatment with β-galactosidase, as in Fig. 6 (Burgert and Thilo, 1983; reprinted with the permission of Academic Press, Inc.).

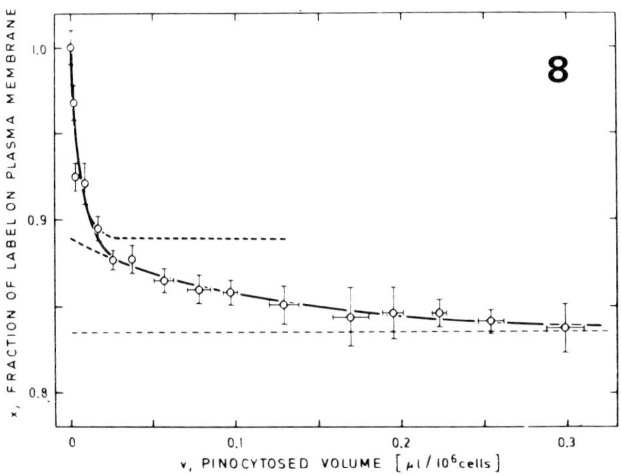

Figure 8. Redistribution of labelled membrane glycoconjugates between the plasma membrane and intracellular membranes as a function of endocytosed volume. The data from Fig. 6 were averaged with those of 4 similar independent experiments (error bars, mean variation, n = 5). The curve can be described by the sum of two decaying exponentials, as indicated by the dashed lines (Burgert and Thilo, 1983; reprinted with the permission of Academic Press, Inc.).

When the redistribution of label is analysed as a function of pinocytosed volume (instead of time, cf. Fig. 6), the data from independent experiments (with slightly varying rates of pinocytosis) can be compared (Fig. 8). Redistribution occurs with biphasic kinetics. This strongly suggests that internalized cell-surface glycoproteins are transported to, and recycle from, at least two kinetically-defined intracellular membrane compartments.

From the kinetics of redistribution of labelled glycoproteins, the rate of membrane traffic between, and the relative sizes of, these compartments has been estimated (Fig. 9). Plasma membrane glycoproteins (pool size 100 at steady state) are internalized once every 20 min (kl = 100/20 = 5 min^{-1}) into an early compartment, m2, containing 13% of the cell-surface pool of labelled glycoproteins at steady state. The bulk of internalized label (kl/(kl + k2) ~ 97%) is recycled directly from m2 after a residence time of about 3 min (m2/kl). Relative pool size and residence time correspond to values found for pinosomal membranes by a morphometric analysis carried out by Steinman et al. (1976). Accordingly, this early compartment is likely to

9

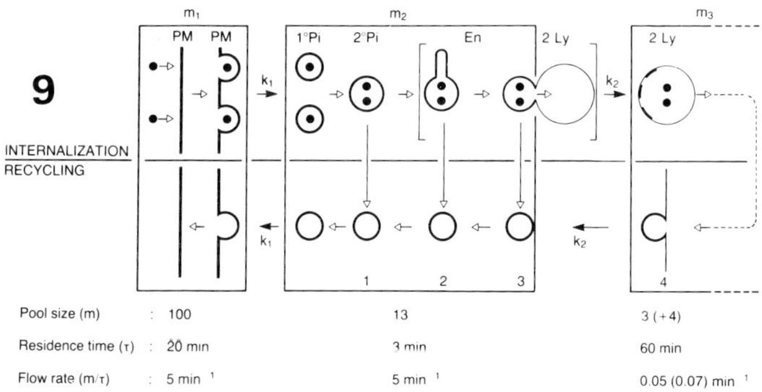

Figure 9. Schematic presentation of intracellular membrane flow and compartmentation. Cell surface derived membrane is indicated by thick lines. PM: plasma membrane; (1°Pi) primary and (2°PI) secondary pinosomes; En: endosomes; 2°Ly: secondary lysosomes. The square brackets indicate the uncertainty concerning the source of endosome membrane and its interaction with 2°Ly. The numbers 1 to 4 refer to probable recycling sites (cf. text) (Haylett and Thilo, 1986; reprinted with the permission of Rockefeller University Press).

represent the first sorting station in endocytosis from where membrane proteins are recycled back to the plasma membrane. The small complementary fraction of internalized label (k2/(k1 + k2) ~ 3%) subsequently enters a second intracellular membrane pool (m3 : pool size 7%) from where it is recycled after a residence time of about 60 min. This compartment represents cell surface-derived glycoproteins in the membrane of secondary lysosomes as has been shown by direct analysis after cell fractionation (cf. next section, Haylett and Thilo, 1986). The above model of membrane traffic has also been confirmed by autoradiography, using the same label in bone marrow-derived macrophages (De Chastellier et al., 1987).

COMPARTMENTS OF MEMBRANE RECYCLING

A first major compartment is the aggregate of incoming vesicles produced by fusion among compartments designated as primary (1°Pi) and secondary (2°Pi) pinosomes (= early endosomes of others). A rapid increase of the average pinosome size has been observed, occurring within less than 1 min after formation of 1°Pi (Fig. 10). Due to the lower surface-to-volume ratio of the larger fusion products, about 50% of internalized membrane could become available for recycling at this point (1 in Fig. 9). From the initial slope of the curve in Fig. 10, the size increase resulting from fusion between two vesicles of equal size, i.e. 1.26-fold, is reached after about 15 seconds. This value could be an indication for the short lifetime of primary pinosomes. Such a single fusion event could already release 20% of the membrane involved. Rapid fusion of newly formed endocytic vesicles with endosome or latex-containing phagosomes was observed in _Dictyostelium discoideum_ by autoradiography (De Chastellier et al., 1983). Recently, Gruenberg and Howell, this volume, have shown with a cell-free system that maximal fusion of vesicles occurs at an early stage of endocytosis.

A second site for membrane recycling could be during delivery of pinocytic contents into endosomes (step 2 in Fig. 9), provided the endosome is an independent organelle not solely derived from the membrane of incoming pinosomes. However, it is still unclear to what extent plasma membrane and endosome share common proteins: the resident proteins of the endosomal membrane remain to be

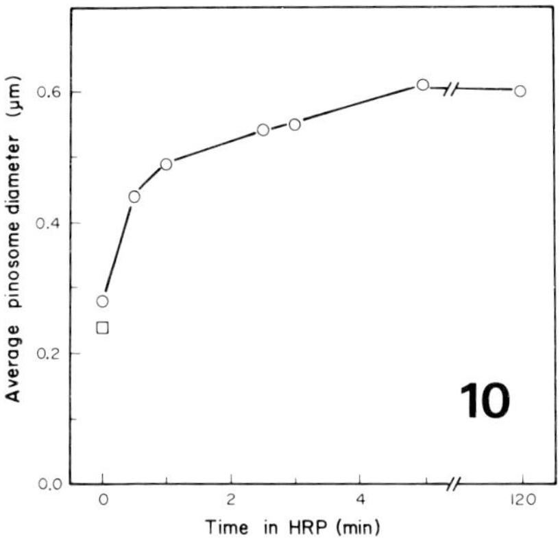

Figure 10. Volume weighted average diameter of endocytic vesicles at various times after formation in macrophages. Calculated from surface-to-volume ratios of cytochemically stained pinosomes after increasing times of horseradish peroxidase uptake (Table IV in Steinman et al., 1976). At t = 0, the size of primary pinosomes was calculated by comparing the rate of membrane internalization (in absolute values: eg. ml = 100 ≙ 1300 μm^2, per 20 min, i.e. 65 μm^2/min) with the rate of fluid-phase uptake (eg. 2.5 μm^3/min.cell, cf. Fig. 6). Assuming spheres, the volume weighted average diameter of 1°Pi becomes, $\phi = 6 \times V/S = 6 \times 25/65 = 0.23$ m) (□; Burgert and Thilo, 1983). The same calculation was made using the data from Steinman et al., 1976: $\phi = 6 \times 12\ (\mu m^3/min)/26\ (\mu m^2/min) = 0.28\ \mu m$(○, at t = 0). (Thilo 1985b, reprinted with the permission of Elsevier Science Publishers B.V.).

identified. In terms of labelled cell-surface glycoproteins at steady-state distribution, the number of autoradiographic grains per morphologically observed membrane area of endosomes is only half the value observed for the cell surface (De Chastellier et al., 1987).

A third compartment for membrane traffic is the lysosome. For labelled cell-surface glycoproteins, this pathway involves only about 1% of the glycoproteins internalized at any one moment. The steady-state contingent of cell surface-derived label in lysosome membrane amounts to about 3% that of the plasma membrane. These directly measured values (rate 1%, pool 3%, cf. Haylett and Thilo, 1986) are somewhat lower than the corresponding values calculated from the redistribution kinetics of label (rate 3%, pool 7%, cf. Burgert and Thilo, 1983). The difference is most likely due to a partial degradation of label after entry into the lysosomal membrane pool. The half-life of labelled glycoproteins has been measured as about 20 h, being the same for all Mr-species. The kinetic analysis indicates that the entire contingent of label passes through the slow recycling pool (m3 in Fig. 9) once about every 12 h. Therefore, less than about 1/3 (< 12.50%/20 = 30%) of label could be degradated during transit through the lysosome membrane pool. However, it is not clear to what extent degradation of label occurs in the lysosome instead of elsewhere along the pathway (cf. Haylett and Thilo, 1986). A point of special interest is the observation that not all labelled Mr-species enter the lysosome membrane to the same extent. Whereas a representative sample of molecules labelled at the cell surface is internalized (Burgert and Thilo, 1983), Mr-species of about 160-190 kD are largely excluded and of 100-120 kD are strongly enriched in the lysosome membrane (Haylett and Thilo, 1986).

For membrane recycling during and after transfer of endocytic contents to secondary lysosomes, two pathways must be considered (steps 3 and 4 in Fig. 9). It would not make sense to assume that only 1% of internalized membrane could suffice to carry endocytic contents to lysosomes (even if an independent source of endosome membrane serves as an intermediate, this membrane still associates with label to 50% the area density as on the plasma membrane, cf. Fig. 11, Haylett and Thilo, 1986). Therefore, whatever fraction of internalized membrane is involved in fusion with secondary lysosomes, only 1% will become trapped as part of the lysosomal membrane, to be recycled only after about 60 min (step 4 in Fig. 9). The remaining membrane, beyond the 1%, will remain available for rapid recycling even <u>after</u> fusion with lysosomes (step 3 in Fig. 9). Such a process would require that during fusion between cell surface-derived membrane and secondary lysosomes, mixing of membrane constituents be restricted (cf. Haylett and Thilo, 1986).

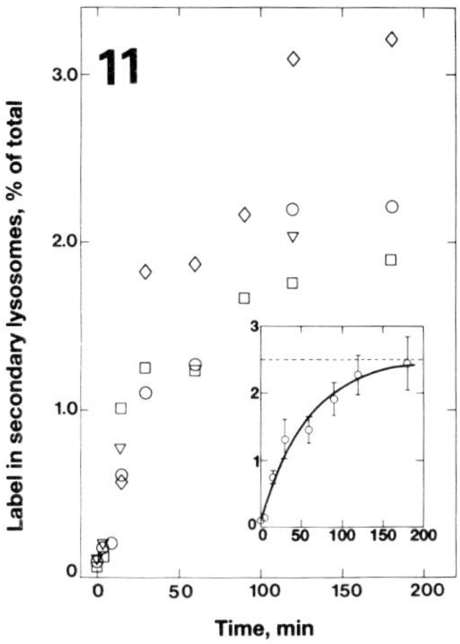

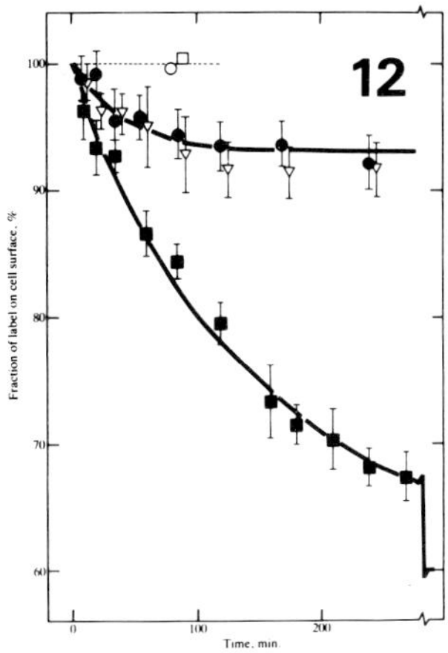

Figure 11. Plasma membrane-derived ^{3}H-galactose in membrane of secondary lysosomes at various times after labelling at the cell surface, followed by membrane internalization during pinocytosis. Different symbols refer to independent experiments. The curve indicates first-order kinetics with k = (56 min)$^{-1}$. (Haylett and Thilo, 1986, reprinted with the permission of Rockefeller University Press.)

Figure 12. Redistribution of labelled membrane glycoconjugates between cell-surface and intracellular membranes in the presence and absence of retrieval of granule membrane. The fraction of label remaining on the cell surface was determined by enzymatic removal. ●, Cells labelled on the plasma membrane without subsequent secretion; ○, same as ● but cells were kept on ice; ▽, cells labelled on plasma membrane followed by secretion; ■, cells labelled on cell surface after secretion; □, same as ■ but cells were kept on ice. Points represent the mean of 3-5 independent experiments with the error bars indicating the average variation (Thilo, 1985a).

The notion of non-mixing of membrane constituents during fusion between membrane of different composition has been tested where the fused membrane complex is more easily accessible, viz. after insertion of granule membrane into the cell surface of mast cells during secretion (Thilo, 1985a). The data in Fig. 12 indicate that granule membrane is retrieved from the cell surface by selective internalization, not affecting the normal rate of internalization observed for plasma membrane constituents in the absence of membrane retrieval.

CONCLUDING REMARKS

When membrane traffic is analysed in terms of labelled cell-surface glycoproteins, the label distributes into at least two consecutive intracellular membrane compartments. These have been distinguished on a kinetic basis, as the average residence time spent by labelled glycoproteins (and by their untagged equivalents) in each, viz. 3 min in a first and 60 min in a second intracellular compartment. In comparison, the residence time on the cell surface is 20 min, which implies a rate of internalization such that the entire contingent of cell-surface glycoprotein is internalized once during this period. The use of a covalent membrane marker has for the first time allowed an estimate of the rate and extent to which internalized plasma-membrane constituents enter and recycle from the lysosome membrane.

Receptor Recycling

The first intracellular compartment (3 min; cf. m2 in Fig. 9) is the site from where major recycling of glycoproteins occurs (endosomes). How does this relate to the kinetics of receptor recycling? For a number of rather different receptors and cell types, surprisingly similar values have been observed for the intracellular residence time of receptors, viz. (13 ± 8) min, as compared to (5 ± 2) min which receptors spend on the cell surface (round trip duration (18 ± 7) min, review: Table III in Thilo, 1985b). Residence time is given by the quotient of pool size and flow rate. Therefore, with the intracellular receptor pool being 2-3 fold that on the cell surface (cf. Table III in Thilo, 1985b), as compared to only 0.13-fold for glycoproteins (m2 vs. m1 in Fig. 9), the flow rate of receptors must still be about 4.5-fold higher than that of the average glycoprotein on the cell surface (2.5/13 vs. 0.13/3; consistently, this comparison also holds when viewed from the side of the cell surface, viz. 1/5 vs. 1/20). This is consistent with observations that receptors are enriched in coated pits prior to internalization (cf. review by Goldstein et al., 1985). Does rapid replacement (note: not equivalent to rapid recycling) of internalized receptors occur from a randomised intracellular pool, as seems to be the case for glycoproteins? For the transferrin receptor in HeLa cells this is clearly not so: receptors constitutively move through a non-randomised intracellular pool in a linear fashion (Ajioka and Kaplan, 1986).

Relative Membrane Composition of Early and Late Endosomes

To the extent that endosomes can be classified into an early and a late stage, according to the sequential processing steps of endocytic contents or receptor-ligand complexes and a related change in their physical properties (cf. this volume), such a differentiation is not reflected by the kinetic behaviour of glycoproteins. During receptor-mediated endocytosis in macrophages, the entire endosome compartment, including the membranes of the tubular network, acquires its steady-state contingent of cell surface-derived glycoproteins within less than 10 min (observed by autoradiography, De Chastellier et al., 1987). An extreme instance of a late endosome would be a long-lived phagosome containing indigestible latex (provided that such a phagosome is not prevented from becoming a late endosome due to the indigestibility of its contents). However, its membrane acquires the label at the same rate and to the same extent as the average pinosome, i.e. the membrane of a late endosome remains linked to the rapid shuttle of membrane to and from the cell surface (De Chastellier et al., 1983). This result again suggests that early and late endosomes must have the same membrane composition in terms of cell surface-derived glycoproteins. Interestingly, this composition is quite distinct from the cell surface, implying that the endosome membrane is not entirely cell surface-derived but exists as an independent organelle (De Chastellier et al., 1987).

REFERENCES

Ajioka RS, Kaplan J (1986). Intracellular pools of transferrin receptors result from constitutive internalization of unoccupied receptors. Proc Natl Acad Sci USA 83:6445-6449.

Burgert HG, Thilo L (1983). Internalization and recycling of plasma membrane glycoconjugates during pinocytosis in the macrophage cell line, P388D1. Exp Cell Res 144:127-142.

De Chastellier C, Ryter A, Thilo L (1983). Membrane shuttle between plasma membrane, phagosomes, and pinosomes in _Dictyostelium_ _discoideum_ amoeboid cells. Eur J Cell Biol 30:233-243.

De Chastellier C, Lang T, Ryter A, Thilo L (1987). Exchange kinetics and composition of endocytic membranes in terms of plasma membrane constituents: a morphometric study in macrophages. Eur J Cell Biol 44:112-123.

Goldstein JL, Brown MS, Anderson RGW, Russel DW, Schneider WC (1985). Receptor-mediated endocytosis: concepts emerging from the LDL receptor system. Ann Rev Cell Biol 1:1-39.

Haylett T, Thilo L (1986). Limited and selective transfer of plasma membrane glycoproteins to membrane of secondary lysosomes. J Cell Biol 103:1249-1256.

Mellman IS, Steinman RM, Unkeless JC, Cohn ZA (1980). Selective iodination and polypeptide composition of pinocytic vesicles. J Cell Biol 86:712-722.

Morré DJ, Kartenbeck J, Franke WW (1979). Membrane flow and interconversions among endomembranes. Biochim Biophys Acta 559:71-152.

Schwarz H, Thilo L (1983). Membrane traffic in _Dictyostelium_ _discoideum_: plasma membrane glycoconjugates internalized and recycled during fluid phase pinocytosis enter the Golgi complex. Eur J Cell Biol 31:212-219.

Steinman RM, Brodie SE, Cohn ZA (1976). Membrane flow during pinocytosis. A stereologic analysis. J Cell Biol 68:665-687.

Steinman RM, Mellman IS, Muller WA, Cohn ZA (1983). Endocytosis and the recycling of plasma membrane. J Cell Biol 96:1-27.

Thilo L, Vogel G (1980). Kinetics of membrane internalization and recycling during pinocytosis in _Dictyostelium_ _discoideum_. Proc Natl Acad Sci USA 77:1015-1019.

Thilo L (1983). Labeling of plasma membrane glycoconjugates by terminal glycosylation (galactosyltransferase and glycosidase). Meth Enzymol 98:415-421.

Thilo L (1985a). Selective internalization of granule membrane after secretion in mast cells. Proc Natl Acad Sci USA 82:1711-1715.

Thilo L (1985b). Quantification of endocytosis-derived membrane traffic. Biochim Biophys Acta 822:243-266.

Cell-Free Analysis of Membrane Traffic, pages 391–409

FUNCTIONAL CHARACTERISTICS OF THE ENDOSOMAL APPARATUS OF RAT LIVER PARENCHYMA

J.J.M. Bergeron, D.G. Kay, W.H. Lai, J.J. Doherty II, C.E. Smith, M.N. Khan, B.I. Posner

Departments of Anatomy and Medicine, McGill University, Montreal, Quebec, Canada. H3A 2B2

INTRODUCTION

The internalization of the peptide hormones insulin and prolactin as well as epidermal growth factor (EGF) and their receptors into liver parenchymal cells has been extensively studied. A major riddle of contemporary cell biology concerns the physiological significance of internalization. Very recent studies from our laboratory and others indicate that peptide hormone and growth factor internalization may be related to ligand action since transmembrane signalling of the insulin and EGF receptors have been demonstrated with isolated endosomes.

The Endosomal Apparatus

The techniques of subcellular fractionation, quantitative electron microscope radioautography and ultrastructural immunocytochemistry have been instrumental in defining a new group of organelles involved in endocytosis. Collectively, these may be considered as components of the endosomal apparatus (reviewed in Bergeron et al., 1985 and schematized in Fig. 1). The endosomal apparatus functions as a locus of ligand-receptor sorting, selective ligand degradation and, we believe, of transmembrane signalling.

Endosome Heterogeneity as Defined by Subcellular Fractionation

Early studies by the in vivo radioautographic technique defined the internalization of ^{125}I-insulin into an heterogenous series of vesicular and tubular components often filled with lipoprotein particles and often associated with stacked Golgi saccules near centrioles (Bergeron et al., 1979). These structures were not lysosomal as defined by acid phosphatase cytochemistry but their intracellular location and lipoprotein content at first indicated that they might be Golgi vacuoles (Bergeron et al., 1979). Subcellular fractionation confirmed the receptor mediated uptake of insulin, prolactin, and ligand-mediated uptake of their receptors into Golgi fractions which were themselves non-lysosomal (e.g. Fig. 2 and refs. Josefsberg et al., 1979; Posner et al., 1980; Khan et al., 1985a). Various attempts were made to resolve the endocytic components from true Golgi components. Percoll gradient centrifugation of hepatic Golgi fractions defined two additional endosomal components to a third (earlier component) previously identified by discontinuous gradient centrifugation (Fig. 3, Posner et al., 1980; Khan et al., 1982; Khan et al., 1985a; Bergeron et al., 1982). These two endosomal components differed to a small extent in density, size, and in their temporal accumulation of ligand (Khan et al., 1982). The latter property led to their designation as early and late endosomes (Fig. 4, Bergeron et al., 1985). Of noteworthy significance was the relative lack of accumulation of the acidotropic agent chloroquine in the 'early' endosomes as compared to the 'late' endosomes (Fig. 3, Khan et al., 1985a).

It was clearly apparent from subcellular fractionation that the early endosomes of peak density ca. 1.042 on Percoll gradients were difficult to resolve from bonafide Golgi components. The innovation of the 'DAB-shift' protocol of Courtoy, Quintart and Baudhuin, (Courtoy et al., 1984; Quintart et al., 1984) was instrumental in clarifying whether internalized insulin entered true Golgi components. Our own application of the technique was conclusive (Figs. 5,6; Kay et al., 1984). Internalized insulin could be resolved from the majority of the Golgi marker enzyme activity, galactosyl transferase. We have used very minor modifications of the original method. These include the use of column chromatography to purify

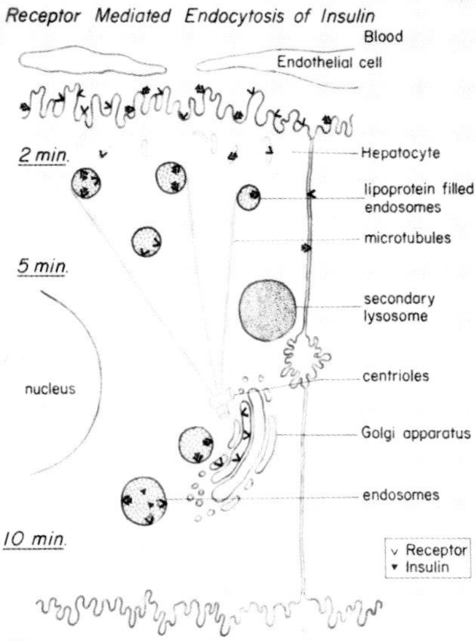

Figure 1. Schematic illustration of receptor and ligand distribution during endocytosis of insulin by liver parenchyma showing the following: 1) Initial internalization (2 min) into a peripheral and chloroquine insensitive tubulovesicular compartment, (Posner et al. 1982b) identified as 100-150 nm vesicles in heavy Golgi fractions (Josefsberg et al. 1979; Posner et al. 1980) and in situ by immunocytochemistry (Dunn et al. 1986); 2) Appearance (5 min) in lipoprotein-filled vesicular compartments of ca. 200-300 nm diameter. These are enriched in all hepatic Golgi fractions and equilibrate on Percoll gradients at $\rho = 1.042$ (Khan et al. 1982, Khan et al. 1986b). Separated as "early" endosomes from galactosyltransferase-containing Golgi elements by DAB shift (Kay et al. 1984), they accumulate little chloroquine (Khan et al. 1985a), but are chloroquine-sensitive for insulin degradation (Khan et al. 1985a; Bergeron et al. 1986; Kahn et al. 1986a), they may represent slightly acidic compartments where insulin degradation is initiated. Interruption by colchicine suggested their transport along microtubules (Bergeron et al. 1983; de Brabander et al. 1986; Carpentier et al. 1986). 3) Endosome enlargement (by transformation via vesicle fusion with primary lysosomes?) during transport into 300-400 nm late endosomes (10 min) (Kahn et al. 1982). The late endosomes accumulate chloroquine (Kahn et al. 1985c) but less than secondary lysosomes (Posner et al. 1982b) and sediment in light mitochondrial fractions that can subsequently be resolved on sucrose gradients with final purification (100-fold enrichment of ligand) on Percoll ($\rho = 1.059$) (Kahn et al. 1986b). Late endosomes occur in hepatic Golgi apparatus intermediate and heavy fractions (Kahn et al. 1985). As selective sites of insulin degradation, they represent a likely locus of recycling of insulin receptors back to the cell surface (via the stacked Golgi cisternae?) as suggested for other receptors (Breitfeld et al. 1985, Snider et al. 1985, Woods et al. 1986, Herzog et al. 1977).

TABLE 1. Properties of Endosomes and Lysosomes of Rat Liver as Evaluated by Subcellular Fractionation

Subcellular Fraction	Fractionation Properties	^{125}I-Insulin Binding	^{125}I-Insulin Peak Uptake		^{125}I-Prolactin Binding	^{125}I-Prolactin Peak Uptake		Chloroquine Uptake	Morphology
			Time (min)	RSA		Time (min)	RSA	RSA	
Small vesicular/ tubular component	In Gh/sv fractions floated from microsomes	8%	2	23	7%	5	31	2.5	100-150 nm diameter empty luminal content
Early endosome	p - 1.042 (Percoll) In: Gl, Gi, Gh fractions	6%	5	90	31%	5	116	1.0	200-300 nm diameter lipoprotein particle filled
Late endosome	p - 1.059 (Percoll) In: Gi, Gh, Li, Lh fractions	13%	10	182	28%	10	250	6.0	300-400 nm diameter lipoprotein particle filled
Lysosome	L2 fraction	14%	2	1	17%	30	2	18.0	540 nm diameter, matrix, often with an associated halo

the galactose-bovine serum albumin-horseradish peroxidase (gal-BSA-HRP) ligand and the use of high resolution Percoll gradients rather than sucrose gradients to separate the 'shifted' organelles (Fig. 6). One positive point of difference between our observations and those of other groups using this technique (e.g. Ajioka and Kaplan, 1987) is that we experience little difficulty in carrying out sodium dodecyl sulfate polyacrylamide gel electrophoresis (SDS-PAGE) of 'DAB-shifted' fractions (Fig. 5). Quantitative analyses of ligand and galactosyl transferase recoveries from the Percoll gradients (Fig. 6) justified two conclusions, (i) that the majority of insulin is not internalized into galactosyl transferase containing Golgi components and (ii) that internalized ligands for the asialoglycoprotein receptor (gal-BSA-HRP) and the insulin receptor are within the same endosomal component.

Selective Ligand Processing in Endosomes

Comparisons of prolactin and insulin clearances from liver have always shown consistently more rapid kinetics for insulin (Josefsberg et al.,1979; Posner et al.,1980;

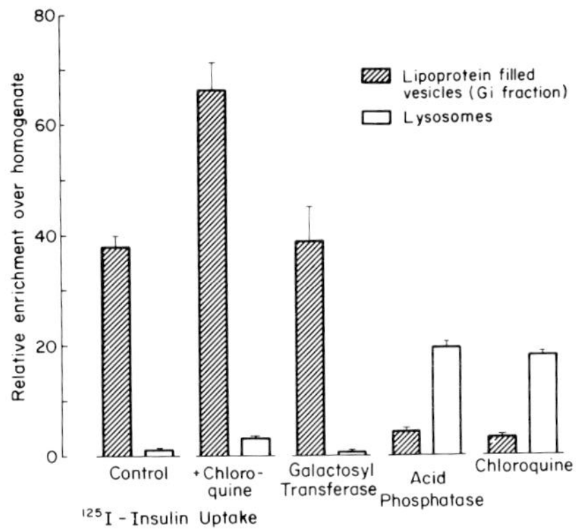

Figure 2. Comparison of insulin uptake, enzyme activities and chloroquine uptake of Gi and lysosome fractions. Standard protocols for the isolation of the Gi fraction and lysosomes were followed (Posner et al., 1982b) after the injection of ^{125}I-insulin into control and chloroquine-treated rats. Fractions were also assessed for their enrichment in marker enzyme activities for the Golgi apparatus (galactosyl transferase) and lysosomes (acid phosphatase) as well as their uptake of the acidotropic agent chloroquine. Negligible ^{125}I-insulin uptake was observed in lysosomes from control or chloroquine-treated rats as defined by acid phosphatase and chloroquine enrichment. Ligand was enriched in lipoprotein-filled endosomal components of the Gi fraction which were non-lysosomal (taken from refs. Posner et al., 1982b; Posner et al., 1979).

Bergeron et al., 1986; Khan et al., 1986a; Posner et al., 1981; Posner et al., 1982a). Thus following the injection of ^{125}I-insulin, no more than a 2-5 min lag was observed before loss of ligand was detected in liver with T½ of ca. 5-6 min (Posner et al., 1980; Bergeron et al., 1986; Khan et al., 1986a; Savoie et al., 1986). By contrast a 15 min lag time was observed before ^{125}I-prolactin was lost with a slower T½ of ca. 10-15 min (Josefsberg et al., 1979; Bergeron et al., 1986; Khan et al., 1986a; Savoie et al., 1986). Furthermore, the kinetics of uptake of insulin into endosomal components was more rapid than that of prolactin (Bergeron et al., 1986). As detailed studies provided no evidence for distinct endocytic organelles harboring insulin as opposed to prolactin (Bergeron et al., 1986; Khan et al., 1986a; Khan et al., 1986b) then the hypothesis that insulin degradation is initiated and substantially effected in endosomes must be considered a likely possibility.

Evaluation of past data clearly showed that the in vivo administration of the acidotropic agent chloroquine

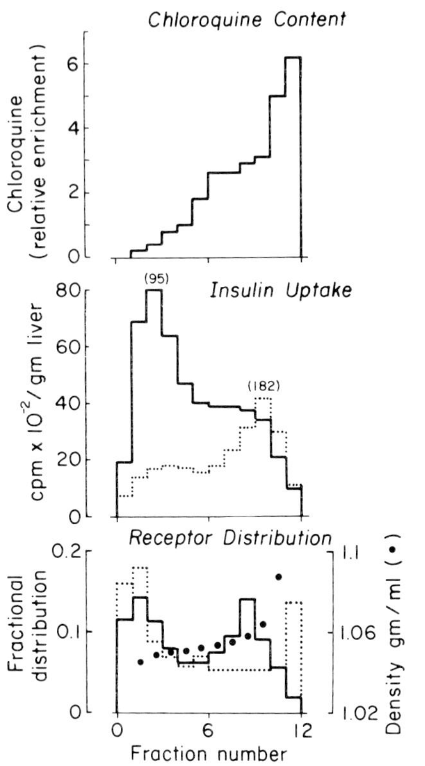

Figure 3. Subfractionation of the Golgi intermediate (Gi) fraction on isoosmolar Percoll gradients. Upper panel: Distribution of ^{14}C-chloroquine in the Gi fraction isolated at 60 min following drug injection. The data are expressed as relative enrichment over the homogenate. Middle panel: Distribution of radioactivity across the Gi fraction. At 5 min (——) or 10 min (········) following the in vivo intravenous injection of ^{125}I-insulin, the Gi fraction was isolated and subfractionated on a Percoll gradient, and the content of radioactivity determined in each fraction. Lower panel: Distribution of specific binding of ^{125}I-insulin (——) and protein content (········) of the Gi fraction. The data are expressed as fractional distribution (left hand ordinate) where the activity (or protein content) of a given fraction is divided by the total recovered from the gradient. The density profile of one of the gradients (●●●●) is on the right hand ordinate. At 5 min after ^{125}I-insulin injection, peak accumulation of radiolabel was observed at a density ca. 1.042 with an enrichment about 95 fold over the homogenate. Peak activity at 10 min after ^{125}I-insulin injection was observed at a density ca. 1.059 with relative enrichment of 182 fold over the homogenate. The lower height of the peak at 10 min after injection (·······) is due to the extensive loss of ^{125}I-insulin during transit from endosomes of ρ 1.042 to those of ρ 1.059 ($T_{1/2}$ of ^{125}I-insulin loss in liver is ca. 5-6 min) (Bergeron et al., 1986). The data are taken from Khan et al., (1985a) (upper panel) and Khan et al., (1982) (middle and lower panels).

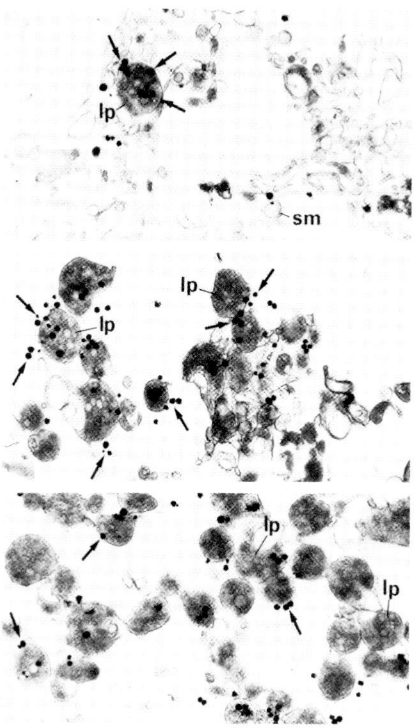

Figure 4. Electron microscope radioautography of low density (upper panel) region of the Percoll gradient (Fractions 1-4 of Fig. 3) enriched in 'early' endosomes. Silver grains (arrows) denote ^{125}I-prolactin and are found associated with lipoprotein-filled components (lp). The major component of this region of the Percoll gradient consists of smooth membranous profiles (sm) presumably derived from the Golgi apparatus. The middle and lower panels denote fractions 5-8 and 9-12 respectively of the Percoll gradient which are highly enriched in lipoprotein-filled vesicles (lp) labeled by ^{125}I-prolactin (arrows) especially near the periphery of the vesicles. The lower panel defines the morphology of the kinetically 'late' endosomes of ca. 300-400 nm in diameter. The random sampling filtration protocol ensured a representative view of the fraction as described more extensively in Khan et al., (1982) and Khan et al., (1985b). Similar morphology for components containing internalized ^{125}I-insulin has been previously documented (Khan et al., 1982). Magnification, x 17,000; exposure, 71 d.

led to an accumulation of insulin but not prolactin in 'early' endosomal components. The earliest small vesicular component was unaffected (Posner et al., 1982b; Bergeron et al., 1986; Khan et al., 1985a; Khan et al., 1986a; Khan et al., 1985b; and Fig 7).

One scenario currently being examined is that the more acid-sensitive pH profile of insulin binding to endosomal insulin receptors, compared to prolactin (Fig. 8) is related to a greater degree of insulin dissociation in endosomes hence easing access to intraluminal degradative enzymes. Diment and Stahl (1985), as well as Geuze and coworkers (Geuze et al.,1985) have independently identified cathepsin D in endosomes. Furthermore Stahl's group has initially proposed that selective ligand processing in macrophage endosomes is regulated by the pH sensitivity of ligand receptor dissociations (Diment and Stahl, 1985). Finally Peters' group in London has independently concluded that liver endosomes serve as a selective locus for insulin degradation (Pease et al., 1985). This group has invoked an insulin specific protease of optimum pH 6-7 but does not believe that ligand dissociation is responsible for the selectivity of insulin processing.

Our own view is that prolactin is internalized into liver parenchyma slower than insulin. Once inside the hepatocyte, however, both ligands enter the same endosomal components but are handled differently. Insulin is selectively degraded perhaps as a consequence of dissociation (Fig. 8). Our findings indicate that little insulin ever makes its way to the lysosome even in chloroquine-treated rats (Posner et al., 1982b). Clearly, the endosome and not the lysosome is the major site of insulin processing in liver.

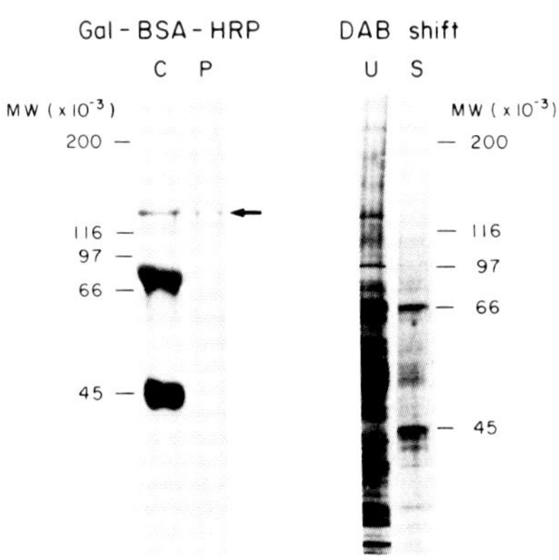

Figure 5. SDS-PAGE of purified Gal-BSA-HRP and of the Gi fraction following the DAB shift protocol. The left hand side of the figure shows the crude reaction mixture (c) and the conjugate of Gal-BSA-HRP following purification on Sephadex G-100 (P). The arrow denotes the purified Gal-BSA-HRP of Mr 130,000. The SDS-PAGE profiles of silver stained proteins of the parent Gi fraction (U) or the Gi endosomal component after the DAB shift (S) is illustrated on the right.

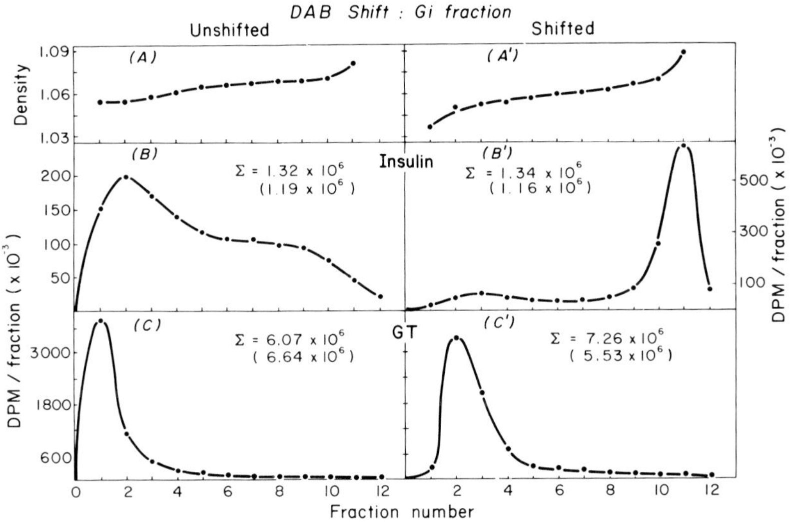

Figure 6. Application of the DAB shift protocol: Distribution of ^{125}I-insulin containing endosomes and galactosyl transferase enriched Golgi apparatus in the Gi fraction: At 5 min after the injection of ^{125}I-insulin (20 x 10^6 cpm) and Gal-BSA-HRP (1 μg per gm body weight), the Gi fraction was isolated then incubated in the absence (A,B,C) or presence (A', B', C') of DAB and H_2O_2 followed by Percoll gradient centrifugation. The majority or 77.02± 1.7% (n=5) of the insulin originally present in fractions 1-4 of the gradient was shifted to the high density region of the gradient (fractions 9-12) while virtually none of the galactosyl transferase activity was shifted. The amount of radiolabelled ^{125}I-insulin or galactosyl transferase activity (dpm ^{3}H-gal transferred to ovomucoid) in the Gi fraction before Percoll gradient subfractionation is indicated in the brackets. The amount of recovered ^{125}I-insulin or of galactosyl transferase activity is indicated following the Σ notation. An average of 90±25% (n=3, ±SD) of input galactosyl transferase activity was recovered from the gradients attesting to the lack of effect of the DAB polymerization reaction on this enzyme activity as previously documented by Quintart et al., (1984).

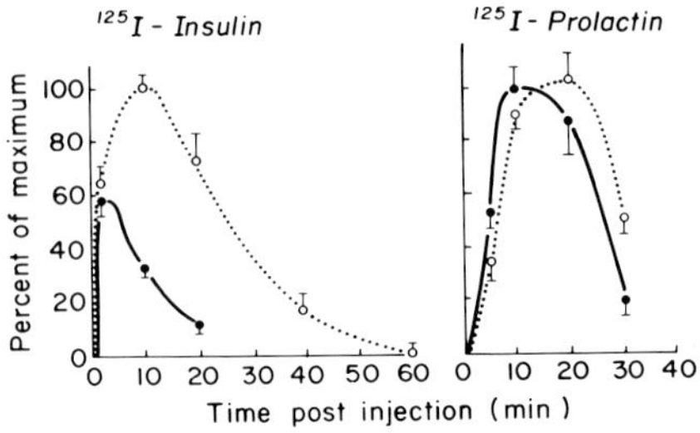

Figure 7. Effect of chloroquine treatment on the time course of ^{125}I-insulin and prolactin uptake into the Gi fraction. At various times following the injection of ^{125}I-insulin or prolactin into control (——) or chloroquine-treated (······) animals, the Gi fraction was prepared and its content of radiolabelled ligand determined. Results are expressed as a percent of the radioactive content of the Gi fraction at the peak time of uptake. The data are taken from Posner et al., (1982b) and unpublished data of the authors. Chloroquine treatment increases selectively the accumulation and time course of ^{125}I-insulin into endosomal components of the Gl and Gi fraction (not shown). By contrast, chloroquine shows no similar effect on ^{125}I-prolactin uptake.

Endosome Isolation and Heterogeneity

In attempting to correlate our in vivo radioautographic studies with subcellular fractionation studies (Bergeron et al., 1979; Bergeron et al., 1983; Josefsberg et al., 1979; Posner et al., 1980; Khan et al., 1982; Khan et al., 1986b; Posner et al., 1982c; Kay et al., 1986) we have developed what we believe are reliable methods based on established protocols for isolating and identifying liver endosomal components. Total endosomal fractions are isolated rapidly from homogenates in one of two ways. In the first method we use a continuous gradient of 0.25 M to 1.07 M STKM (Sucrose, 50 mM Tris pH 7.4, 25 mM KCl, 5 mM $MgCl_2$) over a heavier sucrose layer (1.1 M STKM) containing the homogenate (Bergeron et al., 1982; Kay et al., 1986). Subsequent centrifugation (1.5h x 200,000 gav) results in a combined Golgi-endosome fraction which is enriched ca. 40-60 fold in internalized ligand (unpublished data). Alternatively using a modified de Duve (de Duve et al., 1955) method of homogenization and fractionation, a combined light mitochondrial (L) and microsomal (P) fraction (i.e. an LP fraction) is subjected to discontinuous gradient centrifugation with endosomes collected at the 0.6/0.86 M sucrose interface (Khan et al., 1986b). With both these methods little plasmalemma or lysosomal contamination is evident by electron microscope and marker enzyme analysis. Golgi components however, remain a major contaminant.

More refined fractionation of endosomes is effected in our laboratory currently by taking parent light mitochondrial (L) and microsomal (P) fractions and floating up endosomes by discontinuous sucrose gradient flotation (Khan et al., 1986b). The endosomal fractions at the 0.6/0.86 M sucrose interface have been designated as Li or Gi depending on whether parent L or P fractions are used. The Li fraction is enriched greater than 150 fold in ligand and by subsequent Percoll gradient centrifugation has been shown to consist mainly of 'late' endosomes with little acid phosphatase activity (Fig. 9). The Gi fraction on the other hand is still contaminated with Golgi components and consists of both early and late endosomes which can be resolved by Percoll gradient centrifugation. More details of the fractionation procedures may be found in refs. of Khan et al., (1982); Khan et al., (1986b) and Bergeron et al., (1986).

These more recent subcellular fractionation studies complement our past studies in which lysosomes were concluded to be deficient in insulin and prolactin receptors and internalized insulin (Posner et al., 1982b; Khan et al., 1981; Ferland et al., 1984 and Fig. 10). These studies enable the tabulation of some biochemical features which help distinguish endosomes from lysosomes (Table 1). Recent studies from Fuchs et al., (1986) have demonstrated a novel Na^+K^+ATPase regulation of the ATP dependent H^+-pump of endosomes. The former enzyme activity is apparently absent from lysosomes. Finally, our own laboratory as well as that of Wilson et al., (1986) has provided preliminary evidence for endosome specific antigens (Ahluwalia et al., 1986).

An area of current controversy concerns the lipoprotein nature of the endosomal components (Table 1). Intravesicular lipoproteins of identical morphology to those described here have been previously attributed to Golgi components; indeed they have been employed as a morphological marker for hepatic Golgi fractions (Ehrenreich et al., 1973; Howell and Palade, 1982). Alternately, the intraluminal particles of endosomal components have been viewed as vesicular, therefore these structures have been termed multivesicular bodies (Chao et al., 1981) or multivesicular endosomes (Dunn et al., 1986). However due to the extensive endocytosis of serum lipoprotein by liver parenchyma (Brown and Goldstein, 1986), we believe that the intraluminal particles represent internalized lipoprotein (Chao et al., 1981; Handley et al., 1981; Hornick et al., 1984; Attie et al., 1982) which biases both the morphology and isolation of liver endosomes.

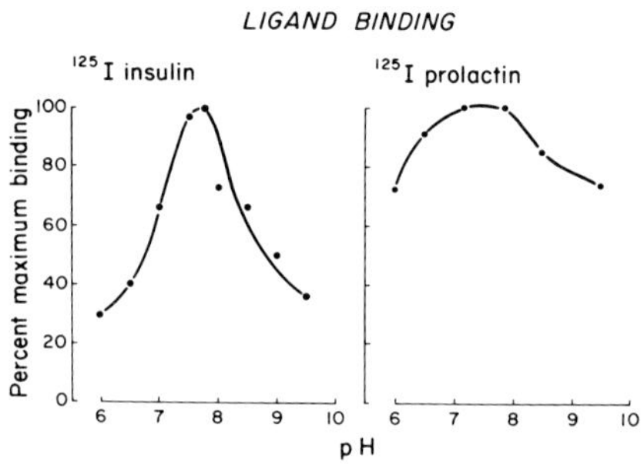

Figure 8. Specific binding of ^{125}I-insulin and prolactin in the Gi fraction expressed as a percent of the maximum specific binding at different pH. Gi fractions were prepared as described (Posner et al., 1982b) and specific binding of ^{125}I-insulin or ^{125}I-prolactin determined as previously described. Taken from refs. Posner et al., (1978); Posner et al. (1979).

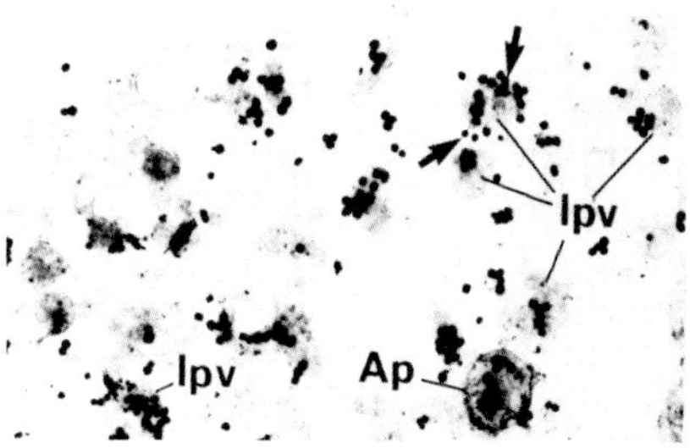

Figure 9. Combined cytochemical and electron microscope radioautographic analysis of Li fraction isolated at 10 min after the injection of ^{125}I-prolactin. After 20 min fixation in 1% glutaraldehyde at pH 7.4 at 4°C, samples were filtered and washed in 100 mM sodium acetate, (pH 5.0) followed by a 30 min incubation in a modified acid phosphatase cytochemical medium which consisted of 8.7 mM β-glycerophosphate in 20 mM sodium acetate, pH 5 and 2.1 mM lead acetate. After incubation for 30 min at 37°C, filters were rinsed in buffer and processed for EM radioautography. Silver grains (arrows) were found over lipoprotein-filled vesicles (lpv) only a minority of which were marked by acid phosphatase (AP). Quantitation revealed that 80.3% of all grains were over acid phosphatase negative components. Magnification, x 12,500.

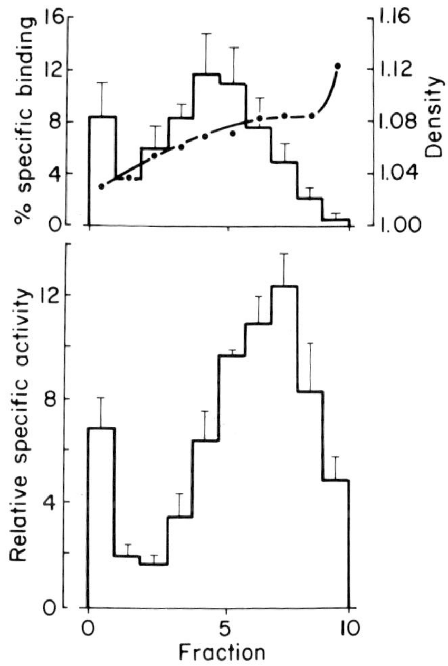

Figure 10. Distribution of hormone binding sites and acid phosphatase activity after subfractionation of the rat liver L fraction on Percoll gradients. Upper panel: The distribution of insulin binding sites expressed as percent specific binding per 50 μg protein of the fraction is plotted against the left hand ordinate. Density in gm/ml (—·—·—·) is plotted against the right hand ordinate. Lower panel: Distribution of the lysosomal marker enzyme, acid phosphatase across the gradient. Activity is expressed as fold enrichment relative to the homogenate. Data are taken from Khan et al. (1981).

The Endosomal Apparatus and Transmembrane Signalling

Following binding in vivo, insulin, prolactin, epidermal growth factor and their receptors are rapidly internalized into liver endosomes (Khan et al., 1985a; Khan et al., 1986a; Kay et al., 1986). Both the insulin receptor and the EGF receptor are units of transmembrane signal generation. Hence upon insulin or EGF binding to their receptors, a cytosolically exposed tyrosine kinase of the receptor is stimulated to phosphorylate its own receptor as well as exogenous substrate(s) of presently unknown significance (Schlessinger, 1986; Khan, 1985; Hunter and Cooper, 1985; Hunter and Cooper, 1981; Erikson et al., 1981; Fava and Cohen, 1984; Sawyer and Cohen, 1985; Giugni et al., 1985; Glenney, 1986; Pepinsky and Sinclair, 1986; Huang et al., 1986; Accili et al., 1986; White et al., 1985; Sadoul et al., 1985; Cohen and Fava, 1985). It has been documented recently by both our group (Kay et al., 1986) and the one of Stanley Cohen (Cohen and Fava, 1985) that isolated intact endosomes function in transmembrane signalling of the EGF receptor (Kay et al., 1986). We have also recently extended this observation to the insulin receptor.

The regulation of transmembrane signalling is expected to be a consequence not only of ligand-receptor stimulation but also of receptor activation due to its phosphorylation state. Recent studies have shown that both these events (stimulation and activation) may be occurring in the endosome. Isolated endosomes show a high proportion of EGF binding to their receptors (unpublished data). Such an observation is consistent with the small volumes and near neutral pH of 'early' endosomes (Khan et al., 1985a; Kielian et al., 1986) as well as clearly documented observations of EGF associated with its receptor in endocytic components in vivo (Lai et al., 1986; McKanna et al., 1979). For insulin, however, it appears that receptor activation via prior phosphorylation may be responsible for insulin receptor transmembrane signalling in the endosome (Khan et al., 1986b). These initial experiments should be helpful in identifying whether unique substrates to receptor kinases of endosomes exist. Hence it is proposed that a major physiological role for receptor internalization is related to the biological actions of insulin and EGF.

REFERENCES

Accili D, Perrotti N, Rees-Jones R, Taylor SI (1986). Tissue distribution and subcellular localization of an endogenous substrate (pp120) for the insulin receptor-associated tyrosine kinase. Endocrinology 119:1274-1280.

Ahluwalia JP, Levine G, Posner BI, Bergeron JJM (1986). Identification of endosome specific antigen(s). J Cell Biol 103:356a.

Ajioka RS, Kaplan J (1987). Characterization of endocytic compartments using the horseradish peroxidase/diaminobenzidine shift technique. J Cell Biol 104:77-85.

Attie AD, Pittman RC, Steinberg D (1982). Hepatic catabolism of low density lipoprotein: mechanisms and metabolic consequences. Hepatology 2:269-281.

Bergeron JJM, Levine G, Sikstrom R, O'Shaughnessy D, Kopriwa B, Nadler NJ, Posner BI (1977). Polypeptide hormone binding sites in vivo: Initial localization of ^{125}I-labelled insulin to hepatocyte plasmalemma as visualized by electron microscope radioautography. Proc Natl Acad Sci USA 74:5051-5055.

Bergeron JJM, Sikstrom R, Hand AR, Posner BI (1979). Binding and uptake of ^{125}I-insulin into rat liver hepatocytes and endothelium. J Cell Biol 80:427-443.

Bergeron JJM, Rachubinski RA, Sikstrom RA, Posner BI, Paiement J (1982). Galactose transfer to endogenous acceptors within Golgi fractions of rat liver. J Cell Biol 92:139-146.

Bergeron JJM, Resch L, Rachubinski R, Patel BA, Posner BI (1983). Effect of colchicine on internalization of prolactin in female rat liver: an in vivo radioautographic study. J Cell Biol 96:875-886.

Bergeron JJM, Cruz J, Khan MN, Posner BI (1985). Uptake of insulin and other ligands into receptor rich endocytic components of target cells: the endosomal apparatus. Ann Rev Physiol 47:383-403.

Bergeron JJM, Searle N, Khan MN, Posner BI (1986). Differential and analytical subfractionation of rat liver components internalizing insulin and prolactin. Biochemistry 25:1756-1764.

Breitfeld PP, Simmons,Jr CE, Strous Ger JAM, Geuze HJ, Schwartz AL (1985). Cell Biology of the asialoglycoprotein receptor system: a model of receptor mediated endocytosis. Int Rev Cytol 97:47-95.

Brown MS, Goldstein JL (1986). A receptor mediated pathway for cholesterol homeostasis. Science 232:34-47.

Carpentier JL, Rees AR, Gregoriou M, Kris R, Schlessinger J, Orci L (1986). Subcellular distribution of the external and internal domains of the EGF receptor in A-431 cells. Exp Cell Res 166:312-326.

Chao YS, Jones AL, Hradek GT, Windler EET, Havel RJ (1981). Autoradiographic localization of the sites of density lipoproteins in the liver in normal and estrogen treated rats. Proc Natl Acad Sci USA 78:597 601.

Cohen S, Fava RA (1985). Internalization of functional epidermal growth factor: receptor/kinase complexes in A-431 cells. J Biol Chem 260:12351-12358.

Courtoy PJ, Quintart J, Baudhuin P (1984). Shift of equilibrium density induced by 3,3' - diaminobenzidine cytochemistry: A new procedure for the analysis and purification of peroxidase-containing organelles. J Cell Biol 98:870-876.

de Brabander M, Nagdens R, Gierts H, Moeremans M, Hopkins C (1986). Intracellular routing of receptors observed in living cells using Nanovid microscopy. J Cell Biol 103:61a.

de Duve C, Pressman BC, Gianetto R, Wattiaux R, Applemans F (1955). Tissue fractionation studies. 6. Intracellular distribution pattern of enzymes in rat liver tissue. Biochem J 60:604-617.

Diment S, Stahl P (1985). Macrophage endosomes contain proteases which degrade endocytosed protein ligands. J Biol Chem 260:15311-15317.

Dunn WA, Connolly TP, Hubbard AL (1986). Receptor-mediated endocytosis of epidermal growth factor by rat hepatocytes: receptor pathway. J Cell Biol 102:24-36.

Ehrenreich JH, Bergeron JJM, Siekevitz P, Palade GE (1973). Golgi fractions prepared from rat liver homogenates I. Isolation procedure and morphological characterization. J Cell Biol 59:45-72.

Erikson E, Shealy DT, Erikson RL (1981). Evidence that viral transforming gene products and epidermal growth factor stimulate phosphorylation of the same cellular protein with similar specificity. J Biol Chem 256:11381-11384.

Fava RA, Cohen S (1984). Isolation of a calcium-dependent 35-kilodalton substrate for the epidermal growth factor receptor/kinase from A-431 cells. J Biol Chem 259: 2636-2645.

Ferland LH, Djiane J, Houdebine LM, Kelly PA (1984). The effect of chloroquine on lysosomal prolactin receptors in rat liver. Endocrinology 115:1842-1849.

Fuchs R, Schmid S, Male P, Hellenius A, Mellman I (1986). Isolation and ATP-dependent acidification of endosome subcompartments from CHO cells. J Cell Biol 103:439a.

Geuze H, Slot JW, Strous Ger JAM, Hasilik A, von Figura, K (1985). Possible pathways for lysosomal enzyme delivery. J Cell Biol 101:2253-2262.

Giugni TO, James LC, Haigler HT (1985). Epidermal growth factor stimulates tyrosine phosphorylation of specific proteins in permeabilized human fibroblasts. J Biol Chem 260:15081-15090.

Glenney J (1986). Two related but distinct forms of the M_r 36,000 tyrosine kinase substrate (calpactin) that interact with phospholipid and actin in a Ca^{2+}-dependent manner. Proc Natl Acad Sci USA 83:4258-4262.

Handley DA, Arbeeny CM, Eder HA, Chien S (1981). Hepatic binding and internalization of low density lipoprotein-gold conjugates in rats treated with 17α-ethinylestradiol. J Cell Biol 90:778-787.

Hornick CA, Jones AL, Renaud G, Hradek G, Havel RJ (1984). Effect of chloroquine on low density lipoprotein catabolic pathway in rat hepatocytes. Am J Physiol 246:6187-6194.

Howell KE, Palade GE (1982). Heterogeneity of lipoprotein particles in hepatic Golgi fractions. J Cell Biol 92:833-845.

Herzog V, Farquhar MG (1977). Luminal membrane retrieved after exocytosis reaches most Golgi cisternae in secretory cells. Proc Natl Acad Sci USA 74:5073-5077.

Huang KS, Wallner BP, Mattaliano RJ, Tizard R, Burne C, Frey A, Hession C, McGray P, Sinclair LK, Chow EP, Browning JL, Ramachandran KL, Tang J, Smart JE, Pepinsky RB (1986). Two human 35 kd inhibitors of phospholipase A_2 are related to substrates of $pp60^{v\text{-}src}$ and of the epidermal growth factor receptor/kinase. Cell 46:191-199.

Hunter T, Cooper JA (1981). Epidermal growth factor induces rapid tyrosine phosphorylation of proteins in A-431 human tumor cells. Cell 24:741-752.

Hunter T, Cooper JA (1985). Protein-tyrosine kinases. Ann Rev Biochem 54:897-930.

Josefsberg Z, Posner BI, Patel B, Bergeron JJM (1979). The uptake of prolactin into female rat liver. J Biol Chem 254:209-214.

Kay DG, Kahn MN, Posner BI, Bergeron JJM (1984). ^{125}I-Insulin in hepatic Golgi fractions: Application of the diaminobenzidine (DAB)-shift protocol. Biochem Biophys Res Commun 123:1144-1148.

Kay DG, Lai WH, Uchihashi M, Kahn MN, Posner BI, Bergeron JJM (1986). Epidermal growth factor receptor kinase translocation and activation in vivo. J Biol Chem 261:8473-8480.

Khan CR (1985). The molecular mechanism of insulin action. Ann Rev Med 36:429-451.

Khan MN, Posner BI, Verma AK, Khan RJ, Bergeron JJM (1981). Intracellular hormone receptors: evidence for insulin and lactogen receptors in a unique vesicle sedimenting in lysosome fractions of rat liver. Proc Natl Acad Sci USA 78:4980-4984.

Khan MN, Posner BI, Khan RJ, Bergeron JJM (1982). Internalization of insulin into rat liver Golgi elements: evidence for vesicle heterogeneity and the path of intracellular processing. J Biol Chem 257:5969-5976.

Khan MN, Savoie S, Khan RJ, Bergeron JJM, Posner BI (1985a). Insulin and insulin receptor uptake into rat liver: chloroquine action on receptor recycling. Diabetes 34:1025-1030.

Khan MN, Savoie S, Bergeron JJM, Posner BI (1986a). Differential kinetics and sensitivity to chloroquine of receptor-mediated insulin and prolactin endocytosis in liver parenchymal cells. Biochem Biophys Acta 888:100-106.

Khan MN, Savoie S, Bergeron JJM, Posner BI (1986b). Characterization of rat liver endosomal fractions: in vivo activation of insulin-stimulable receptor kinase in these structures. J Biol Chem 261:8462-8472.

Khan RJ, Khan MN, Bergeron JJM, Posner BI (1985b). Prolactin uptake into liver endocytic components: reduced sensitivity to chloroquine. Biochim Biophys Acta 838:77-83.

Kielian MC, Marsh M, Helenius A (1986). Kinetics of endosomal acidification detected by mutant and wild type Semliki Forest Virus. EMBO J 5:3103-3109.

Lai WH, Guyda HJ, Bergeron JJM (1986). Binding and internalization of epidermal growth factor in human term placental cells in culture. Endocrinology 118:413-423.

McKanna JA, Haigler HT, Cohen S (1979). Hormone receptor topology and dynamics: morphological analysis using ferritin-labelled epidermal growth factor. Proc Natl Acad Sci USA 76:5689-5693.

Pease RJ, Smith GD, Peters TJ (1985). Degradation of endocytosed insulin in rat liver is mediated by low density vesicles. Biochem J 228:137-146.

Pepinsky RB, Sinclair LK (1986). Epidermal growth factor-dependent phosphorylation of lipocortin. Nature 321:81-84.

Posner BI, Josefsberg Z, Bergeron JJM (1978). Intracellular polypeptide hormone receptors: characterization of insulin binding sites in Golgi fractions from the liver of female rats. J Biol Chem 253:4067-4073.

Posner BI, Josefsberg Z, Bergeron JJM (1979). Intracellular polypeptide hormone receptors: characterization and induction of lactogen receptors in the Golgi apparatus of rat liver. J Biol Chem 254:12494-12499.

Posner BI, Patel BA, Verma AK, Bergeron JJM (1980). Uptake of insulin by plasmalemma and Golgi subcellular fractions of rat liver. J Biol Chem 255:735-741.

Posner BI, Bergeron JJM, Josefsberg Z, Khan MN, Khan RJ, Patel BA, Sikstrom RA, Verma AK (1981). Polypeptide hormones: intracellular receptors and internalization. Rec Prog Hormone Res 37:539-582.

Posner BI, Khan MN, Bergeron JJM (1982a). Endocytosis of peptide hormones and other ligands. Endocr Rev 3:280-298.

Posner BI, Patel BA, Khan MN, Bergeron JJM (1982b). Effect of chloroquine on the internalization of ^{125}I-insulin into subcellular fractions of rat liver. J Biol Chem 257:5789-5799.

Posner BI, Verma AK, Patel BA, Bergeron JJM (1982c). Effect of colchicine on the uptake of prolactin and insulin into Golgi fractions of rat liver. J Cell Biol 93:560-567.

Quintart J, Courtoy PJ, Baudhuin P (1984). Receptor-mediated endocytosis in rat liver: purification and enzymatic characterization of low density organelles involved in the uptake of galactose exposing proteins. J Cell Biol 98:877-884.

Sadoul JD, Peyron JF, Ballotti R, Debant A, Fehlmann M, Van Obberghen E (1985). Identification of a cellular 110,000 -Da protein substrate for the insulin receptor kinase. Biochem J 227:887-892.

Savoie S, Rindress D, Posner BI, Bergeron JJM (1986). Tunicamycin sensitivity of prolactin, insulin and epidermal growth factor receptors in rat liver parenchyma. Mol Cell Endo 45:241-246.

Sawyer ST, Cohen S (1985). Epidermal growth factor stimulates the phosphorylation of the calcium-dependent 35,000-Dalton substrate in intact A-431 cells. J Biol Chem 260:8233-8236.

Schlessinger J (1986). Allosteric regulation of the epidermal growth factor receptor kinase. J Cell Biol 103:2067-2072.

Snider MD, Rogers OD (1985). Intracellular movement of cell surface receptors after endocytosis: Resialylation of asialotransferrin receptor in human erythroleukemia cells. J Cell Biol 100:826-834.

Wattiaux R, Connick S, Ronveaux DuPal JW, Dubois MF (1978). Isolation of rat liver lysosomes by isopycnic centrifugation in a metrizamide gradient. J Cell Biol 78:349-368.

White MF, Maron R, Khan CR (1985). Insulin rapidly stimulates tyrosine phosphorylation of a Mr - 185,000 protein in intact cells. Nature 318:183-186.

Wilson JM, Whitney JA, Neutra MR (1986). An endosome-specific antigen in absorptive cells of suckling rat ileum. J Cell Biol 103:148a.

Woods JW, Doriaux M, Farquhar MG (1986). Transferrin receptors recycle to cis and middle as well as trans Golgi cisternae in IgG-secreting myeloma cells. J Cell Biol 103:277-286.

Cell-Free Analysis of Membrane Traffic, pages 411–415

IDENTIFICATION OF ANTIGEN(S) SPECIFIC TO A 'LATE' ENDOSOME FRACTION

J.P. Ahluwalia, J.J. Doherty, II, M. Troulis, B.I. Posner and J.J.M. Bergeron

Departments of Anatomy, (J.P.A., J.J.D., M.T., J.J.M.), and Medicine, (B.I.P.), McGill University, Montreal, Quebec, Canada. H3A 2B2.

INTRODUCTION

A current focus of research in endocytosis has been an attempt to identify antigen(s) specific to components of the endocytic pathway. Although antibodies specific to proteins that are predominantly localized to lysosomes (Lewis et al., 1985; Lippincott-Schwartz and Fambrough, 1986; Reggio et al., 1984) have been characterized, to our knowledge, no endosome-specific antibodies have yet been described. The availability of such an antibody would not only be useful in dissecting the endocytic pathway, but would also help in establishing endosomes as distinct entities.

We describe here preliminary data on the characterization of a rabbit antiserum raised to 'late' endosomes using an established protocol of immunization with subcellular fractions (Lippincott-Schwartz and Fambrough, 1986). The antiserum recognizes an immunologically distinct set of proteins in the parent endosome fraction.

EXPERIMENTAL PROCEDURES

Cell Fractionation and Immunization

'Late' endosome fractions were obtained by Percoll gradient centrifugation of rat liver Golgi intermediate (Gi) and equivalent fractions from light-mitochondrial pellets (Li) as described by Khan et al. (1986). After isolation, 'late' endosomes were further separated into membrane and content subfractions by treatment with 100 mM Na_2CO_3, pH 11.3 (Howell and Palade, 1982). Rabbit polyclonal antibodies were raised to the endosomal membrane subfractions by injection into the popliteal lymph node, essentially as described by Louvard et al. (1982).

Preparation of Other Subcellular Membrane Fractions

Rat liver lysosomes were purified by isopycnic centrifugation in a metrizamide gradient according to the method of Wattiaux et al. (1978). Only fraction L2 was used. Plasmalemma was isolated from rat hepatocytes by a modification of the method of Hubbard et al., as described by Kay et al. (1986). Golgi-endosome (GE) fraction containing endocytic elements was prepared as described by Kay et al. (1986). Stripped rough microsomes were isolated according to Paiement and Bergeron (1983).

SDS-PAGE and Immunoblotting

Rat serum as well as other membrane fractions were subjected to electrophoresis on 5-15% linear gradient polyacrylamide gels. Blotting of SDS polyacrylamide gels onto 0.45 μm nitrocellulose paper was essentially according to Burnette (1981). After the transfer procedure, nitrocellulose blots were incubated for 1 h at room temperature in 150 mM NaCl, 10 mM Tris, pH 7.4, containing 0.5% Tween 20 (Buffer TWA). The blots were subsequently incubated for 2 h at room temperature or overnight at 4°C in first antibody diluted appropriately in Buffer TWA, washed five times for 5 min each with Buffer TWA, incubated for 2 h at 37°C with alkaline phosphatase-conjugated goat anti-rabbit IgG diluted 1:1000 in Buffer TWA, and washed again with Buffer TWA as above. After the last wash, blots were stained for alkaline phosphatase as described by Smith and Fisher (1984).

RESULTS AND DISCUSSION

The antiserum raised to Na_2CO_3 washed 'late' endosomal membranes, using the Louvard method of immunization with subcellular fractions, reacted with various bands in all of the fractions tested (Fig. 1). However, a 100,000 dalton band was stained fairly intensely by the antiserum to near equal intensity in lysosomes, L intermediate and Golgi intermediate fractions. This band was absent from other subcellular fractions except the Golgi endosome fraction where it was stained to a much lesser extent. A lower molecular weight band (Mr ≃30-35 kDa) was identified by the antiserum in L intermediate and Golgi intermediate fractions, but not in any of the other fractions except the Golgi endosome fraction where it was stained very lightly. An intense band due to rat albumin also was stained by the antiserum in lanes corresponding to rat serum, L intermediate, Golgi intermediate and Golgi endosome fractions.

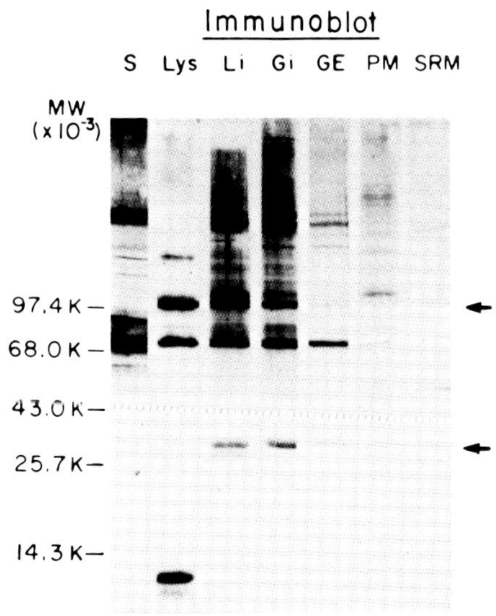

Figure 1. Identification of antigens recognized by antiserum to rat liver endosomal membranes by immunoblotting. Membrane fractions prepared as described in Experimental Procedures were separated by SDS-PAGE, transferred to nitrocellulose paper, probed with 1:1000 diluted immune serum and calf alkaline phosphatase-conjugated goat anti-rabbit IgG antibody respectively, and then stained for alkaline phosphatase activity (10 min, room temperature). Approximately 20 μg protein was loaded per lane. Molecular weight markers are shown on the left and arrows indicate the positions of 100 kDa and 30-35 kDa bands. S, rat serum; Lys, lysosomal fraction L2; Li, L intermediate

fraction; Gi, Golgi intermediate fraction; GE, Golgi endosome fraction; PM, plasmalemma; SRM, stripped rough microsomes.

Since most of the albumin was washed away from membrane fractions by Na_2CO_3 wash as judged by silver stained gels of Na_2CO_3 endosomal membranes (unpublished data), the presence of rat serum albumin-specific antibodies in the antiserum indicates the high immunogenic nature of rat albumin. The lightly stained bands in other subcellular fractions that are not shared by the Percoll purified 'late' endosomes (for review, see Bergeron et al., 1985) used for immunization may be due to slight contamination of this fraction by elements of the other fractions. The 100 kDa band shared by lysosomal and endosomal fractions may be due to the 100,000 mol-wt. protein reported by Reggio et al. (1984). On the other hand, the 30-35 kDa band seems, by this method of detection, to be restricted to the L intermediate and Golgi intermediate parent endosome and Golgi endosome fractions. The presence of this band as well as the 100 kDa band in Golgi endosome fraction reflects the well established presence of endocytic elements in this fraction (Kay et al., 1986).

Therefore, the polyclonal antibodies raised to purified 'late' endosomal membranes recognize a protein that is restricted to the two parent endosome fractions. Immunoadsorption of this native antiserum with cross-reacting fractions will further aid in establishing the unique protein composition of the endosomal fractions.

ACKNOWLEDGEMENTS

This work was supported by grants from the Medical Research Council and the National Cancer Institute of Canada. J.P.A. is a recipient of a predoctoral fellowship from Fonds de la Recherche en Santé du Québec.

REFERENCES

Bergeron JJM, Cruz J, Khan MN, Posner BI (1985). Uptake of insulin and other ligands into receptor-rich endocytic components of target cells: The endosomal apparatus. Ann Rev Physiol 47:383-403.

Burnette WN (1981). "Western Blotting": Electrophoretic

transfer of proteins from Sodium Dodecyl Sulfate-Polyacrylamide Gels to unmodified nitrocellulose and radiographic detection with Ab and radioiodinated Protein A. Anal Biochem 112:195-203.

Howell KE, Palade GE (1982). Hepatic Golgi fractions resolved into membrane and content subfractions. J Cell Biol 92:822-832.

Hubbard AL, Wall DA, Ma A (1983). Isolation of rat hepatocyte plasma membranes. I. Presence of three major domains. J Cell Biol 95:217-229.

Kay DG, Lai WH, Uchihashi M, Khan MN, Posner BI, Bergeron JJM (1986). Epidermal growth factor receptor kinase translocation and activation *in vivo*. J Biol Chem 261:8473-8480.

Khan MN, Savoie S, Bergeron JJM, Posner BI (1986). Characterization of rat liver endosomal fractions. *In vivo* activation of insulin-stimulable receptor kinase in these structures. J Biol Chem 261:8462-8472.

Lewis V, Green SA, Marsh M, Vihko P, Helenius A, Mellman I (1985). Glycoproteins of the lysosomal membrane. J Cell Biol 100:1839-1847.

Lippincott-Schwartz J, Fambrough DM (1986). Lysosomal membrane dynamics: Structure and interorganellar movement of a major lysosomal membrane glycoprotein. J Cell Biol 102:1593-1605.

Louvard D, Reggio H, Warren G (1982). Antibodies to the Golgi complex and the rough endoplasmic reticulum. J Cell Biol 92:92-107.

Paiement J, Bergeron JJM (1983). Localization of GTP-stimulated core glycosylation of fused microsomes. J Cell Biol 96:1791-1796.

Reggio H, Bainton D, Harms E, Coudrier E, Louvard D (1984). Antibodies against lysosomal membranes reveal a 100,000-mol-wt protein that cross-reacts with purified H^+,K^+ ATPase from gastric mucosa. J Cell Biol 99:1511-1526.

Smith DE, Fisher PA (1984). Identification, developmental regulation, and response to heat shock of two antigenically related forms of a major nuclear envelope protein in *Drosophila* embryos: Application of an improved method for affinity purification of antibodies using polypeptides immobilized on nitrocellulose blots. J Cell Biol 99:20-28.

Wattiaux R, Wattiaux-De Connick S, Ronveaux-Dupal M-F, Dubois F (1978). Isolation of rat liver lysosomes by isopycnic centrifugation in a metrizamide gradient. J Cell Biol 78:349-368.

Cell-Free Analysis of Membrane Traffic, pages 417–419

PHOSPHOMANNOSYL RECEPTOR BINDING IS LOCALIZED IN VESICLES SEPARATED BY FREE-FLOW ELECTROPHORESIS FROM ROUGH ENDOPLASMIC RETICULUM PREPARATIONS OF RAT LIVER

Nita Minnifield, Keri Safranski, and D. James Morré

Department of Biological Sciences and Department of Medicinal Chemistry and Pharmacognosy, Purdue University, West Lafayette, IN 47907

Transport of lysosomal enzymes (LE) into lysosomes, localization of the phosphomannosyl receptor (PMR), and binding of PMR to LE are currently active areas of investigation. PMR has been found in cis and trans Golgi apparatus (GA) regions as well as in endosomes and coated vesicles by both biochemical and immunocytochemical methods (von Figura et al., 1986; Farquhar, 1985). Studies by Fischer et al. (1980) on PMR binding to LE suggested that β-hexosaminidase might bind at low levels to PMR in rough endoplasmic reticulum (RER) as well. These experiments were to analyze further the nature of LE binding in membranes of highly purified RER fractions from rat liver.

RER vesicles prepared from rat liver by sucrose density gradient centrifugation were subsequently separated by preparative free-flow electrophoresis (FFE) (Morré et al., 1984). The electrophoretic separation (Fig. 1) revealed a peak of glucose-6-phosphatase activity that coincided with RER fractions having attached polyribosomes, and determined to be >98% RER by morphometry. Except for fractions 36-38, galactosyltransferase was relatively constant among fractions suggesting the activity to be endogenous to RER. β-Glucuronidase also displayed activity across the separation. Across the main part of the separation all fractions contained glucose-6-phosphatase and β-glucuronidase.

Binding of LE to PMR was determined using ^{125}I-β-glucuronidase (Minnifield et al., 1986). β-Glucuronidase (bovine source, Sigma) was iodinated with ^{125}I with Iodo-beads (Pierce) according to the manufacturers' recommendations. Binding assays contained in a final volume of 300 µl, 50 mM Tris, pH 7.5, 0.1% saponin, 1.0 mg/ml BSA, 10 µg/ml aprotinin, [^{125}I]-β-glucuronidase (4000 units, 500,000 cpm), and membrane fraction. Samples were incubated at room temperature for 1 h and stopped by the addition of 0.4 ml of 25 mM Tris, pH 7.5. Membranes were pelleted by ultracentrifugation in an SW 50.1 rotor at 30 K for 20 min. The pellet was washed a second time with 0.5 ml of 25 mM Tris, pH 7.5, pelleted as above, and radioactivity determined in a gamma counter. Radioactivity

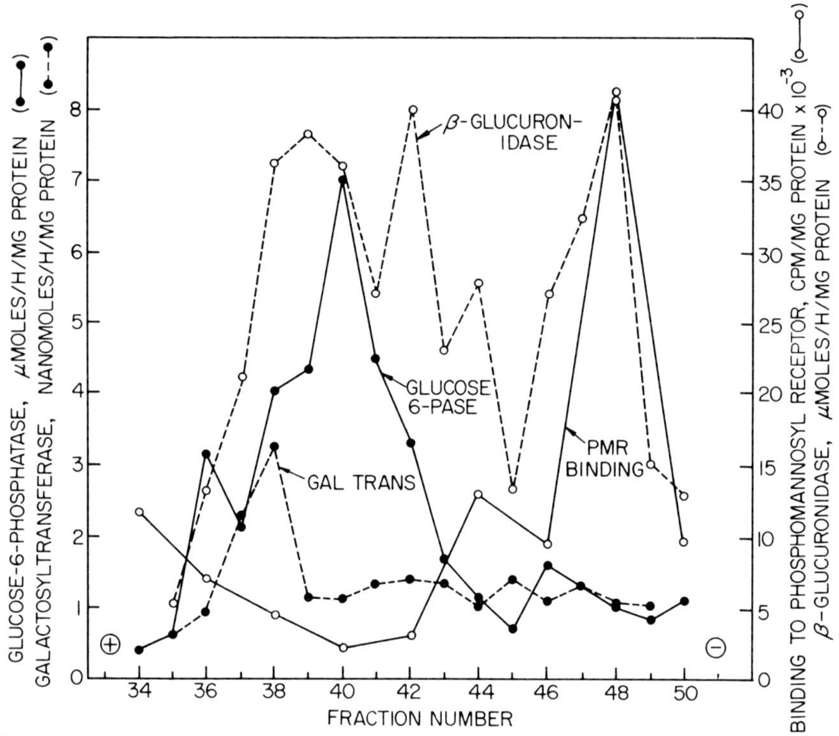

Figure. Electrophoretic separation of sucrose gradient purified rough endoplasmic reticulum from rat liver. Fractions of increasing electronegativity were obtained from right to left. Contaminating Golgi apparatus elements, when present, were found in fraction 36. The bulk of the endoplasmic reticulum was at fractions 37-42. Transition elements and vesicles were in fraction 48.

associated with membranes in duplicate tubes containing 100 mM mannose-6-phosphate was subtracted from the membrane-bound radioactivity in the absence of mannose-6-phosphate and this value was taken as specific binding of unoccupied receptor. Controls included a measure of nonspecific binding to plastic tubes and a demonstration of reversibility from assays both in the presence and in the absence of mannose-6-phosphate during the binding assay. For PMR, low specific activity binding was observed in all RER fractions, whereas high specific activity binding was detected in a subfraction less electronegative than the bulk RER. This subfraction contained high β-glucuronidase but low glucose-6-phosphatase specific activity. By electron microscopy it was determined to contain part rough-part smooth transitional regions of ER as well as small vesicles of undetermined origin. The findings support the presence of low levels of activity of PMR binding as well as galactosyltransferase endogenous to RER membranes. The separation from RER membranes by FFE of a fraction enriched in both transitional elements and small vesicles and PMR activity is of interest. Anticipated roles of such ER elements include the transfer of LE and PMR from sites of translation on polysomes attached to RER to sites of concentration in cis GA elements (Brown et al. 1984).

REFERENCES

Brown WJ, Farquhar MG (1984). The mannose-6-phosphate receptor for lysosomal enzymes is concentrated in cis Golgi cisternae. Cell 36:295-307.

Farquhar MG (1985). Progress in unraveling pathways of Golgi traffic. Ann Rev Biol 1:447-489.

Fischer HD, Gonzalez-Noriega S, Sly WS, Morré DJ (1980). Phosphomannosyl-enzyme receptors in rat liver. J Biol Chem 255:9608-9615.

Minnifield N, Creek KE, Navas P, Morré DJ (1986). Involvement of cis and trans Golgi apparatus elements in the intracellular sorting and targeting of acid hydrolases of lysosomes. Eur J Cell Biol 43:92-100.

Morré DJ, Creek KE, Matyas GR, Minnifield N, Sun I, Baudoin P, Morré DM, Crane FL (1984). Free-flow electrophoresis for subfractionation of rat liver Golgi apparatus. BioTechniques 2:224-233.

von Figura K, Hasilik A (1986). Lysosomal enzymes and their receptors. Ann Rev Biochem 55:167-193.

Cell-Free Analysis of Membrane Traffic, pages 421–425

IN VITRO FUSION OF MICROLIPID DROPLETS FROM MAMMARY GLAND

H. M. Valivullah, D. R. Bevan, *A. Peat and T. W. Keenan

Department of Biochemistry, Virginia Polytechnic Institute and State University, Blacksburg, Virginia 24061, U.S.A., and *Sunderland Polytechnic, Sunderland, England

INTRODUCTION

Milk lipids exist in the form of globules composed of a triacylglycerol-rich core surrounded by a membrane derived from apical plasma membrane. Milk lipid globules range in size from less than 1 μm to more than 10 μm in diameter. Very small globules are most numerous but globules in the 1 to 4 m size range account for most of the volume of milk lipids. The process by which milk lipid globules are secreted appears to be unique. Within the cell, precursors of lipid droplets of milk occur in the cytoplasm. Here they are surrounded by a protein and polar lipid surface coat material, but not by bilayer membrane. The droplets grow in volume as they migrate toward the apical region of the cell. During secretion they become progressively enveloped in apical plasma membrane. As they are ultimately pinched off into the alveolar spaces, they are completely surrounded by a differentiated region of apical plasma membrane (Mather and Keenan, 1983).

Within mammary epithelial cells the first morphologically recognizable precursors of milk lipid globules are microlipid droplets, structures of less than 0.5 μm diameter. These appear to originate from endoplasmic reticulum and are composed of a triacylglycerol-rich core surrounded by a non-bilayer protein and polar lipid coat. Observations to date suggest that microlipid droplets may (1) be secreted directly as the very small lipid globules of milk, (2) fuse to form larger structures known as cytoplasmic lipid droplets (Dylewski et al., 1984; Deeney et al., 1985). To begin to elucidate those factors which promote microlipid droplet fusion, a process which ultimately controls size distribution of milk lipid globules, an *in vitro* system to study microlipid droplet fusion was developed.

METHODS

Microlipid droplets, isolated from homogenates of rat mammary glands (Dylewski et al., 1984), were labeled with the fluorescent dye octadecyl rhodamine B chloride (Hoekstra et al., 1984) by adding 10 μl of a 4.3 mM solution of the dye in ethanol to 1 ml of a suspension of microlipid droplets in 50 mM Tris, pH 7.5, and incubating for 1 h at 21°C. Unincorporated dye was removed by gel filtration on Sephadex G-75 eluted with 50 mM Tris, pH 7.5. Fusion was monitored by measuring the increase in fluorescence caused by release from self-quenching when surface density of octadecyl rhodamine B was decreased by fusion of labeled with unlabeled microlipid droplets. Incubations were in a final volume of 1 ml of 50 mM Tris, pH 7.5, containing dye-labeled and unlabeled microlipid droplets (ratio of 1: 2 or greater) and other constituents as indicated in Table 1. Excitation and emission were at 525 and 585 nm, respectively. After recording emission for 30 min at 21° C, Triton X-100 was added to a final concentration of 1%, and fusion was calculated as percent increase in fluorescence emission in the presence of Triton X-100. Cytosol was prepared by homogenization of mammary tissue in 0.25 M sucrose in 50 mM Tris, pH 7.5, and centrifugation of the homogenate for 60 min at 120,000 *g*.

RESULTS

Small, gradual increases in fluorescence were observed with microlipid droplets suspended in buffer alone. Fusion was markedly stimulated by addition of cytosol (100 μl) to the reaction mixture (Table 1). Chelex treatment of cytosol reduced fusogenic activity, but addition of calcium or magnesium restored activity. Calcium added to buffer alone stimulated fusion, and addition of cytosol and calcium together increased fusion over that obtained with cytosol alone. Protease treatment of cytosol reduced fusogenic activity, suggesting an involvement of proteins. On fractionation of cytosol on Sephacryl S-200, fusogenic activity was found in fractions corresponding to the ascending portion of the first protein peak eluted. Fusogenic activity of this fraction was enhanced by addition of calcium. On electrophoresis in nondenaturing gels, this fusogenic fraction from cytosol had three components, with M_r of about 290, 360 and 370 kD. In denaturing gels four component polypeptides, of about 15, 16, 16.5 and 18 kD, were detected. As noted by others studying fusion of liposomes, homopolymers of lysine, arginine and histidine promoted fusion of microlipid droplets. Addition of calcium with polyarginine or polyhistidine resulted in a

TABLE 1. Factors Affecting Fusion of Microlipid Droplets

Addition	% Fusion
Cytosol	11.9
Chelex-treated cytosol	7.4
Chelex-treated cytosol + 2.5 mM calcium	16.8
Chelex-treated cytosol + 2.5 mM magnesium	12.2
Trypsin-treated cytosol	5.9
Trypsin-treated cytosol + 2.5 mM calcium	7.5
Pronase-treated cytosol	3.8
Cytosol preincubated 1 h at 37°C	5.2
2.5 mM Calcium	13.3
Sephacryl fraction of cytosol + 2.5 mM calcium	33.0
Polyarginine (M_r 40kD), 100 μg/ml	18.0
Polyarginine, 100 μg/ml + 2.5 mM calcium	21.4
Polylysine (M_r 8kD), 10 μg/ml	22.8
Polylysine, 10 μg/ml + 2.5 mM calcium	14.2
Polyhistidine (M_r 11kD), 100 μg/ml	4.0
Polyhistidine, 100 μg/ml + 2.5 mM calcium	11.8
Mixed brain gangliosides, 50 nmol	51.0
GM3, 50 nmol	23.9
GM2, 50 nmol	16.6
GM1, 50 nmol	17.6
GD3, 50 nmol	13.1
GD1a, 50 nmol	8.0

further stimulation of fusion, but addition of calcium was inhibitory to polylysine-stimulated fusion.

Mixed gangliosides from bovine brain, when added as an aqueous solution, strongly promoted fusion, with maximum effect being obtained near 50 nmol/ml (sialic acid basis). Individual gangliosides GM3, GM2, GM1, GD3 and GD1a all promoted fusion, but none was as effective as mixed brain gangliosides. Control experiments (see below) showed that gangliosides promoted fusion of microlipid droplets without causing lysis. Gangliosides enhanced fusion at concentrations below their critical micelle concentrations, suggesting that fusion was not due to detergent properties of gangliosides. The physiological relevance of this effect of gangliosides remains to be assessed. On analysis, microlipid and cytoplasmic lipid droplets from rat mammary gland were found to contain about 0.2 and 2.0 nmol of ganglioside sialic acid/mg protein, respectively. With cytoplasmic lipid droplets, this level of ganglioside is much higher than can be ascribed to plasma membrane or endomembrane contamination of preparations (Deeney et al., 1985). Since these lipid droplets do not traffic through the Golgi apparatus where glycosyltransferases of ganglioside biosynthesis are localized (Keenan et al., 1974), how gangliosides come to be present in intracellular lipid droplets remains to be established. When nonionic detergent extracts of microlipid droplets were applied onto affinity columns of immobilized brain gangliosides (Parikh and Cuatrecasas, 1974) a protein fraction was strongly retained. This fraction was not eluted with 1% Triton X-100 or 1 M guanidine in 1% Triton, but was eluted with 4 M guanidine in 1% Triton. Polypeptides in this fraction had mobilities similar to those in the fusogenic fraction of cytosol on electrophoresis in denaturing gels. These observations suggest that gangliosides or proteins with affinities for gangliosides may have a functional role in microlipid droplet fusion.

Two approaches were taken to show that results obtained were due to fusion rather than lysis of microlipid droplets. Microlipid droplets were isolated from an animal which had received intraperitoneal injections of radioactive leucine and glycerol to label proteins and lipids. After incubation with various fusogens, larger droplets were separated from smaller droplets by centrifugal flotation and the ratio of lipid to protein radioactivity in small and large droplets was measured. Protein was maintained on the surface of larger droplets in the amounts expected based on surface area of these droplets. This suggests that larger droplets originated by fusion. If lysis had occurred, the proteinaceous surface coat material would have been lost and the large droplets would have been formed by coalescence of triacylglycerols released from microlipid droplet interiors. Alternatively, portions of fusogenic mixtures were taken after incubation, spread onto grids and stained with phosphotungstate for electron microscopic examination. Both large and small droplets were observed to have morphologically distinct plate-like coat material over their surfaces, and many fusion figures were observed. Morphometric analysis showed that there were many more larger droplets in incubation mixtures containing calcium or gangliosides than in control incubations with buffer alone suggesting that true fusion occurs. This system is expected to help identify those factors that ultimately determine the size distribution of milk lipid globules.

REFERENCES

Deeney JT, Valivullah HM, Dapper CH, Dylewski DP, Keenan TW (1985). Microlipid droplets in milk secreting mammary epithelial cells: evidence that they originate from endoplasmic reticulum and are precursors of milk lipid globules. Eur J Cell Biol 38:16-26.

Dylewski DP, Dapper CH, Valivullah HM, Deeney JT, Keenan TW (1984). Morphological and biochemical characterization of possible intracellular precursors of milk lipid globules. Eur J Cell Biol 35:99-111.

Hoekstra D, de Boer T, Klappe K, Wilschut J (1984). Fluorescence method for measuring the kinetics of fusion between biological membranes. Biochem 23:5675-5681.

Keenan TW, Morré DJ, Basu S (1974). Ganglioside biosynthesis: concentration of glycosphingolipid glycosyltransferases in Golgi apparatus from rat liver. J. Biol Chem 249:310-315.

Mather IH, Keenan TW (1983). Function of endomembranes and the cell surface in the secretion of organic milk constituents. In Mepham TB (ed): "Biochemistry of Lactation," Amsterdam, Elsevier, pp 231-283.

Parikh I, Cuatrecasas P (1974). Ganglioside-agarose and cholera toxin. Meth Enzymol 34:610-619.

Cell-Free Analysis of Membrane Traffic, pages 427–429

ENDOCYTOSIS AND SUBCELLULAR DISTRIBUTION OF TRANSFERRIN IN RAT LIVER

H. Goldenberg, E. Wallner, R. Pumm and M. Hüttinger

Department of Medical Chemistry, University of Vienna, Vienna, Austria, A-1090 Vienna, Waehringerstrasse 10

Transferrin-iron uptake by the liver has been studied extensively. Mechanisms include a specific receptor mediated uptake primarily by hepatocytes (Sibille et al., 1986, Morgan et al. 1986) as well as unspecific uptake by fluid-phase pinocytosis (Cole and Glass, 1983, Sibille et al., 1982). Uptake is also mediated by liver endothelial cells which have a much higher transferrin receptor density than hepatocytes (Kishimoto and Tavassoli, 1985, Tavassoli et al., 1986). Reductive release of iron at the cell membrane without endocytosis also is possible (Morley and Bezkorovainy, 1985).

In the experiments described here transferrin uptake was examined in livers of rats in situ. Radioactive transferrin was injected intravenously and the liver subcellular distribution studied by differential and density gradient fractionation.

After icecold perfusion, the radioactivity in the liver reached a constant level after 5 to 10 min and remained so for at least 30 min. Beginning after 2 min, the subcellular distribution of transferrin was constant. Sucrose density gradient centrifugation (15 to 35% w/w, two steps with 37% and 55% w/w, two hours at 55000 rpm in a Z-60 zonal rotor, Fig. 1) show a distribution of radioactivity from transferrin distinct from marker enzymes for plasma membrane, lysosomes and Golgi apparatus. The fraction containing the bulk of the radioactivity from transferrin showed an appearance typical of endosomes upon electron microscope examination. These membranes also exhibited high N-ethylmaleimide sensitive ATPase activity and ATP-dependent fluorescence quenching of acridine orange (a measure of vesicle acidification). The findings demonstrate that radioactive transferrin is taken up by liver cells in situ and that the radioactivity enters a compartment with sedimentation density and morphological properties characteristic of the endosomal compartment of liver (Saermark et al., 1985). This fraction can provide an excellent source of starting material for further studies using more decisive fractionation procedures necessary to determine the nature and heterogeneity of the compartment as well as the kinetics of transfer from one subcompartment to another within the endosomal systems.

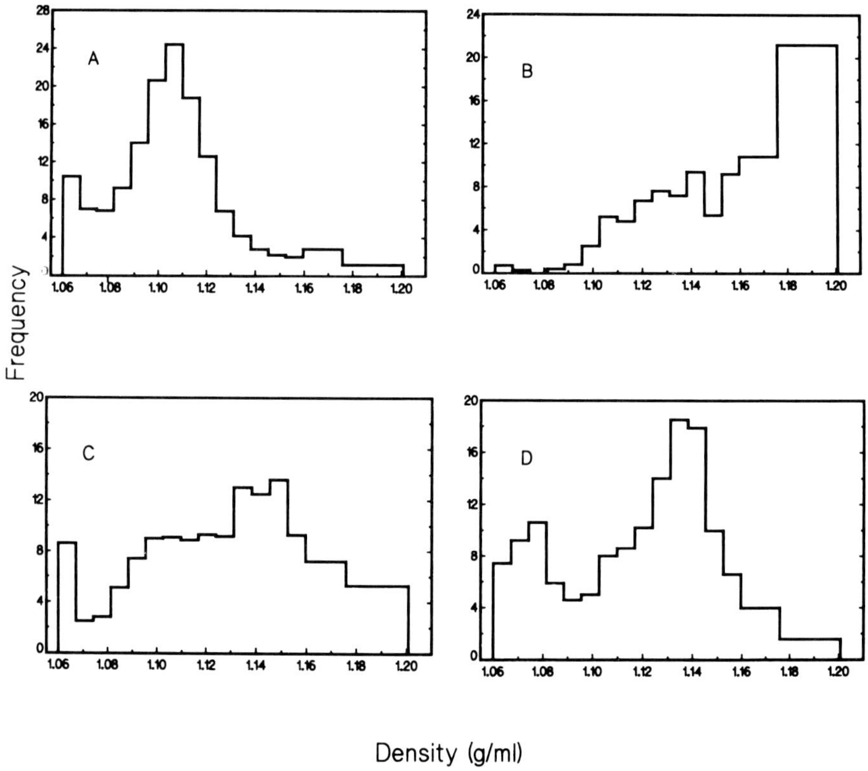

Figure 1: Sucrose density gradient fractionation of a liver homogenate 30 min after intravenous injection of ^{125}I-transferrin. A: radioactivity, B: β-glucosaminidase, C: 5'-nucleotidase, D: galactosyl transferase.

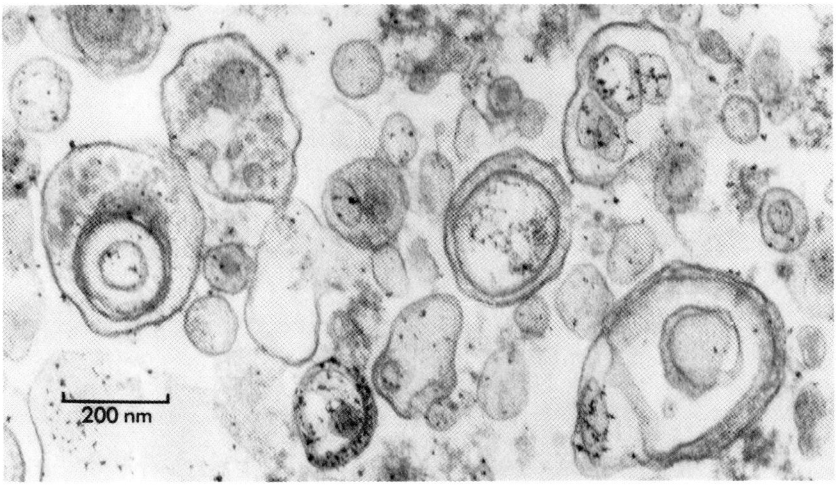

Figure 2. Electron micrograph of a low density fraction.

REFERENCES

Cole EJ, Glass J (1983). Transferrin binding and iron uptake in mouse hepatocytes. Biochim Biophys Acta 762:102-110.

Kishimoto T, Tavassoli M (1985). Endothelial binding of transferrin in fractionated liver cell suspensions. Biochim Biophys Acta 846:14-20.

Morgan EJ, Smith GD, Peters TJ (1986). Uptake and subcellular processing of ^{59}Fe-^{125}I-labelled transferrin in rat liver. Biochem J 237:163-173.

Morley CGD, Bezkorovainy A (1985). Cellular iron uptake from transferrin: is endocytosis the only mechanism? Int J Biochem 17:553-564.

Saermark T, Flint N, Evans WH (1985). Hepatic endosome fractions contain an ATP-driven proton pump. Biochem J 225:51-58.

Sibille JC, Octave JN, Trouet A, Crichton RR (1982). Transferrin protein and iron uptake by cultured hepatocytes. FEBS Lett. 150:365-369.

Sibille JC, Octave JN, Schneider YJ, Trouet A, Crichton RR (1986). Subcellular distribution of transferrin protein and iron in the perfused rat liver: effect of Triton WR-1339, digitonin and temperature. Eur J Biochem 155:47-55.

Tavassoli M, Kishimoto T, Soda, R, Kataoka M, Harjes K (1986). Liver endothelium mediates the uptake of iron-transferrin complex by hepatocytes. Exp Cell Res 165:369-379.

Cell-Free Analysis of Membrane Traffic, pages 431–440

A MITOTIC FORM OF THE GOLGI APPARATUS IDENTIFIED USING IMMUNO-ELECTRONMICROSCOPY

John M. Lucocq, James G. Pryde, Eric G. Berger[1] and Graham Warren
Department of Biochemistry, University of Dundee, Dundee DD1 4HN, Scotland, U.K. and (1) the Department of Pharmacology, Biocenter, University of Basle, CH 4056 Switzerland.

Introduction

Membrane-bound organelles do not appear to be made *de novo* so at least one copy of every organelle must be distributed to each daughter cell during the division process (Birky, 1983). If there is more than one copy of the organelle at the onset of division then the problem is to partition them equally. If there is but a single-copy then there must also be a mechanism which divides the organelle before partitioning can take place. The Golgi complex in animal cells appears to be a single copy organelle (Rambourg et al., 1981) and both cytochemical (Ludford, 1924) and immunofluorescence (Burke et al., 1982; Hiller & Weber, 1982) studies at the light microscopic level have shown that it divides by a process involving fragmentation, dispersal and re-assembly. The large number and small size of the fragments appears to ensure random dispersion throughout the mitotic cell cytoplasm so that equal partitioning is effected by division of the cell into two equal halves, a process known to be a function of the mitotic spindle poles (Rappaport, 1961 and 1969).

Little is known about the detailed mechanism of fragmentation and re-assembly because, at the electron microscopic level, the Golgi apparatus has been shown to lose its characteristic morphology - the distinctive stacks of closely-apposed and flattened cisternae cannot be found in mitotic cells (see Zeligs & Wollman, 1979) and attempts to label the products of fragmentation have been unsuccessful. A prerequisite for a complete morphological description of Golgi division is that Golgi membranes can be reliably identified and quantitated throughout mitosis. We have therefore used immunoelectron microscopy to localise galactosyltransferase, a resident protein of Golgi apparatus membranes.

Results

Sections of interphase HeLa cells embedded in Lowicryl K4M were incubated with antibodies to galactosyltransferase followed by protein A gold. As shown previously in interphase cells (Roth & Berger, 1982), labelling was found only over the *trans* cisterna(e) of the Golgi stack (Fig 1A) and nowhere else confirming the specificity of the antibody for Golgi membranes.

Previous studies on Golgi division by both light and electron microscopy suggested that the products of fragmentation were small vesicles (see Zeligs & Wollman, 1979). This, combined with the profound inhibition of membrane traffic during mitosis led to the hypothesis (Warren, 1985) that the two are causally related and that inhibition of membrane traffic at the level of vesicle fusion causes fragmentation of the Golgi stack into small vesicles - in essence the organelle is homogenised thereby ensuring equal partitioning at cytokinesis. We therefore expected to find dispersed vesicles containing galactosyltransferase in mitotic cells.

To our surprise we found that galactosyltransferase was almost exclusively to be found in distinct globular clusters of vesicles and tubules which we term **Golgi clusters** (Lucocq et al., 1987). Examples are shown in figures 1B and C and a schematic summarising the main features to date is shown in figure 2. They comprise two classes of vesicle segregated in different regions of the cluster. One class comprises electron-lucent vesicles up to 250nm in diameter and the other smaller vesiculo-tubular vesicles with a denser content. Both are set in a dense matrix which excludes cytoplasmic structures such as ribosomes and the cluster is bounded by elements of the rough endoplasmic reticulum, sometimes on all sides. Each cluster is usually 1µm in diameter and is not connected to any of the others. There are around a score in each mitotic cell, dispersed apparently at random.

Most of the labelling for galactosyltransferase was located over the smaller vesiculo-tubular structures although a few of the electron-lucent ones were also positive (Figures 1B and C). We could not find qualitative or quantitive evidence for galactosyltransferase outside the clusters. The lectin, *Ricinus communis* I, binds to galactosylated oligosaccharide chains (Baenziger & Fiete, 1979; Debray et al., 1981), the products of the galactosyltransferase reaction, and labels the *trans* Golgi cisterna(e) of interphase BHK (Griffiths et al.,

(Text continued on page 434.)

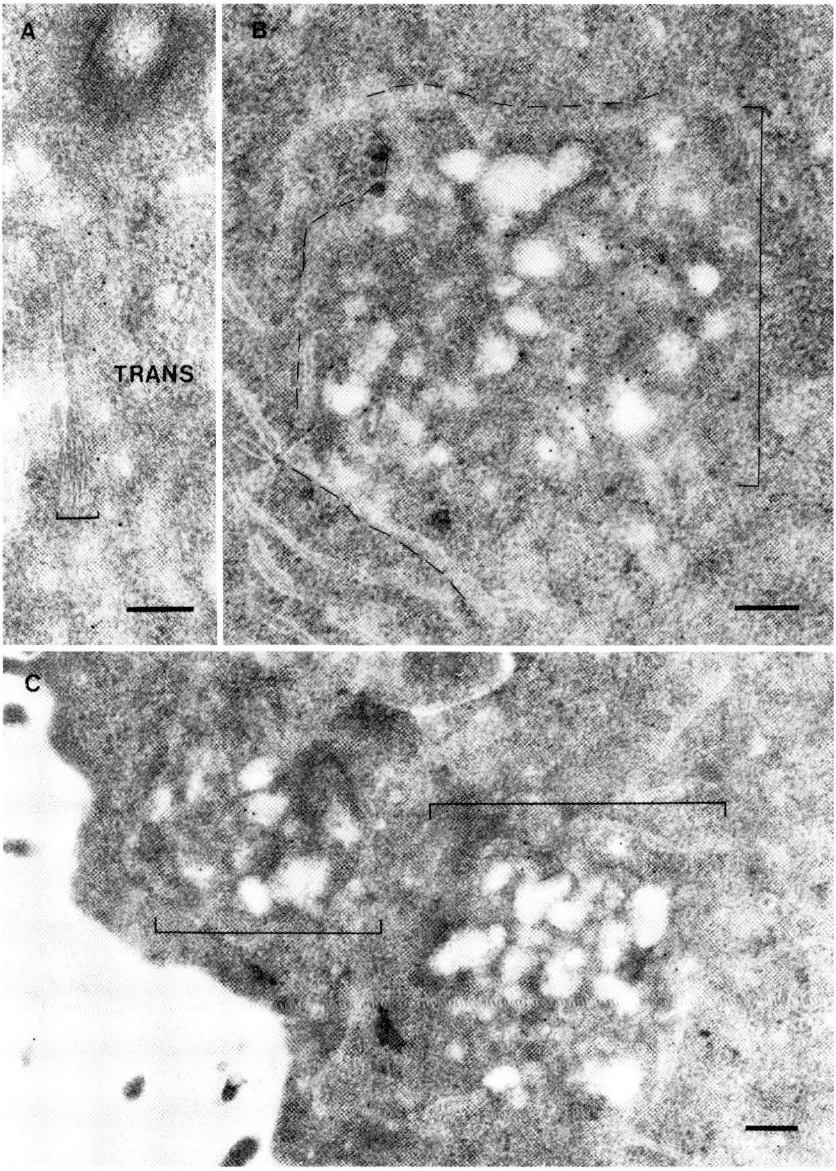

Figure 1. Galactosyltransferase labelling of interphase and mitotic cells. Sections of HeLa cells embedded in Lowicryl K4M, labelled with antiserum against galactosyltransferase followed by 7nm protein A-gold. (A) Interphase. Note labelling of *trans* (see over)

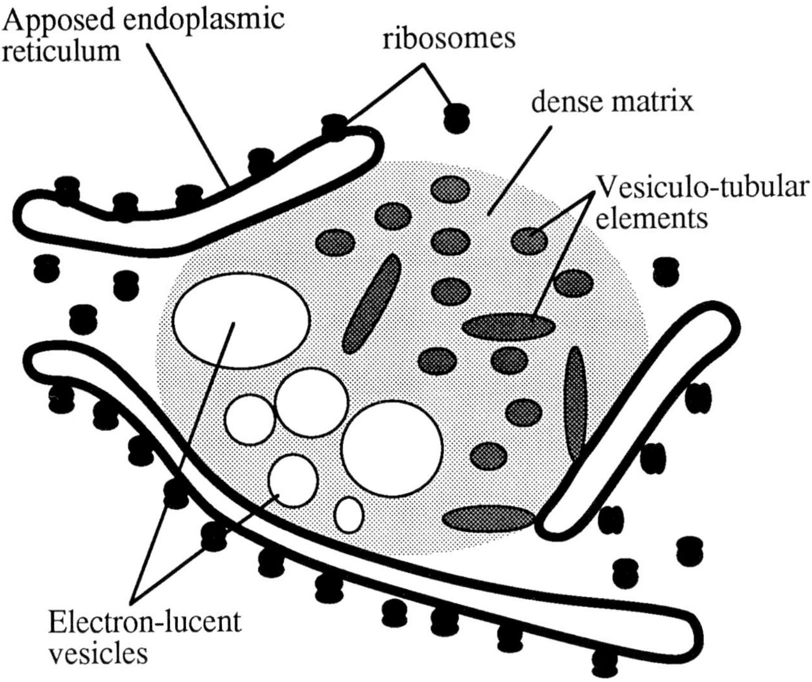

Figure 2. Schematic illustrating main features of Golgi clusters.

(Text continued from page 432.)

1982) and HeLa cells (Figure 3A) as well as lysosomes and the plasma membrane. This lectin labelled Golgi clusters in mitotic HeLa cells (Figure 3B and C) giving the same distribution of labelling as galactosyltransferase. This provides further evidence that elements of the *trans* Golgi apparatus are located among the small vesiculo-tubular structures.

Figure 1 (contd.) cisterna(e) of stack (bracketed). (B,C) Mitosis. Clusters (bracketed) contain electron-lucent vesicles and smaller, vesiculo-tubular vesicles over which most of the labelling for galactosyltransferase is found. Note the apposed endoplasmic reticulum (outline dashes in B). Bars 0.2μm.

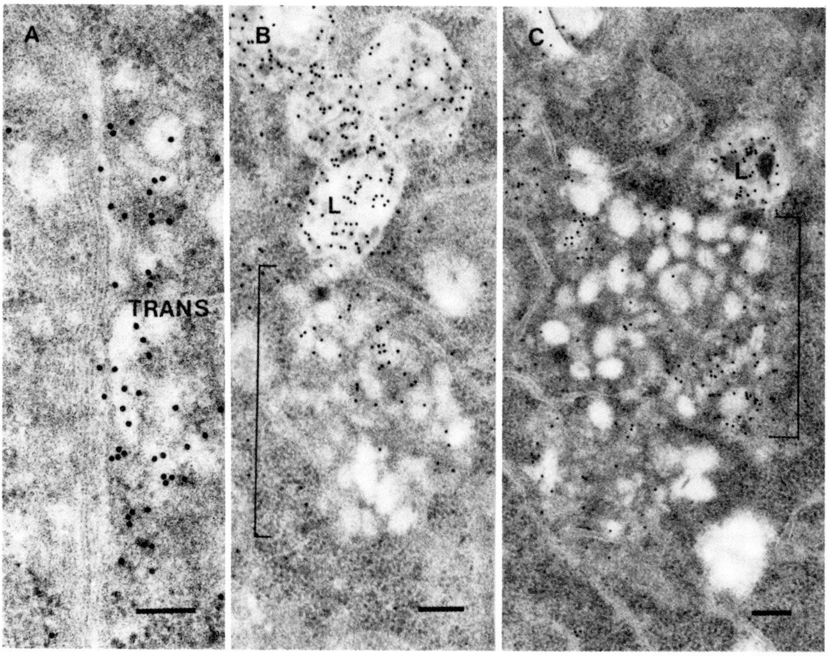

Figure 3. ***Ricinus communis*** **I labelling of interphase and mitotic cells.** Lowicryl K4M sections were labelled with the lectin conjugated to 22nm gold. (A) Interphase. Labeling restricted to *trans* cisterna(e). (B,C) Mitosis. Labelling found mainly over the smaller vesiculo-tubular elements of the Golgi clusters (bracketed). Two lysosomes (L), near to the clusters, are also labelled. Bars 0.2μm.

The morphology of clusters was poor in Lowicryl K4M so we looked for clusters in HeLa cells postfixed in osmium tetroxide. Galactosyltransferase cannot be localised using antibodies after osmium tetroxide treatment so we used the thiamine pyrophosphatase (TPPase) reaction (Novikoff & Goldfisher, 1961), which stains *trans* Golgi cisterna(e) (Figure 4A) and is known to co-localise with galactosyltransferase in interphase HeLa cells (Roth & Berger, 1982). Clusters identified using the morphological criteria described above proved to be positive for TPPase (Figures 4B and C). Some small vesicles and tubules and some larger structures were stained. Very rarely, clusters contained cisternal elements positive for TPPase (Figure 4C) indicating that in a few cells Golgi disassembly does not go to completion.

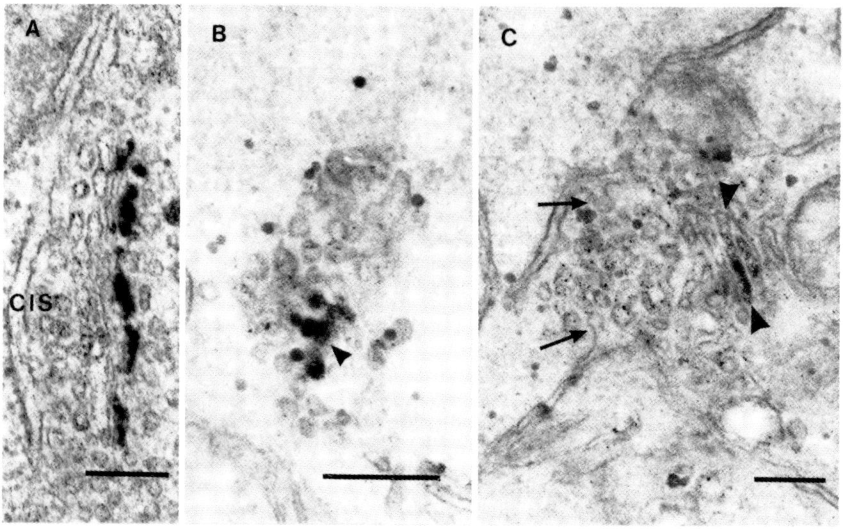

Figure 4. Thiamine pyrophosphatase staining of osmified HeLa cells. (A) Interphase. Labeling restricted to *trans* cisterna(e). (B) Mitosis. A few of the vesiculo-tubular structures are stained (arrowhead) in this small cluster. (C) Metaphase. A rare cluster showing remnants of the *trans* Golgi stack stained for TPPase (arrowheads). Note budding (arrows) in the apposed endoplasmic reticulum suggestive of transitional elements. Bars 0.2μm.

Not all of the membranes in the Golgi clusters are labelled by the available Golgi markers which opens up the possibility that other membranous organelles are present and might be partitioned along with Golgi membranes. HeLa cells were incubated with horseradish peroxidase (HRP) for 90min., sufficient time to fill all of the compartments along the endocytic pathway (Marsh et al., 1986). Mitotic cells were then isolated by shake-off (Klevecz, 1975). The enzyme was found in tubules, often single, in the peripheral cytoplasm and collections of tubules close to the plasma membrane (Figure 5A and B). Structures resembling multivesicular bodies and secondary lysosomes were also positive for HRP and most of these structures were also labelled in separate experiments by the lysosomal marker acid phosphatase (Novikoff, 1976; Pavelka & Ellinger, 1983)(Figure 6). However, we did not find any convincing evidence for HRP or acid phosphatase inside the cluster vesicles (Figures 5 and 6) though occasionally we found HRP-positive tubules nearby (Figure 5B). Interestingly, bodies positive for acid phophatase were frequently found next to the clusters (Figure 6) mostly in the region labelled by *ricinus*

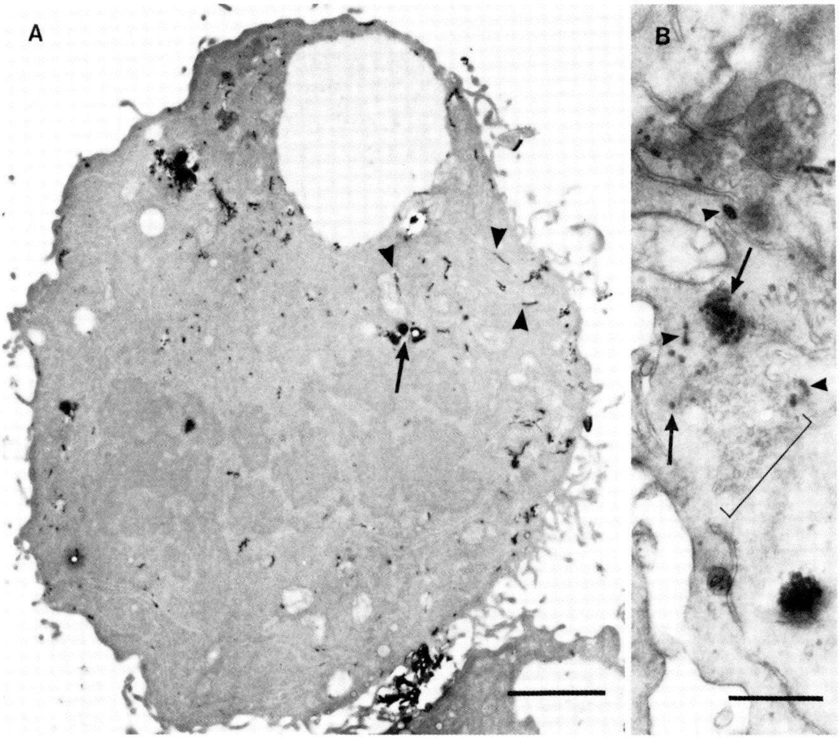

Figure 5. HRP staining of the endocytic pathway in mitotic cells. HRP was visualised using diamino-benzidine. (A) Low magnification of metaphase cell showing stained tubules (arrowheads) throughout cytoplasm and some vacuoles (arrow). (B) Higher magnification of unstained Golgi cluster (bracketed) showing juxtaposed stained tubules (arrowheads). The granular structures (arrows) are glycogen. Bars (A) 2μm and (B) 0.5μm.

communis I and containing galactosyltransferase. We conclude that the compartments of the endocytic pathway are not part of the Golgi clusters but the proximity of lysosomes and other endocytic compartments suggests that some organelles may be partitioned as a group.

Discussion

We do not yet know the role played by the Golgi clusters in the division process but their presence at all stages of mitosis that we have been able to examine argues that they are an end product of

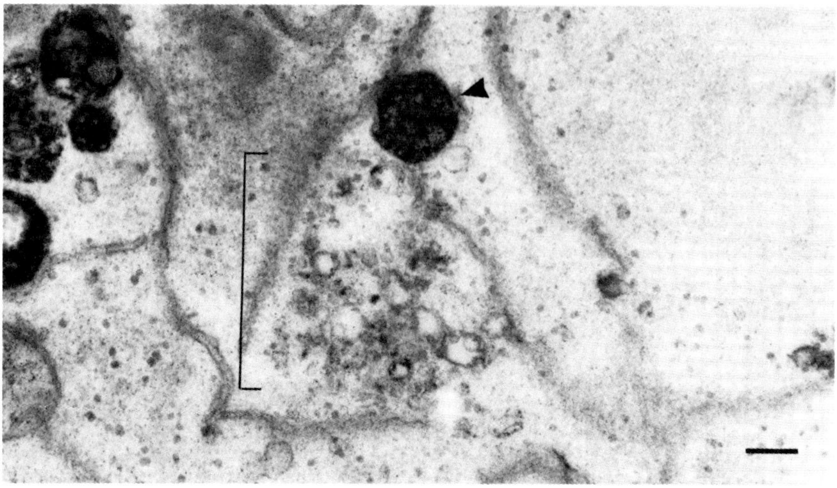

Figure 6. Acid phosphatase staining of lysosomes in mitotic cells. A stained lysosome (arrowhead) lies close to the unstained Golgi cluster (bracketed). Bar 0.2µm.

fragmentation and not an intermediate. This poses a problem because there are so few of them and random partitioning alone will not ensure equal numbers in each daughter cell. A mechanism for equal partitioning would have to exist (Birky, 1983). The presence of a small number of clusters is at variance with certain immunofluorescence experiments (Burke et al., 1982) which led in part to the idea that thousands of small vesicles are produced. But these experiments were performed using antibodies to different parts of the Golgi apparatus and we do not yet know whether all Golgi membrane proteins are present in the Golgi clusters. Only further detailed studies of the mitotic Golgi apparatus using other antibodies will resolve this anomaly and enable us to elucidate the division mechanism.

Finally it is worth pointing out that the Golgi clusters were only discovered because an immunocytochemical technique was used. Antibodies against resident membrane proteins are the only reliable means of following the fate of an organelle which undergoes major changes in morphology. This point is highlighted by the fact that Golgi clusters have been described before in other cell types (Melmed et al., 1973; Zeligs & Wollman, 1979) but their origin was not determined because they bore no resemblance to the classical Golgi apparatus. In one of the first papers showing the disappearance of the Golgi apparatus during mitosis (Robbins & Gonatos, 1964) there is even a clear example

(Fig. 20) of a Golgi cluster which the authors did not comment upon. This oversight shows that the importance of using an immunocytochemical technique cannot be overstated.

Acknowledgements

This work was supported by grant number SP1757 from the Cancer Research Campaign.

References

Baenziger J, Fiete D (1979). Structural determinants of Ricinus communis agglutinin and toxin specificity for oligosaccharides. J Biol Chem 254:9795-9799.

Birky C W (1983). The partitioning of cytoplasmic organelles at cell division. Int Rev Cyt 15:49-89.

Burke B, Griffiths G, Reggio H, Louvard D, Warren G (1982). A monoclonal antibody against a 135K Golgi membrane protein. EMBO J 1:1621-1628.

Debray H, Decout D, Strecker G, Spik G, Montreuil J (1981). Specificity of twelve lectins towards oligosaccharides and glycopeptides related to N-glycosyl proteins. Eur J Biochem 117:41-45.

Griffiths G, Brands R, Burke B, Louvard D, Warren G (1982). Viral membrane proteins acquire galactose in trans Golgi cisternae during intracellular transport. J Cell Biol 95:781-792.

Hiller G, Weber K (1982). Golgi detection in mitotic and interphase cells by antibodies to secreted galactosyltransferase. Exp Cell Res 142:85-94.

Klevecz R R (1975). Automated cell cycle analysis. Methods Cell Biol 10:157-173.

Lucocq J M, Pryde J G, Berger E G, Warren G (1987). A mitotic form of the Golgi apparatus in HeLa cells. J Cell Biol in press.

Ludford R J (1924). Thc distribution of the cytoplasmic organs in transplantable tumour cells with special reference to dictyokinesis. Proc Roy Soc Lond 97:50-60.

Marsh M, Griffiths G, Dean G E, Millman I, Helenius A (1986). Three-dimensional structure of endosomes in BHK-21 cells. Proc Natl Acad Sci USA 83:2899-2903.

Melmed R N, Benitez C J, Holt S J (1973). An ultrastructural study of the pancreatic cell in mitosis with special reference to changes in the Golgi complex. J Cell Sci 12: 163-173.

Novikoff A B, Goldfisher R (1961). Nucleosidediphosphatase activity in the Golgi apparatus and its usefulness in cytological studies. Proc Natl Acad Sci USA. 47: 802-810.
Novikoff A B (1976). The endoplasmic reticulum: A cytochemist's view (A Review). Proc Natl Acad Sci USA 73:2781-2787.
Pavelka M, Ellinger A (1983). The trans Golgi face in rat small intestinal absorptive cells. Eur J Cell Biol 29:253-261.
Rambourg A, Clermont Y, Hermo L (1981) Three-dimensional structure of the Golgi apparatus. Meth Cell Biol 23: 155-166.
Rappaport R (1961). Experiments concerning the cleavage stimulus in sand dollar eggs. J Exp Zool 148:81-89.
Rappaport R (1969). Aster-equatorial surface relations and furrow establishment. J Exp Zool 171:59-67.
Robbins E, Gonatas N K (1964). The ultrastructure of a mammalian cell during the mitotic cycle. J Cell Biol 21:429-463.
Roth J, Berger E G (1982). Immunocytochemical localization of galactosyltransferase in HeLa cells: codistribution with thiamine pyrophosphatase in trans Golgi cisternae. J Cell Biol 93: 223-229.
Warren G (1985). Membrane traffic and organelle division. TIBS 10:439-443.
Zeligs J D, Wollman S H (1979). Mitosis in rat thyroid epithelial cells in vivo. I. Ultrastructural changes in cytoplasmic organelles during the mitotic cycle. J Ultrastruc Res 66: 53-77.

Cell-Free Analysis of Membrane Traffic, pages 441–446

APPLICATIONS OF ENDOCYTOSIS RESEARCH TO DRUG DELIVERY

Glynn Wilson, Clive J. Dix and Ian F. Hassan

Advanced Drug Delivery Research, Ciba-Geigy Pharmaceuticals, Horsham, West Sussex, RH12 4AB, United Kingdom.

INTRODUCTION

From the early considerations of lysosomotropic carriers (Trouet et al., 1972) drug delivery systems, with the potential to utilise a variety of receptor-mediated endocytic processes, have been described (Gregoriadis et al., 1984). Few of these systems have yet advanced to the stage of human clinical trials. For human use drug carrier systems should ideally have the characteristics shown in Table 1. Recent progress in understanding the morphological and biochemical details of endocytic compartments, in a variety of cell types, is aiding the design of soluble and particulate drug carriers with optimal selectivity, stability and biocompatibility.

CARRIER SYSTEMS

Much attention has focused on the use of liposomes as carriers for different drugs (Poste, 1984). A major limitation of such particles is that they do not extravasate the circulation, however, they are rapidly internalised by fixed and free phagocytic cells. This property has been used advantageously to deliver a variety of drugs, including immunomodulators, to phagocytic cells of the reticuloendothelial system, resulting in activation of macrophages to destroy bacteria, fungi, viruses and tumours (Fidler, 1986). The internalisation of drug delivery systems by the asialoglycoprotein receptor of the liver best exemplifies the utilisation of a receptor-mediated pinocytic pathway. Antitumour drugs such as

TABLE 1. Idealised Characteristics of an Endocytosable Drug Delivery System

Selectivity:

1. Able to penetrate anatomical barriers between administration and target sites.
2. Specific for target cell surface in disease state.
3. Efficiently endocytosed and specifically routed along a defined intracellular pathway.

Stability:

1. Stable in blood stream and extracellular spaces.
2. Sensitive to defined intracellular trigger e.g. pH and lysosomal enzymes, to release active drug.
3. Drug stable during passage through intracellular compartments and membranes.

Biocompatibility:

1. Pharmacologically inactive during transport to final site.
2. Non-immunogenic.
3. Carrier degraded and excreted.

Pharmaceutics:

1. Chemically and physically stable in dosage form.
2. Convenient and cost effective to prepare and store.
3. Sufficient drug loading can be achieved.

daunorubicin are coupled to galactosylated human serum albumin via peptidic linkages that allow the selective release of active drug in the lysosome. The system shows specificity for hepatoma cells *in vivo*, is biodegradable and poorly immunogenic. One application of this specific carrier system in the human clinic is in the treatment of hepatocarcinoma and their metastases (Schneider et al., 1984). In this example details on the binding and internalisation of a range of natural and neo-glycoproteins have permitted the construction of carrier systems that satisfy many of the criteria given in Table 1. These systems show potential for the delivery of a variety of pharmacological agents in the treatment of a range of liver diseases (Table 2).

TABLE 2. Utilisation of the Hepatocyte Asialoglycoprotein Receptor (cf. Fallon and Schwartz, 1985; Trouet et al., 1984).

Disease	Drug
Cancer	Daunorubicin Methotrexate Diptheria Toxin
Viral	Arabinoside A Trifluorothymidine
Drug Toxicity	N-Acetylcysteine Folinic Acid
Malaria	Primaquine

EXPLOITABLE RECEPTORS IN MUCOSAL EPITHELIAL CELLS

For many cellular targets much basic information is required on the morphology and biochemistry of cell or tissue specific receptors to establish how they can be exploited for drug delivery. Much current interest is directed towards epithelial and endothelial cell barriers that delivery systems need to traverse on route to their final destination. Studies on endocytic pathways in cells of the GI-tract have been difficult owing to the lack of *in vitro* cell systems that retain the anatomical and biochemical features of the differentiated cells *in vivo*. We have begun to study the endocytic pathways in a human colon carcinoma line, CaCo-2, which in culture shows many of the characteristics of differentiated epithelial cells (Rousset, 1986). When grown on 0.45 μ filters in Millicell chambers (Millipore), the cells form a functionally tight monolayer and develop a high transcellular electrical resistance of 300-400 ohms/cm^2, 10-15 days after seeding. The cells have well developed brush borders which express high levels of sucrase. The cells will transcellularly transport fluid-phase markers such as inulin and lucifer yellow from apical to basolateral membrane and vice versa. The cells are functional in the polarised transport of vitamin B12 (cobalamin) from apical to basolateral membrane. They express and internalise receptors (200-400 per cell) for the intrinsic factor-cobalamin complex, and secrete a cobalamin-binding protein exclusively from the basolateral membrane (Dix et al., 1987). These results demonstrate the potential of this *in vitro* cellular barrier for defining and characterising endocytic pathways that may be utilised for drug transport.

In defining exploitable cell surface receptors and membrane pathways for drug delivery, determination of the cellular selectivity and mechanism of viral entry and tropism *in vivo* (cf. Tomlinson, 1986), may provide interesting leads. Developing methodologies for the isolation of viral proteins and their functional reconstitution into lipid vesicles are exemplified by the reconstruction of the pH dependent fusogenic activity of vesicular stomatitis virus G protein (Metsikkö et al., 1986). The systems that are currently being used to insert molecules into target cells *in vitro* (Loyter et al., 1986), may have future applications for the delivery of drugs *in vivo*. The use of replication-defective retroviral envelopes for the delivery of replacement genes to cells *ex vivo* is close to human clinical trials (Wilson, 1986).

REFERENCES

Dix CJ, Obray HY, Hassan IF, Wilson G (1987). Vitamin B12 transport through polarised monolayers of a colon carcinoma cell line. Biochem Soc Trans 15: 439-440.

Fallon RJ, Schwartz AL (1985). Receptor-mediated endocytosis and targeted drug delivery. Hepatology 5:899-901.

Fidler IJ (1986). Immunomodulation of macrophages for cancer and antiviral therapy. In Tomlinson E, Davis SS (eds): "Site-Specific Drug Delivery," Chichester: John Wiley & Sons Ltd, pp 111-134.

Gregoriadis G, Poste G, Senior J, Trouet A (1984). "Receptor-Mediated Targeting of Drugs," New York and London: Plenum Press.

Loyter A, Gitman AG, Chejanovsky N, Nussbaum O (1986). Fusion-mediated microinjection of macromolecules into living cultured cells by 'targeted' sendai virus envelopes. In Celis JE, Graessman A, Loyter A (eds): "Microinjection and Organelle Transplantation Techniques," New York and London: Academic Press, pp 179-197.

Metsikkö K, Van Meer G, Simons K (1986). Reconstitution of the fusogenic activity of vesicular stomatitis virus. EMBOJ 5:3429-3435.

Poste G (1984). Drug targeting in cancer therapy. In Gregoriadis G, Poste G, Senior J, Trouet A (eds): "Receptor-Mediated Targeting of Drugs," New York and London: Plenum Press, pp 427-474.

Rousset M (1986). The human colon carcinoma cell lines HT-29 and CaCo-2: two *in vitro* models for the study of intestinal differentiation. Biochimie 68:1035-1040.

Schneider YJ, Abarca J, Aboud-Pirak E, Baurani R, Ceulemans F, Deprez-de Campaneere D, Lesur B, Masquelier M, Offe-Slachmuylder C, Rolin-van Swiefen R, Trouet A (1984). Drug targeting in human cancer chemotherapy. In Gregoriadis G, Poste G, Senior J, Trouet A (eds): "Receptor-Mediated Targeting of Drugs," New York and London: Plenum Press, pp 1-25.

Tomlinson E (1986). Pathophysiology and the temporal and spatial aspects of drug delivery. In Tomlinson E, Davis SS (eds): "Site-Specific Drug Delivery," Chichester: John Wiley & Sons Ltd, pp 1-26.

Trouet A, Deprez-de Campaneere D, DeDuve C (1972). Chemotherapy through lysosomes with a DNA-daunorubicin complex. Nature 239:110-112.

Trouet A, Pirson P, Baurain R, Masquelier M (1984). Cell targeting of primaquine. In Peters W, Richards WHG (eds): "Handbook of Experimental Pharmacology," Heidelberg: Springer Verlag, pp 253-264.

Wilson G (1986). Gene therapy: rationale and realisation. In Tomlinson E, Davis SS (eds): "Site-Specific Drug Delivery," Chichester: John Wiley & Sons Ltd, pp 149-164.

Index

Acer pseudoplatanus, 257–258, 260–261
Acetate, 239, 303
Acetic acid, 8, 23, 36, 270
Acetylcholine, 97–98
N-Acetyl-ethylmaleimide (NEM), 314, 424
N-Acetyl-glucosamine, 318, 377
N-Acetylglucosaminidase, 86, 105
N-Acetyl-β-glucosaminidase, 175, 178, 181
N-Acetylglucosaminyltransferase I, 107–108, 129, 131
N-Acetyl-glucosaminyl phosphotransferase, 365
N-Acetyl-*D*-mannosamine, 344
N-Acetyl-neuraminic acid, 309, 344
Acid phosphatase, 392, 395, 401–403, 436, 438
Acidification
 assays, 29, 37–38
 endosomal, 43, 45–46, 161–162
Acridine orange, 428
Acyl-coenzyme A, 314
Adenosine
 diphosphate (ADP), 98
 monophosphate (AMP), 98, 160
  cyclic, 161
 triphosphate (ATP), 29–30, 40–41, 45, 98, 162, 260–261, 270, 277, 307, 312–313, 326–327, 334, 338, 345–347, 350 352, 401, 428; *see also* ATPase
Adenylate cyclase, 97, 160–161
Aldehyde, 135
Alkaline phosphatase, 160, 412–413
Alkaline phosphodiesterase I, 86, 161, 175, 178, 180–181, 216–218, 219, 222
Aminoacridine, 161–162
Aminopeptidase N (APN), 196–197, 199, 206
α-Amylase, 9
Analytical differential centrifugation, 247–249
Antibodies
 affinity purified, 125
 anti-58 kD IgG, 136
 anti-140 K, 199–201
 anti-AP-FITC, 202
 anti-APN, 199, 201, 203, 206
 anti-clathrin, 5, 81
 anti-cytochrome P_{450} reductase, 81, 83
 anti-G, 108, 319
 -antigen interactions, 4
 -antigen recognition system, 77
 anti-Golgi, 4, 131–132
 anti-GSA, 201–202
 anti-Ig, 92–94
 anti-organelle, 91, 93
 anti-RLA, 199, 201, 203
 anti-RLA-TRITC, 202
 anti-viral protein, 94
 concentrations of determined by A_{280}, 116
 domain specific, 77–78
 endosome-specific, 164, 364
 goat anti-rabbit IgG, 104, 109, 412–413
 monoclonal, 92, 196, 199, 319
  anti-aminopeptidase N, 198
  anti-AP, 202
  anti-clathrin, 5, 141–149
  anti-G, 107, 320, 322
  anti-Golgi, 102, 104, 131
  anti-heavy chain, 144, 146–148
  anti-human blood group A, 198–202
  anti-light chain, 146–148
  anti-rabbit histocompatibility class I, 197–198, 200, 203
  B3/25, 185–191
  colloidal gold complexes, 5
  flow cytometry and, 206
  HA 4 mouse, 117, 122
  ^{125}I-HA 4 mouse, 123
  high affinity, 94
  LC_a-specific, 147

LC_b-specific, 147
organelle specific, 129–139
immuno-isolation and development of, 75–88, 91–99, 101–112, 115–126, 129–139, 141–149
polyclonal, 4–5, 92–93, 135, 198
affinity purified, 94
anti-AP, 201–202
anti-APN, 200
anti-G, 102, 104, 108, 321–322
anti-heavy (cis) Golgi subfractions, 133–134
pre-immune, 78
preparation and characterization of, 115
production of, 46–47, 129–139
rabbit anti-mouse IgG, 117, 123
rat serum albumin-specific, 414
rhodaminated goat anti-mouse IgG, 104
selection of, 93–94
-vesicle complexes, 117
see also Antigen(s); Antiserum; Serum, immune
Antigen(s)
58 kD, 136
-antibody interactions, 4
-antibody recognition system, 77
anti-NADPH cytochrome P_{450} reductase, 84–85
BHK cell surface, 327
endocytic pathway specific, 401, 411–414
endogenous, 84–87
foreign, 84, 317
histocompatibility, 195
membrane vesicle, 199
peptide, 144
RLA class I, 206
selection of for immuno-isolation, 84
see also Antibodies; Antiserum; Serum, immune
Antipain, 213
Antiserum, 124
anti-cell, 96
polyclonal, 93
rabbit 411, 413
see also Antibodies; Antigen(s); Serum, immune
Apamin, 158
Aprotinin, 22, 213
Arginine, 423
Argon, 204
Arthritis, rheumatoid, 55
Aryl sulfatase, 86
Ascorbate, 285
Asialoglycoprotein(s), 115, 156–157, 159–160; *see also* Protein(s)
Asialoorosomucoid (ASOR), 232, 366
-LPO, 368–370
^{125}I-Asialotransferrin(s), 157
Aspartate, 52
Aspergillus oryzae, 9
Assays
acidification, 24, 29, 37–38
cell-free, 307, 328, 333, 344, 346–347, 353, 356, 358–359
colorimetric, 175
complementation, 308, 312
immuno-fluorometric, 319
ATPase, 21, 46, 160–161, 195, 265, 424
Na_2VO_4 inhibited, 282
nitrate inhibited, 270–271, 276–277, 282
vanadate inhibited, 270–271, 277
Autoimmune disorders, 55
Autoradiography, 110, 380, 384, 389
Auxins, 282
Avidin, 6, 244; *see also* Streptavidin
Azide, 293

Bacteria, 441
Bicarbonate, 237
Bile, canaliculi, 177, 232
Biotin, 6, 244
Biotin-ϵ-aminocaproyl concanavalin A, 243
Bone marrow, 56, 59, 384
Bovine serum albumin (BSA), 108, 123, 146, 176–180, 187, 216, 285, 292, 359, 394
Buffer
chamber, 8–9, 270
HEPES/KCl, 29
homogenization, 105
phosphate saline, 197, 204
sodium ion containing, 45–46
STEA, 187
TD, 337
TEA, 23, 338
Tris, 400, 412, 422
-Cl, 359

-Mes, 258, 285
TWA, 412

Calcium, 364, 422–424
chloride, 294
ion, 23, 60, 160, 290, 356, 358
Carbon, 188
Carbonylcyanide *p*-trifluoromethoxy phenylhydrazone (FCCP), 24, 29–30, 326
Carboxyl groups, 70
Carbowax, 187
Catalase, 217–218, 220, 222
Cell lines
colon carcinoma, 444
epithelial, 185, 194
fibroblast, 24
Hep 2, 190
KB, 190
WISH, 190
YB2/0, rat myeloma, 144
Cell(s)
15B, 335
A431, 185–187, 189–190
absorptive, 195
-free systems
reconstitution of membrane traffic in, 305–314, 317–329, 333–340, 343–353, 355–360
reconstruction of organelle function in, 311–314
BHK, 101–112, 309, 317, 327, 355
BHK-21, 3, 21, 22–24, 28–30, 308, 343, 344, 347, 349
CHO, 3, 31, 36–38, 42, 44–45, 60
CV1, 149
Daucus carota L., 292
endothelial barriers, 443
electrokinetic property of. *See* Zeta potential
enterocytes, 206
epithelial, 352–353, 421
mucosal, 443–444
erythrocytes, 54
ghosts, 148–149
fibroblasts, 174
V79, 291
germ, 17
goblet, 198
granulocytes, 55–56
green, 231
growth and synchronization of, 337
HeLa, 182, 389, 432–434, 436
Hep G_2, 174, 178, 366, 368–369, 372
hepatocytes, 10, 13, 60, 98, 115–119, 121–125, 156, 161, 176, 178, 412, 424, 443
hepatoma, 58, 443
homogenization of, 23
interphase, 432–433, 435
keratinized, 69
leukocytes, polymorphonuclear, 55
lymphoblastoid, 24
lymphocytes, 142, 144
B, 55
T, 55
macrophages, 56, 385, 398, 441
bone marrow-derived, 384
P388D1, 379–380
Madine Darby canine kidney (MDCK), 101, 309, 352, 356
metaphase, 436
mitotic, 308, 335, 337, 364, 431–438
monocytes, 56
myeloma, 131, 136, 144
nerve, 234
neutrophils, 57, 61
NRK, 135–136
nuclei, 25, 77
pancreatic exocrine, 135
perinuclear region of, 35
phagocytic, 441
plant, 253–265, 269–277, 281–287, 289–294, 297–300, 303–304
red, 55
RLT-28, 238, 240
RPC 5.4 mouse myeloma, 135
separation of
by continuous flow electrophoresis, 51–63
by phase partition, 227–234
Staphylococcus aureus, 79, 115–126, 367
surface labelling of with SFV, 22
tumor, 57
see also Cell lines; *specific organelles*
Cellulose, 95
fiber meshwork, 78–83
immunoadsorbents, 96

Chelex, 423
Chemotaxis, 56
Chloroplasts, 231, 233–234, 286
Chloroquine 356, 357, 392, 393, 395, 398, 400
Cholesterol, 160
  -7α-hydroxylase, 212
Chondroitin sulfate, 56
Chromatography
  affinity, 91, 367–369
  column, 392
Chromosomes
  flow cytometry and, 196
Chymostatin, 213
Cisternae
  free-flow electrophoresis and, 10–13, 17
  Golgi, 13, 130–131, 139, 393, 431–432, 435–436
  model, 129–130
  plant cell, 269
  rough endoplasmic reticulum, 138, 211
  trans-Golgi, 131
Clathrin, 5, 13, 17, 79, 141–142, 145–149, 160, 299, 313, 355
  heavy chain, 147–148
  homologues, 313
  light chain, 145–148
  -minus yeast mutants, 141
  triskelions, 145–147
Clostridium perfringens, 345
Clostripain, 147
CMPase, 129–130
Coagulopathy, platelet consumptive, 59
Cobalamin, 444
Colchicine, 393
Collagenase, 96
Colloids, 52
Columns
  gel filtration, 261
  pentomannosyl-6-phosphate Sepharose, 368–369
  Sephacryl
    S-200, 423
    S-300, 366
  Sephadex
    G-75, 422
    G-100, 398
Compartments
  endocytotic, 7, 17, 163–165, 364, 437
  endosomal, 176, 178–180, 193
  exocytotic, 7, 17
  membrane, recycling of, 384–388
  prelysosomal, 373
Concanavalin A, 243–244
Coomassie blue staining, 164–165
Cores, peroxisomal, 212, 217, 220
Cotyledons, 289–290
o-Coumaric acid, 260–262
Creatine
  phosphate, 338
  phosphokinase, 338, 345
Cucurbita pepo L., -var. Cocozelle, 270
Cycloheximide, 104–105, 110, 369–372
Cytochrome
  b_5, 212
  c reductase, 23, 25
  oxidase, 97, 217–220, 222
  P_{450}, 212
Cytoplasm, 85, 149, 211, 318, 335, 351, 359, 437
  filament proteins, 190
  peripheral, 436
  plant cell, 253, 265
Cytoskeleton, 13, 56, 62
Cytosol, 43, 52, 98, 263–264, 307, 312–314, 340, 352, 355, 357–360, 364, 422–423

DAB
  -induced density shift, 5, 169–182, 392, 394, 398–399
  polymerization reaction, 399
Daucus carota L., 292
Daunorubicin, 443
DD-NANA, 334–345; *see also* NANA
DEAE, 256–257
Density gradient centrifugation, 98
Density modification, 5, 153–165, 169–182, 185–207, 211–222, 227–234, 237–241, 243–244, 247–249
Deoxycholate, 367
Depolarization, 98
Development, and immuno-isolation of organelle-specific antibodies, 75–88, 91–94, 96–99, 101–112, 115–126, 129–139, 142–149

Dextran, 9, 22–30, 36–37, 40–41, 44–45, 227–228, 239–240, 259, 276, 381–382
-polyethylene glycol system, 231
sulfate, 256–257
Diamino-benzidine, 437
O-Dianisidine, 23
o-Dichlorobenzene, 215
Dictyosomes. *See* Cisternae
Dictyostelium discoideum, 384
Dimethyl sulfoxide (DMSO), 24, 337
Diphenylhexatriene (DPH), 203
Duodenum, 197

Ecto-phosphohydrolase, 98
EDTA, 23–24, 60, 105, 368
Elastase, 147
Electrophoresis
continuous flow. *See* free-flow
free-flow, 1–17, 21, 23–32, 35–47, 51–63, 240–241, 254, 257, 269, 289–290, 417–419
tonoplast isolation by, 269–277
preparative, for preparation of plant cell membranes, 281–284
one-dimensional gel, 110
SDS-PAGE, 26, 31, 41–42, 104, 106, 108, 110–111, 117, 119, 126, 164–165, 190–191, 328, 335–336, 338, 369, 394, 398, 412
two-dimensional gel, 109, 111, 212, 243
Endocytic
compartments, 163–165, 364, 437
fusion event in cell-free system, 355–360
membrane cycling, 21
pathway, 30, 84, 87, 98, 292, 303, 317–329, 355–356, 370, 436–437, 444
antigens specific to, 411–414
tracers, 36, 45
vesicle proteins, 368
vesicles, 358, 384–385
Endocytosis, 59, 126, 142, 317, 324, 377, 391
constitutive, 30
drug delivery and, 441–444
fluid phase, 368
G protein and, 320–321
insulin, 393
kinetics of membrane traffic in, 380–384
plant protoplast, 253–254, 291–294
of prebound SFV, 37
receptor-mediated, 21, 141, 149, 169, 171, 312, 355, 368, 389
regulation of, 30
subcellular distribution of transferrin in rat liver and, 427–428
see also Endocytic
Endoglycosidase-H, 101, 104
Endomembrane flow hypothesis, 297
Endoplasmic reticulum, 27, 60, 84, 119, 131, 186, 216, 232, 239, 312, 333–335, 339, 340, 434, 436
adsorption, 120
cytochrome *c* as marker for, 25
free-flow electrophoresis and, 15, 17, 51, 69
immuno-isolation and, 85
isolations, 17
lysosomal enzyme transfer and, 363
markers, 86, 117
plant cell, 283, 286
purification of, 13, 17
rough, 5, 102, 130, 138, 240, 333, 340, 417–419, 432
DAB and, 177
plant cell, 269
purification of membranes enriched in, 211–222
smooth, 81
flow cytometry and, 197
Endosomal
apparatus, characteristic functions of in rat liver parenchyma, 391–404
compartments, 176–180, 193
receptors, 175, 371
Endosome(s), 3, 77, 98, 130, 309, 317, 320–321, 355, 383–385, 399, 404
acidification of, 30–31, 161–162
CHO cell, 37
co-fractionation of BHK-21 cell, 21–32
DAB-induced density shift and, 169–182
early, 158–161, 384, 392–393, 397
fractions, labelling of, 187–193
hepatic subfractions of, 156–165
Hep G_2 cell, 369, 372
heterogeneity of, 392–393, 400–403
isolation of functionally different subpopulations, 35–39, 43–44

isolation of, using colloidal gold, 185–194
late, 39, 45–46, 159–162, 177, 364, 392, 397, 401, 411–412
latent, 27
Man-6-P receptor located in, 366
membrane, 164
multivesicular, 322–323
pathway, 193–194
plant cell, 293
purification of, 176
rat liver, 162, 176, 181–182, 363
receptor
containing, 126
negative, 120–121
positive, 120–121
selective ligand processing in, 394
separation of dependent on trypsinization, 27–29
subcompartments, 328
subfractions, 161, 164
transferrin receptor, 370
vesicles, origin and biochemical properties of 159–161
Enzyme(s), 88, 117, 164, 387, 395
assays, 217
free- (continuous) flow electrophoresis and, 58
glycoprotein processing and, 132
immuno-isolation and, 85
lysosomal, 101, 141, 363, 403, 417, 419
markers, 9, 11, 25, 60, 69–71, 86, 179, 192, 239–240, 270
see also specific enzymes
Epidermal growth factor (EGF), 115, 391, 403–404
receptors 163, 185, 193, 364
Epoxide hydrolase, 212
Esterase, 212, 220, 222
Ethanolamine, 270
Etioplasts, 283
Exocytosis, 59
green algae and, 297
pathway for, 322
plant, 253
Exoskeleton, plant, 253
Extra-corporeal circuitry, 59

Fab fragments, 147, 149
Fetal calf serum (FCS), 103
Fatty acids
in higher plants, 303–304
very long chain, 303–304
Filipin, 253, 271, 273, 275, 276–277
Flow cytometry, 5, 195–207
Flow fluorometry. *See* Flow cytometry
Fluorescein isothiocyanate (FITC), 22, 45, 200
-dextran, 23–30, 36–37, 40–41, 44–45, 381–382
Fluorography, 42
Formalin, 54
Formol, 58
Forskolin, 160
Fowl plague virus (FPV), 355–359
Fractions
endosome, labelling of, 187–194
immuno-isolation of time-defined, 318–326
Free-flow. *See* Electrophoresis
Fucosyl transferase, 283
Fungi, 441
Fusion
event, endocytic, in cell-free system, 355–360
in vitro, 355–360
of microlipid droplets from mammary gland, 421–425
protease-sensitivity of, 351–352
specificity of, 346–350

Galactose, 378, 394
β-Galactosidase, 378–379, 381–382
Galactosyltransferase, 14, 10, 17, 23–25, 45, 70, 86, 104–105, 107–108, 129–131, 134, 175, 178, 181, 216–218, 219, 222, 239–240, 379, 378, 392–395, 399, 417, 428, 431–434, 437
galBSA-HRP, 172–173, 176–178, 180, 394, 398
Gangliosides, mixed brain, 423–424
Gene(s)
replacement, 444
sequencing, 147
GERL structure, 101
Glucagon, 160
Glucose, 8, 105
oxidase, 368
-6-phosphatase, 14, 70, 218, 222, 239–240, 369, 417

Glucosides
  o-coumaric acid, 260–262
β-Glucuronidase, 13, 417–419
Glucuronosyltransferase, 212
Glutamate, 52
Glutaraldehyde, 54, 381, 402
Glycans, 244
^{3}H-Glycerol, 57
Glycerol gradient centrifugation, 274–275
β-Glycerophosphate, 367, 402
Glycine max L., 270
Glycoconjugates, 70, 243–244
Glycophorin A, 87
Glycogen, 437
Glycoprotein(s)
  135 kD membrane, 131
  cell-surface, kinetics of internalization and recycling of, 377–389
  envelope, 352
  extracellular, modified, 30
  insertion of into plasma membrane, 343–344
  Man_5 $GlcNAc_2$ residues, 308
  membrane-bound, 334
  plasma membrane, 329
  Semliki Forest virus, 309
  transport of, 290
  viral, 26–27, 101, 141–142, 149
  see also Protein(s); Sialoglycoprotein(s)
Glycosyltransferase, 11, 424
Glycosylation, 129
Gold, colloidal, 185–194, 327–328, 432–435
Golgi apparatus, 3–4, 7–11, 13–14, 17, 25, 27, 60, 83–86, 120, 216–217, 307, 312, 323, 335, 339–340, 352, 363–364, 373, 395, 397, 411, 413, 424
  CHO membranes, 312, 314
  contaminating elements in, 418
  clusters, 364, 432, 434–439
  flow cytometry and, 197
  DAB and, 169, 177
  free-flow electrophoresis and, 12, 14, 15–17, 21–22, 69
  galactosyltransferase enriched, 399
  in green algae, 297–300
  Man-6-P receptor located in, 366
  membranes, 163, 232
  mitotic form of, 431–439
  plant cell, 283
  saccules, 392
  subcompartments, 77, 129–132, 134–139
  vacuoles, 392
  vesicles, 432
  see also trans-Golgi network
Growth hormone, 58
  plant, 283; *see also* Auxins

Hemagglutinin, 149
Hepatic
  endosome subfractions, 155–165
  Golgi fractions, 401
Hepatoma(s), 58
  cells, 443
  isolation of highly purified plasma membrane from by two-phase partition, 237–241
  RLT-N, 14, 238, 240
HEPES, 24, 29, 239, 337–338
Heterogeneity, endosomal, 400–403
β-Hexosaminidase, 12, 23, 25–28, 39
High pressure liquid chromatography (HPLC), 145–146
Histidine, 423
Homeostasis
  cytoplasmic, 255
  regulators of, 30
Hormones. *See specific hormones*
Horseradish peroxidase (HRP), 22, 44, 169–178, 180, 321–323, 385, 394, 436–437
Hydrogen chloride, 171
Hydrolase, 195–196
  lysosomal, 365–366, 369, 372
Hydroxymethylglutaryl-coenzyme A reductase, 212
Hypocotyls, 274, 284

IgA, 81
pIgA, 174–175, 177–178, 180
IgG, 81–82, 94–95, 118, 120, 122–125, 136, 138, 150, 198, 319
IgM, 131
Imidazole, 116, 171, 213
Immuno-electronmicroscopy, 431–435
Immunoadsorption, 4, 91, 93, 96–97, 115–117, 119, 121–126, 322

Immunoaffinity, purification techniques, 4, 91–99
Immunoblotting technique, 412
Immunodetection, of membrane markers, 197–206
Immunogold labelling, 198
Immuno-isolation
  and development of organelle specific antibodies, 75–88, 91–94, 96–98, 101–111, 115–126, 129–139, 141–149
  of hepatocyte compartments using *S. aureus* cells, 115–126
  of subcellular components, 4, 77–88
  selection of antigens for, 84
  solid supports used for, 78–83
Immunomatrices, 95
Immunoperoxidase, 138
Immunoprecipitation, 6, 131
Influenza virus, 101, 322, 343–344, 346, 349, 351–352
Insulin, 156, 159–160, 364, 391–395, 397–398, 401, 403–404
^{125}I-Insulin, 157, 392, 395–397, 399, 402
Inulin, 261, 444
Iodogen, 36
Ion(s)
  argon, 204
  calcium, 23, 60, 160, 290, 356, 358
  channels, 163–164
  magnesium, 23, 161
  manganese, 290
  potassium, 69, 239
  pumps, 163
  sodium, 195
Iron, 38, 363, 424
Isoelectric focussing, 63
Isolation(s)
  of endocytic compartments, 17
  of endoplasmic reticulum, 17
  endosomal, 7, 35–447, 400–403
  Golgi, 9–10, 17
  lysosomal, 30
  microsomal, 213–216
  mitochondrial, 17
  of plant cell plasma membrane via free-flow electrophoresis, 269–277
  of plasma and vacuole membranes from green leaves by preparative free-flow electrophoresis, 285–287
  of post-mitochondrial supernatants, 213, 214, 216
  of receptors, 233
  of surface glycoconjugates, 243–244
  of tonoplast vesicles, 17, 255–265
  of vacuoles, 255–265
Isoprenaline, 160
Isotachophoresis, 63

Karnovsky fluid, 186
Keratinization, 69
Keyhole limpet hemocyanin, 116
Kinase, 163

Lactate dehydrogenase, 97
Lactoperoxidase, 327, 366, 368
Lanthanum chloride, 293
Lead acetate, 402
Lectins, 71–72, 243–244
  Ricinus communalis, 432, 435–437
Lepidium sativum L., 270
Lesions, filipin-induced, 271, 273
Leucine aminopeptidase, 232
^{14}C-Leucine, 176
Leucyl-leucyl-leucine, 213
Leucylnaphthylamidase, 160
Leupeptin, 213
Ligandosomes, 156
Ligand(s), 21, 44, 120, 157, 317, 364, 366
  binding, 367, 373
  DAB and, 179–181
  foreign, 158
  HRP and, 169–174
  internalized, 115, 118, 163, 394
  -receptor complexes, 35, 38, 62, 88, 389, 391
  selective degradation of, 391
  self-sorting of, 159
  sorting, demonstration and kinetics of, 177–178
  as subcellular markers, 155–156
  synthesis, 372
Lipase, 61
Lipid(s)
  droplets, cytoplasmic, 421–422
  extraction of, from plant cells, 293

flow cytometry and, 206
in higher plants, 276, 303–304
milk, 421
neutral, 303–304
phosphate group, 52
see also Phospholipid(s)
Lipoproteins, 402; *see also* Protein(s)
Liposomes, 423
Lucifer yellow, 444
Lumen, 163, 195, 198, 258
Lycine, 423
Lymph nodes, popliteal, 412
Lymphoproliferative disorders, 55
Lysomotropic carriers, 441
Lysophosphatidylcholine, 291
Lysosome(s), 158–159, 161, 163, 232, 308, 317, 329, 355–356, 365, 369, 372, 388, 398, 400–401, 412–414, 424, 434, 436–438
co-fractionation of BHK-21 cell, 21–32
continuous flow electrophoresis and, 51
DAB and, 169, 174, 175, 177
delivery of endocytosed ligand receptor complexes to, 185
endosome content targeted to, 38
enzymes, 363, 417, 419
late, 46
plant cell, 291
pre-, 367, 373
primary, 393
secondary, 383, 386, 393
separation of dependent on trypsinization, 27–29

Magnesium, 161, 422
acetate, 359
chloride, 8, 338
ion, 23
Mammary gland, in vitro fusion of micro-lipid droplets from, 421–425
Manganese, 290
Mannitol, 258
Mannose, 12, 334
Mannose-6-phosphate, 12, 363, 365–373
α-1,2 Mannose, 365
^{3}H-Mannose, 22, 26, 105
Mannosidase
I, 334, 339
II, 12, 129, 131
Markers, immunodetection of, 197–206
Melilotus alba, 253, 258, 260–262
Membrane(s)
apical, 59–60
basolateral, 205–206, 444
boundary, 61
brush border, 59, 195, 198–199, 201, 205–206
CHO cell, 31
continuous functional domains, 52
depolarization of, 98
ER-like, 10
flux, 11, 13
functional domains, 51, 53–63
intracellular, 61
lysosomal, 31
markers, immunodetection of, 197–206
microsomal, 42
nuclear, 51
pericellular, 175
plant cell, identification of, 269–277
plasma, 3–8, 10–11, 13–16, 21–22, 27, 78, 80, 85–86, 158–159, 164, 186, 191, 217, 231–232, 234, 352, 384, 386–388, 424, 434, 436
absorptive cell, 195
amphibian epidermis, 69–72
basolateral domain, 197–199, 203
BHK cell, 317
cell-free reconstitution of insertion of glycoprotein into, 343–344
donor, 349
fat cell, 95
in vitro intermediate transfer of lipids and fatty acids to in higher plants, 303–304
isolation of from green leaves by preparative free-flow electrophoresis, 285–287
isolation of from hepatomas and cultured cells by two-phase partition, 237–241
plant cell, 282–284, 289, 291, 294
reorganization model for, 61
separation of, using continuous flow electrophoresis, 51- 63
sinusoidal, 160–161
trypsinization of, 38

vesicles, 124
virion, 26
recycling, compartments of, 384–388
RER enriched domain, purification of by density perturbant, 211–222
traffic
analysis and applications of, 361–373, 377–387, 393–404, 411–414, 417–419, 421, 423–425, 427–429, 431–439, 441–444
in plants, 251–277, 281–294, 297–300, 303–304
reconstitution of in cell-free systems, 305, 307–314, 317–329, 333–340, 343–353, 356–360
during secretion of scales in green algae, 297–300
Mercaptoethanol, 9
β-Mercaptoethanol, 237, 258
Methionine, 110, 337–338
^{35}S-Methionine, 41–43, 104, 106, 110–111, 119, 134, 192
Methyl-^{3}H-choline chloride, 187
^{14}C-Methylamine, 326
Metrizamide, 349
Mice
athymic, 55
BALB/c, 145
SJL, immunodeficient, 144
Microlipids, droplets, in vitro fusion of, 421–425
Microperoxisomes, 249; *see also* Peroxisomes
Microsomes, 41–43, 71, 77, 79, 116, 119, 134, 157, 213–221
plant cell, 284
Microtubules, 393
Microvilli, 59
Mitochondria 15–17, 27, 51, 59, 69, 105–106, 216, 219, 231–232, 234, 239–240, 248–249, 393
light, 411
plant cell, 269, 277, 283, 286
Mitogens, 55
Mitosis, 431–438
Monensin, 161–162, 298
Monoamine oxidase (MAO), 216–219, 222
Myeloma, 135, 138

NADH
cytochrome *c* reductase, 11, 62, 97, 124, 212, 218, 220, 222, 239–240
ferricyanide reductase, 119
NADPase, 12, 129–130
NADPH
cytochrome *c* reductase, 70, 212
cytochrome P_{450} reductase, 86
NANA, 344; *see also* DD-NANA
Naphthylphthalamic acid, 282–283, 290
NBD
-caproic acid, 291–294
-PC, 291–294
Nerves, cholinergic terminals, 97
Neuraminidase, 56, 58, 61–62, 309, 343–344, 346–349, 351–352, 356–358, 378–379
Nitrate, 270–271, 276–277, 282
Nitrocellulose, 119
Nitrogen, liquid, 367
p-Nitrophenylphosphatase, 14, 69–71
Nocodazole, 337–339
Nuclear
envelope, 312, 335, 340
fragments, continuous flow electrophoresis and, 51
magnetic resonance, 265
Nucleoside diphosphatase (UDP), 11, 212, 283
5′-Nucleotidase, 24, 63, 86, 97–98, 160–161, 175, 178, 181, 232, 428
Nycodenz gradients, 5, 157–159, 162

Octadecyl rhodamine B chloride, 422
Oligomerization, 322
Oligomycin, 162
Oligosaccharides, 12
$Man_5GlcNAc_2$, 334, 339
N-linked, 372
Orthovanadate, 270
Osmium tetroxide, 186, 435
Ouabain, 14, 69, 239–240
Ovomucoid, 399

Paraformaldehyde, 104
Parasites, intracellular, 30
Parenchyma, functional characteristics of endosomal apparatus in rat liver, 391–404

Partides, submitochondrial, 234
Pathways
endocytic, 30, 84, 87, 98, 193–194, 292, 329, 355–356, 370, 436–437, 444
antigens specific to, 411–414
fusion of in reconstituted cell-free system using immuno-isolated fractions, 317
exocytic, 322
pinocytic, 441
Pentamannosyl-6-phosphate, 367–369
Pepstatin, 213
Peptides, 144, 147
Percoll gradient centrifugation, 3–4, 392, 394, 396–397, 399, 401
Peroxidase, 138, 289–290, 328
Peroxisomes, 216–217, 220
analytical differential centrifugation of, 249
cores, 212, 217, 220
modal densities of, 219
plant cell, 286
see also Microperoxisomes
Petroleum ether, 215
Phagocytosis, 56, 61, 63
Phagosomes, 389
Pharbitis, 289–290
Phase partition, 5–6, 153–165, 169–182, 185–207, 211–222, 227, 237–241, 294
Phenyl methyl sulfonyl fluoride (PMSF), 213
Phosphatase, 61
Phosphate 270, 271, 367
buffered saline (PBS), 22, 36, 108, 123, 237, 239
Phospholipase, 254, 291–294
Phospholipid(s), 53, 160
NBD PC, 254, 291–294
see also Lipid(s)
Phosphomannan, 367
Phosphomannosyl, 12, 363
Phosphotungstic acid, 282, 286
pIgA. *See* IgA
Pinocytosis, 387
fluid-phase, 424
Pinosomes
primary, 383–385
secondary, 383–384
Plants
membrane
traffic in, 251–265, 269–277, 281–287, 289–291
transport in, 292–294, 297–300, 303–304
Plasmalemma, 131, 414
Plasma membrane, 4–5, 13–16, 21–22, 27, 78, 80, 85–86, 158–159, 164, 186, 191, 195, 217, 231–232, 234, 352, 384, 386–388, 424, 434, 436
amphibian epidermis, 69–72
basolateral domain, 197–199, 203
BHK cell, 317
cell-free reconstitution of insertion of glycoprotein into, 343–344
donor, 349
fat cell, 95
in vitro intermediate transfer of lipids and fatty acids to in higher plants, 303–304
isolation of from green leaves by preparative free-flow electrophoresis, 285–287
isolation of from hepatomas and cultured cells by two-phase partition, 237–241
liver, 165
plant cell, 269–277, 282–284, 289, 291, 294
proteins, 365
sinusoidal, 160–161
Plastids, 269
Platelets, 55, 58–59, 61–62
Platinum, 188
Pleurochrysis, 297–298
Polyacrylamide beads, 79–80, 82, 86, 123
Polyarginine, 423
Polydispersity, 4, 78, 169
Polyethylene glycol (PEG), 227–228, 231, 239, 276
Polyhistidine, 423
Polypeptides, 94, 101, 111–112, 143, 158, 182, 191, 423
Polyribosomes, 15, 417
Polysomes, 363

Potassium, 14
chloride, 24, 29
ion, 69, 239
phosphate, 231, 239
Primaquine, 178
Prolactin, 160, 364, 391–392, 394–395, 397–398, 401, 403
^{125}I-Prolactin, 395, 397
Pronase, 423
Protease, 314, 351–352
Proteinaceous halo, 52, 60
Protein(s)
58 kD, 5, 136–139
103–108 kD, 131
A, 92, 94, 117, 122–124, 198, 327, 432–433, 435
assay of, 28, 216
cell-surface, 85
cobalamin-binding, 444
cytoplasm filament, 190
DAB and, 181
determination of, 24
endocytic vesicle, 368
endosomal, 163
envelope, 309
epidermal content of, 69
flow cytometry and, 206
free-flow electrophoresis and, 9, 70–71
G, 43–44, 87, 102, 104, 106–109, 112, 307–308, 312–314, 317–322, 326–329, 333–336, 339–340, 352, 444
Golgi membrane, 438
immuno-isolation and, 85
liver, 172, 217, 221
microsomal, 40, 43, 134
-NH_2 subgroups, 52
peaks, 25, 70, 104
phosphate group, 52
plant cell, 275, 289
plasma membrane, 11, 13, 87, 115, 365
polymers, phase partition and, 227–230, 233
reconstitution of transport from ER to Golgi using a cell-free system, 333–340
recovery, 8
secretory, 101, 134, 141, 132, 211
spike, 322
Staphylococcus aureus, 92, 94
targetting of to lysosomes, 365
see also Glycoprotein(s); Lipoprotein(s); Sialoglycoprotein(s)
Proteolysis, 163
Protonophores, 326
Protoplast(s), 253, 265
endocytosis, 254, 291–294
fluorescent labeling of, 292–293
fusogenic, 293–294
Melilotus alba, 262
nonfusogenic, 293–294
phospholipase activity in, 291–294
Provacuoles, 293; *see also* Vacuole(s)
PTA staining 272, 275

Rabbits, A^+, 197
Rats
Holtzman, male, 8
Wistar, female, 213
Reagents, antibody binding, 4
Receptor(s)
anti-asialoglycoprotein, 119–120
anti-Man-6-P, 370
asialoglycoprotein, 116, 121, 125, 156–157, 368, 394, 441, 443
beta-adrenergic, 158
cell-surface, 156, 318, 377, 444
EGF, 116, 121–122, 125, 185, 193, 364, 391, 403–404
endocytic, 126
endosomal, 175, 371
IgA, 81
insulin, 364, 391–393, 397, 403
internalization mediated by, 60
isolation of, 233
kinases, 163, 404
-ligand complexes, 35, 88, 389, 391
lymphokine, 57
mannose-6-phosphate, 363, 365–373
mediated endocytosis, 169, 171, 312, 355, 368
-mediated pinocytic pathway, 441
opiate, 233
phosphomannosyl, 12, 363, 417–419
prolactin, 364, 392
recycling of, 115, 388–389
SAGP, 118
self-sorting of, 159

transferrin, 46, 193, 369–371
Receptosomes, 51, 156
Reductase, 85
Retroviruses, 444; *see also* Viruses
Rhodamine, 109, 202–203
Ribophorins, 212
Ribosomes, 15, 77, 211–212, 220, 432
plant cell, 269
Ricinus communalis, 432, 435–437
RNA
microsomal, 134, 218
viral, 150

Salicylic acid, 339
Saponin, 418
Sarcoplasmic reticulum, 51
Scherffelia dubia, 297–300
Semliki Forest virus (SFV), 22, 25–26, 36–38, 148, 309, 344–346, 348, 352, 355
Separation techniques, new developments in, 1–17, 21–32, 35–47, 51–63, 69–72
Serum, immune, 413; *see also* Antiserum
Sialic acid, 10–11, 17, 52, 55, 58, 60, 70, 161, 424
^{3}H-Sialic acid, 355–356, 359
Sialoglycoprotein(s), 31, 61, 63; *see also* Protein(s)
Sialyltransferase, 10, 104–105, 121–22, 130–131
Signalling, transmembrane, 391, 403–404
Sodium
azide, 293
chloride, 36
dodecyl sulfate, 338, 394
hydroxide, 8
ion 195
pump, 71
salicylate, 42
Sorbitol, 62, 228, 292
Specificity of fusion, 346–350
Spectrofluorometry, 24
Spectrophotometry, 148
Sphingomyelin, 160
Staphylococcus aureus, 79, 115–122, 367
Stratum corneum, 70
Streptavidin, 6, 243–244; *see also* Avidin
Streptococcus pneumoniae, 379
Succinate-INT-reductase, 23, 25–26, 28, 70, 105, 239–240
Sucrose, 8–9, 16, 23, 27, 40, 83, 105–106, 108, 116–117, 155–156, 162, 172, 176, 187, 192–193, 213–215, 220–221, 231, 237, 276, 285, 304, 359, 394, 400–401, 417–418
Sulfate, 17, 70, 71
Supernatants
microsomal, 16, 41–42
plant cell, 285
post-chromosomal, 335, 338
post-mitochondrial, 158, 213–214, 216
post-nuclear, 23, 25–26, 37, 45, 327, 352, 355
TCA, 348
Synapses, cholinergic, 98
Synaptosomes, 234

Taxol, 59
^{35}S-TCA, 111
Tetraselmis, 298–300
Thiamine pyrophosphatase, 11, 86, 435–436
Thimerosal, 171, 174
Thrombocytopenia, 59
Thylakoids, 230, 233, 286
Thymidine, 337
Tonoplast(s), 17, 253, 289
identification of, 269–277
isolation of via free-flow electrophoresis, 269–277
preparative free-flow electrophoresis and, 281–283, 286
vesicles, isolation of, 255–265
TPCK, 23, 25, 37, 351
TPPase, 129–130, 435–436
Transferase, 85
Transferrin, 40, 44, 46, 193, 364
diferric, 36, 38
FITC conjugated, 45
receptor, 185, 369–371
subcellular distribution of, 424, 427–428
uptake by liver, 363
^{125}I-Transferrin, 193
Transformations, viral, 55
trans-Golgi network, 4, 27, 101–112, 347, 349–350, 352–353; *see also* Golgi apparatus
Transplant rejection, 59
Triacylglycerol, 421
Trichloroacetic acid, 344

Triethanolamine, 8, 23, 270, 338
Tritium, 309
Triton
  X-100, 270, 346–347, 367, 422
  X-114, 105, 132–133, 139
Trypsin, 23, 25, 27–29, 31, 37–40, 45, 187–190, 337, 351
Tubulin, 160
Two phase polymer partition. *See* Phase partition
Tunicamycin, 372

UDP-GlcNAc transferase I, 307, 312
UDPG-sterolglucosyltransferase, 271, 275, 277
Urate oxidase, 217–218, 220, 222

Vacuoles, 253
  condensing, 300
  Golgi, 392, 437
  isolation of, 255–265
  membrane, isolation of from green leaves by preparative free-flow electrophoresis, 285–287
  plant cell, 293–294
Vanadate, 270–271, 277
Vesicle(s), 83–85, 350, 352
  -antibody complexes, 117
  ASGP receptor-positive, 118
  basolateral, 204–205
  brain coated, 125
  brush border, 199, 201, 203–206
  budding, 312, 314
  clathrin coated, 141–149, 355
  EGF receptor-positive, 121
  endocytic, 327–328, 358, 384–385
  endoplasmic reticulum, 15, 220, 231
  endosome, origin and biochemical properties of, 159–161, 363
  fusion, 312, 314
    in plant cells, 293
  Golgi, 364, 432, 438
  immunoadsorption of, 116–117, 119
  immuno-isolated acceptor, 327
  intestinal, characterization of with flow cytometry, 195–207
  lipoprotein-filled, 402
  -mediated intracellular transport, 345
  neural membrane, 232, 234
  neutrophil, phagocytic, 61
  plant cell, 271–274, 282
  post-Golgi, 351
  rough endoplasmic reticulum derived, 211–212
  secretory, 61, 365
    in green algae, 299
    plant cell, 269
  separation of by phase partition, 227–234
  synaptic, 98
  targeting, 312, 314
  TGN, 353
  thylakoid, 231, 233–234
  tonoplast, 253
    isolation of, 255–265
  transition, 13, 17
  vesiculo-tubular, 434
Vesicular stomatitis virus (VSV), 27, 43–44, 101, 110, 307–308, 312, 317, 319–322, 333–335, 337, 352, 444
Virions, 26, 318
Viruses, 101–103, 111, 441, 444; *see also* Retroviruses; *specific viruses*

Western blotting analysis, 145–146

Xenobiotics, lipophilic, 211

Zeta potential, 52